D0268690

Cellular Radio

Analog and Digital Systems

RECEIVED

AUG - 8 1995

MSU - LIBRARY

The Artech House Mobile Communications Series

John Walker, *Series Editor*

Cellular Radio: Analog and Digital Systems, Asha Mehrotra

Mobile Information Systems, John Walker, editor

Narrowband Land-Mobile Radio Networks, Jean-Paul Linnartz

Land-Mobile Radio System Engineering, Garry C. Hess

The Evolution of Mobile Communications in the U.S. and Europe: Regulation, Technology, and Markets, Michael Paetsch

For a complete listing of *The Artech House Telecommunications Library,* turn to the back of this book

Cellular Radio

Analog and Digital Systems

Asha Mehrotra

Artech House
Boston • London

TK
6570
. M6
M44
1994

Library of Congress Cataloging-in-Publication Data
Mehrotra, Asha.
Cellular radio: analog and digital systems/Asa Mehrotra.
Includes bibliographical references and index.
ISBN 0-89006-731-7
1. Cellular radio. I. Title.
TK6570.M6M44   1994
621.3845'6–dc20

94-5944
CIP

A catalogue record for this book is available from the British Library

© 1994 ARTECH HOUSE, INC.
685 Canton Street
Norwood, MA 02062

All rights reserved. Printed and bound in the United States of America. No part of this book may be reproduced or utilized in any form or by any means, electronic or mechanical, including photocopying, recording, or by any information storage and retrieval system, without permission in writing from the publisher.

International Standard Book Number: 0-89006-731-7
Library of Congress Catalog Card Number: 94-5944
10 9 8 7 6 5 4 3 2

Contents

To the memory of my father, Narsingh Dass Khattry

Preface

Cellular radio is a fast-changing field of mobile communication. The whole concept of mobile communication is changing so rapidly that it is impossible to keep pace with the changes. The early 1980s saw the advent of analog systems that had an uncertain future. Subscriber growth in this area has been phenomenal, however, and, as a result, low-capacity analog systems are rapidly being replaced by digital systems throughout the world. Both Europe and the U.S. produced their own systems, based on TDMA technology. It is expected that within the next few years analog systems will be replaced by their digital counterparts in most industrialized countries. It is anticipated that a digital TDMA system will not satisfy the growing capacity requirements of the industrialized world, which will force a change to an even higher-capacity system like CDMA. By the turn of the century, personal communication systems (PCS), based on microcellular technology, will nearly satisfy individuals' speech and data requirements. The FCC has already cleared the path for the fast development of PCS in the U.S. by allocating 220 MHz of spectrum in the 1.85 to 2.2-GHz band.

No single book can cover all the areas of cellular radio. Thus, this book focuses on select system aspects of analog and digital cellular radios. Accordingly, the big picture is presented without going into the mathematical details. Chapter 2 examines the central core of analog systems, discussing all the relevant design requirements and the techniques used in meeting them. After bringing the traffic concept and its applications to cellular radio, Chapter 4 takes up different analog systems of the world. Background material is presented first so that the reader can follow different sections of this chapter without consulting external material. Chapter 5 discusses coverage problems of analog systems and provides solutions. Capacity limitations of analog systems are presented in Chapter 6 along with some suggested techniques for improvement by adapting to digital techniques and microcellular technology. In essence, this chapter provides the necessary link between the analog and the digital cellular radios. Both the American and European systems (GSM) are discussed in detail in Chapter 7, along with some improvements gained by adopting E-TDMA and CDMA. The full utility of cellular cannot be achieved without networking, and so Chapter 8 discusses the techniques of authentication and fraud minimization. Lastly, the multiplexing of data at cell sites and then transporting them to a mobile switching office by using microwave links is becoming popular. I have treated this from the application point of view in Chapter 9.

I believe the best way to learn a subject is to solve problems. With this in mind, each chapter is supplemented with several problem sets. Adequate references have been provided for the interested reader who wants to pursue further a particular topic.

This book is intended for the practicing engineer and final-year undergraduate or graduate student who want working in the cellular field. For practicing engineers, a B.S. degree in electrical engineering and some working knowledge of mobile communication is assumed. For students, it is assumed that they have taken senior-level courses in communication engineering.

I would like to express a few words of acknowledgment. First of all, I would like to thank The Analytic Sciences Corporation (TASC) for granting permission to publish this book. I extend my appreciation to my colleagues Mr. Richard G. Moldt, for helping me review the complete book, and Michael Scheidt, for consultations while this book was being written. I am extremely grateful to Professor R. Pickholtz of George Washington University, who was instrumental in initiating this project. I also want to thank my graduate students at George Washington University, who have contributed by asking the right questions and, in general, by helping me in my thought processes. I also must thank the reviewers of this book for their excellent comments, which have tremendously improved it.

Finally, a project of this type can never be completed without the continuous support of one's family. In this regard, thanks are due to my wife, Nisha, my daughters, Anuja, Sonia, and Vinita, and my son, Neil. Special thanks to my seventeen-year-old daughter, Sonia, for drawing all the figures and graphs and for putting up with this unending project for the last several years.

Chapter 1
Introduction to Cellular Mobile Radio

1.1 BACKGROUND AND HISTORY

What is cellular radio, and how is it different from other forms of mobile radio communication? The Federal Communication Commission (FCC), in Title 47, Part 22 of the Code of Federal regulation, effective June 22, 1981, has defined a cellular system as:

> A high capacity land mobile system in which assigned spectrum is divided into discrete channels which are assigned in groups to geographic cells covering a cellular geographic service area. The discrete channels are capable of being reused in different cells within the service area.

The three basic parameters defining a cellular radio system, from the FCC definition, are high capacity, cells, and frequency reuse.

1. High Capacity: theoretically, a cellular radio system can be configured and expanded to serve a limitless number of subscribers. (There are practical limitations on subscriber traffic that will be discussed in Chapter 3.)
2. Cells are defined as individual service areas, each of which has an assigned group of discrete channels assigned to it from the available spectrum. Subscribers in a particular cell can utilize the channels assigned to that cell. A group of contiguous cells make up the cellular geographic service area (CGSA) served by a specific system. A system can grow geographically by adding new cells.
3. Frequency reuse allows the discrete channels assigned to a specific cell (for example, cell #1) to be used again in any cell which is separated from cell #1 by enough distance to prevent cochannel interference from deteriorating the quality of the service. As a system grows, the discrete channels originally assigned to cell #1 can be continuously reassigned so that the system will never run out of available channels to serve the public. The frequency reuse concept is not only used in cellular radio telephone service but also in the television and radio industries.

A cellular radio system is the culmination of all prior mobile communication systems, starting with the demonstration of land-to-mobile communications by Guglielmo Marconi in 1898. Indeed, the potential uses for cellular radio systems go beyond just mobile communications. Many consider the cellular concept to be a viable alternative to wireline local-loop distribution, vastly expanding the potential for cellular radio telephone systems.

Any history of radio communications must begin with Heinrich Hertz, the discoverer of electromagnetic waves. By 1880, Hertz had demonstrated a practical radio communication system. The world's first commercial radio service was established by Guglielmo Marconi, and his first customer was Lloyd's of London. The first radio link covered 7.5 miles and provided information about incoming shipping. Marconi began the first ship-to-shore communications in 1898. Marconi also accomplished the first long distance transatlantic transmission in 1901. Mandatory 24-hour ship-to-shore communications were established during 1910–1912 by the United States, Britain, and other maritime nations as a direct result of two ships sinking: the *Republic* in 1909 and the *Titanic* in 1912. By 1918, 5,700 ships had wireless telegraphy installations.

The first human voice transmission via radio was accomplished by Reginald Fessenden in December, 1900. The first link was one-mile long. The demonstration took place in Maryland and marked the beginning of radio telephony. Six years later, on Christmas Eve 1906, Reginald Fessenden transmitted the world's first radio broadcast. The transmitter was located at Brant Rock, Massachusetts, and good-quality voice and music was received by ship and shore wireless operators within 15 miles of Brant Rock. In 1915, a team of Bell Telephone engineers, using the giant antennas of the U.S. Navy station NAA at Arlington, Virginia, were the first to span an ocean with the human voice—a milestone of intercontinental radio telephony. The voice radio transmissions were received in France, Panama, and Hawaii.

1.2 MOBILE RADIO

The development of the mobile radio system can be divided into two parts: phase 1 is when the early systems were being developed, and phase 2 begins after the FCC's classification of "Domestic Public Land Mobile Radio Service" (DPLMRS) [1]. See Table 1.1. The first important use of mobile radio using an automobile instead of a ship was in 1921, when the Detroit Police Department instituted a police dispatch system using a frequency band near 2 MHz. The service was so successful that the channels in the band were soon utilized to the limit. In 1932, the New York City Police Department instituted the use of the 2-MHz band for mobile communication. In 1934, the FCC opened four new channels in the 30-MHz to 40-MHz band, and by the early 1940s a significant buildup of police and other public service systems was realized. In the late 1940s, the FCC made mobile radio available to not only police and fire departments, but also to the private sector.

In 1946, Bell Telephone Laboratories (BTL) inaugurated the first mobile system for the public, in St. Louis. Three channels in the range of 150 MHz were put into service.

Table 1.1
Milestones in Cellular Radio Development

Year	Milestone
A. Phase 1: Early Development	
1921	Detroit Police department (2 MHz)
1932	New York Police Department (2 MHz)
1934	FCC authorizes four channels in the 30–40 MHz range
B. Phase 2: Post DPLMR	
1946	First public mobile system from Bell Labs (150 MHz)
1947	Highway mobile system (35–44 MHz, simplex)
1955	11 channels around 150 MHz with 30-kHz channel spacing
1956	12 channels at 450 MHz (manual)
1964	First automatic 150-MHz system, known as MJ (duplex)
1969	First automatic 450-MHz system, known as MK (duplex)
1971	Bell proposal to FCC for cellular radio
1977	FCC authorizes installation and test of AMPS
1981	FCC releases 800–900 MHz band for cellular land-mobile phone service (40-MHz band)
1990	FCC NOI into the establishment of new PCS

Note: DPLMR = domestic public land mobile radio service; FCC = Federal Communications Commission; AMPS = advanced mobile phone service; NOI = notice of inquiry; PCS = personal communication services.

A "highway" system to serve the corridor between New York and Boston began operating in 1947. The system operated at frequencies between 35 and 44 MHz. All of the initial systems were simplex push-to-talk systems. In spite of this, demand for service grew and in some areas the number of prospective customers outpaced the number of channels available to serve them. Initially, six channels with 60-kHz channel spacing were available at 150 MHz. In 1955, the channel spacing was reduced to 30 kHz, and 11 channels became available as a result of this reduced channel bandwidth. In 1956, 12 channels were added near 450 MHz to meet increased public demand. All of the mobile systems in use at that time were manual systems, requiring operator assistance to set up the calls.

The first automatic system was installed in 1964 and operated at 150 MHz. It was a full-duplex system, eliminating the push-to-talk requirement of the older systems. It had automatic channel selection and allowed the mobile user to dial directly, thereby eliminating the need for operator assistance. Automatic mobile systems operating at 450 MHz were authorized and built in 1969. The original simplex push-to-talk system was classified as a DPLMRS, while the automatic systems fall under the improved mobile telephone system (IMTS) designation. Prior to 1949, all mobile service was supplied by the wireline telephone companies. In 1949 the FCC authorized a new type of entity, designated as radio common carriers (RCCs), to provide mobile radio service to the public. As of December, 1977, RCC's had 80,000 mobile units in service compared to 63,000 for wireline common carriers (WCC).

The limited number of channels that were available in any given area coupled with the demand for mobile radio telephone service, particularly in urban areas, resulted in unacceptable service. Users were not able to readily access the network, often requiring three or four attempts (or more in cities like New York) to seize an idle channel. This situation provided the impetus for the development of cellular radio, resulting in FCC Docket 18262 and the Bell System response in December 1971, which explained in detail the system architecture now known as cellular radio.

On June 7, 1982, the FCC started to accept license applications to construct systems and operate in the top 30 cellular markets of the country. The first cellular system was put into operation in the Chicago area on October 13,1983, by American Telephone and Telegraph (AT&T). According to an earlier FCC ruling, each new market will be granted two operational licenses: one to WCC and one to RCC. The goal is to provide continuous, nationwide cellular service within a reasonable time frame. Appendix A lists the first 120 cellular systems operating in the United States. Although analog cellular telephony will continue to satisfy American needs, the big mobile explosion is expected to come from new technologies—personal communication network (PCN) and personal communication services (PCS), based on low-power microcellular technology, which will provide both the voice and data communication and can be made available in the street, home, and office. The development in this direction is rapid in Europe, especially in the United Kingdom. In keeping with the rapid development in cellular industry, the FCC has recently issued a notice of inquiry (NOI) for the establishment of PCS.

1.3 HISTORY OF FCC REGULATION

The first decade of the twentieth century saw the initial development of a useful application for Heinrich Hertz's discovery of radio waves for ship-to-shore radio telegraphy. The value of ship-to-shore radio telegraphy was recognized early on, and the first federal regulation of radio communications began in 1910 when Congress passed legislation mandating that U.S. flag ships carry radio telegraphy equipment.

The proliferation of radio transmitters prompted Congress to pass the Radio Act of 1912, which provided for the first licensing of radio stations. The radio-broadcasting industry then started to grow by leaps and bounds, and, to alleviate the resultant congestion and chaos, the Federal Radio Commission was established in 1927. With the technological advances in the radio industry, it was soon evident that broader regulatory rules would be required. The result was the Communications Act of 1934 that created the FCC, which is operated by seven commissioners appointed by the president for seven-year terms (pending Senate approval). The Communications Act of 1934 centralized regulation of U.S. interstate and foreign communications: radio, wire, and cable.

Along with its other powers, the FCC is responsible for the allocation of radio frequencies, establishing transmitter power limits, and providing broadcast station call letters. It is also responsible for issuing the licenses necessary to provide service to the public and approves (or disapproves) interstate telephone and telegraph rates. The United States now has about 13 million licensed radio stations, utilizing about 50 million transmitters in 70 different categories. Additional jurisdiction was assigned to the FCC when Congress passed the Communications Satellite Act of 1962, reflecting the continuous recognition by the Congress of the growth and expansion of radio technology. The FCC operates through various bureaus and divisions. The one of immediate concern to us is the Common Carrier Bureau, which is responsible for the control of land mobile services, including cellular radio telephone service.

1.3.1 Land Mobile and Cellular Mobile History

While the history of land mobile service began with the Detroit Police Department in 1921, it was not until 1946 that Bell Telephone System planners started looking for a large-scale system that would satisfy mobile customer demands. Proposals for different systems were made from time to time, as described below. These proposals are generally associated with the FCC dockets summarized in Table 1.2. In 1947, Docket 8658 was the first adaptation where the FCC set aside 40 MHz of spectrum for land mobile use. In 1958, in response to an inquiry for land mobile service made by the FCC in Docket 11997, the Bell System proposed allocating 75 MHz of spectrum at 800 MHz for this service.

Upon formal recognition of the congestion on land mobile frequencies, which was approaching unacceptable levels, the FCC instituted Docket 18262 to set aside sufficient spectrum to meet the demand for land mobile communications into the foreseeable future. This action by the FCC can be considered as the starting point for cellular radio telephone communications. In 1970, the FCC tentatively set aside 75 MHz of spectrum (at 850 MHz) for common carrier cellular systems, under Docket 18262.

To this point, only wireline carriers were eligible to provide service under the terms of Docket 18262. In 1971, upon reconsideration the FCC modified its 1970 decision to

Table 1.2
Summary of FCC Regulation

Year	FCC Docket Number	Regulations
1947	8658	Spectrum allocation of 40 MHz
1958	11997	Bell proposal for 75-MHz bandwidth system at 800 MHz
1970	18262	75-MHz tentative allocation to common carriers
1974	18262	40-MHz firm allocation to wire line common carriers
1975	18262	40-MHz bandwidth opened to any common carrier
1977	18262	Authorizes two developmental systems (wire line and nonwireline)
1988		Addition of 10-MHz bandwidth, making a total cellular allocation of 50 MHz

Source: [1]

allow nonwireline as well as wireline carriers access to the 75 MHz allocated for cellular common carrier systems. This decision was not well received by the telephone companies and in 1974 the FCC again restricted eligibility to wireline carriers. The FCC once again adopted a one system per market policy because of the belief (since proven to be false) that technical complexity and expense would not allow competing systems in a market to be viable. At the same time, the FCC revised the common carrier wireline spectrum allocation from 75 MHz to 40 MHz and decided to license developmental systems to enable optimal development of cellular technical capability. Since then the FCC has added an additional 10 MHz of bandwidth, which makes the total cellular allocation 50 MHz.

In 1975, upon reconsideration, the FCC again reversed itself and lifted its restriction on limiting eligibility to wireline carriers. The rationale for this reversal was that if only wireline carriers are technically and financially capable of amassing the resources necessary to develop and implement cellular service, it was therefore unnecessary to mandate this eventuality by rule. It was this forward-looking decision on the part of the FCC more than anything else that gave impetus to the development of cellular radio. Removing wireline cellular systems from under the aegis of the monopoly-protected business, the non-wireline carriers saw expanded interest in their potential service possibilities. This universal interest sped up the process of getting cellular service to the public by many years.

In 1976, the District of Columbia Circuit of the U.S. Court of Appeals reaffirmed the FCC's position, removing the last obstacle to the commercial development of cellular radio systems. The FCC authorized two developmental cellular systems in 1977:one to WCC and one to RCC. The wireline authorization, for the Chicago metropolitan area, was granted to the Illinois Bell Telephone Company, for the advanced mobile phone system (AMPS). American Radio Telephone Service, Inc., (ARTS) was granted the nonwireline authorization to build a developmental system in the Baltimore–Washington, DC. area. A third developmental cellular radio license was granted to Millicom, Inc., another nonwireline company, in 1980, with the system to be built in Raleigh-Durham, North Carolina. The license grant was partly based on Millicom's plan to build a distributed switching capability into its system, thereby helping to minimize landline costs to subscribers.

The FCC initiated Docket 79-318 with the release of a "Notice of Inquiry and Notice of Proposed Rulemaking" in 1980. This was the initial step in the process of defining cellular service requirements so that operating license applications could be submitted and acted upon. Subsequently, the FCC took the next step by releasing a "Report and Order," adopting rules for commercial cellular service in 1981. In 1982, "Memorandum Opinion and Order on Reconsideration" affirmed the 1981 "Report and Order." This was the final step necessary to initiate the rush of applications for cellular operating licenses.

The FCC agreed to accept license applications in a set order. The thirty most densely populated mobile service areas (MSAs) in the country were designated for the first round of applications, followed in descending population order by additional MSAs in groups of thirty until all areas were reached. The FCC will continue to accept license applications, thirty at a time, until the entire country has cellular service.

For the first thirty markets, licenses were granted based on the comparative content of each application. The number and size of the applications for the first round, as well as for subsequent submissions, soon exceeded the resources of the FCC to properly evaluate each application in a timely manner. To resolve this situation, and to prevent delay in granting operating licenses, the FCC instituted a lottery procedure whereby the license for an area was granted to the winner of a lottery. The comparative process was abandoned. Since each service area had to have two systems, one by the WCC (designated by letter A), and the other by the RCC (designated by letter B), service in an area was usually started first by the wireline, followed by the non-wireline company. However, in some cases this order has been reversed.

1.4 CONVENTIONAL MOBILE RADIO VERSUS CELLULAR MOBILE RADIO

An understanding of conventional mobile telephone system will aid in recognizing the advantages of the cellular systems [1–4, 9–10]. In the conventional mobile telephone system, the transmitted signals are strong enough that the channels assigned in one service area can't be reused in nearby service area. This severely limits the number of available channels. Due to high-powered transmitters, the area of coverage of the transceivers can be thousands of square miles. The size of the coverage area varies depending upon the transmitted power, transmission frequency, and the antenna height. Unlike conventional systems, cellular mobile radio systems may use a large number of small power transmitters (100W maximum permitted by FCC per channel), each covering a small area in the range of 75 to 300 square miles. Because of the short distance covered by each transceiver, the particular channel frequency can be reused over and over in multiple nonadjacent cells. Each transceiver is connected to a central switching office, which controls and monitors the overall system and provides the interface to the local telephone company.

Figures 1.1(a) and 1.1(b) depict the conventional mobile radio and cellular radio layout, respectively. A comparative summary of conventional mobile and the cellular radio is provided in Table 1.3, which shows that the current conventional New York City mobile system can serve only twelve callers at any time, while as many as 333 subscribers could theoretically be serviced within the area of one cell. More subscribers can be serviced simultaneously depending upon that cell's output power, separation from other cells, and the number of channels allocated. As shown in Figure 1.1(b), unlike single transmitters for conventional systems serving 10–25 miles radius, the cellular system has multiple transmission sites designated as cells. Cells are connected to the *mobile telephone switching office* (MTSO), which plays the central role in the successful operation of mobile system. It connects to the telephone network, provides maintenance and testing capabilities, and stores the information for billing purposes.

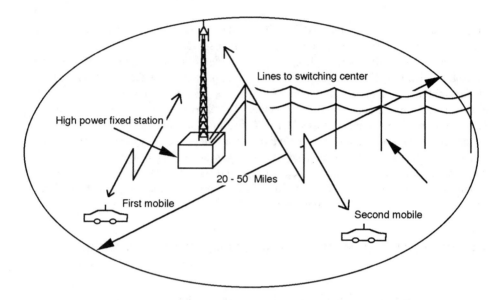

Figure 1.1(a) Conventional mobile radio system.

1.5 OBJECTIVES IN THE DESIGN OF CELLULAR SYSTEMS

Some major design objectives of the cellular systems are as follows [1–4].

- *Large subscriber capability.* The system should be capable of serving many thousands of mobile users within the local serving area with a fixed number of several hundred channels.
- *Spectrum utilization.* The multiple use of the same channels in cells with geographical separation ensures that the radio spectrum is used efficiently.
- *Nationwide compatibility.* The mobile users should be able to use their equipment even though they have drifted from their home base to other areas that are served by different cellular systems. A mobile user in this situation is known as a *roamer.*
- *Service to portable, vehicles, and other specialized services.* A cellular system should provide adequate service to handheld portable phones as well as to regular mobile units in vehicles. The system should provide specialized services, such as dispatch or fleet operation. The system should also provide special features, such as abbreviated dialing.
- *Adaptability to traffic density.* Since the traffic density will differ from one point in a cellular coverage area to another, the capability to cope with different traffic must be designed as an inherent feature of the cellular system.
- *Quality of service and affordability.* The service has to be comparable to regular telephone service. Since cost and economic considerations play a major role, it must be made affordable for general public.

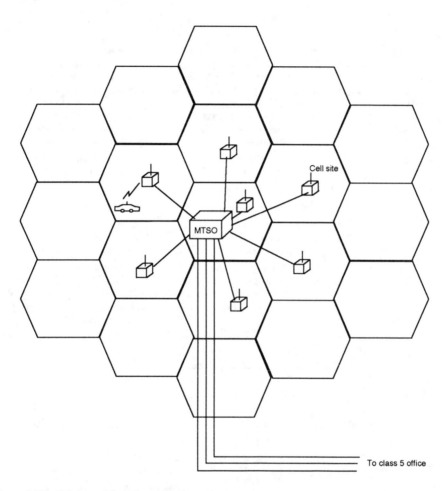

Figure 1.1(b) Cellular mobile radio system.

1.5.1 Frequency Allocation

As described earlier, in 1974 the FCC allocated a 40-MHz bandwidth in the 825–890 MHz frequency range for high-capacity mobile radio telephone use. This 40-MHz bandwidth is divided equally between transmit and receive bands. Mobile transmit channels are in the 825–845 MHz range, while the mobile receive channels are in the 870–890 MHz range. This original spectrum is shown in Figure 1.2. Channel band A is assigned to radio common carrier (RCC), channel band B is assigned to the WCC (local telephone company).

Due to a shortage of channels, the FCC recently (1989) increased the original 40-MHz frequency allocation to 50 MHz. To the original spectrum A, spectra designated as A' and A" were added, and spectra B' and B" were added to B. Both wireline and

Table 1.3
Conventional Mobile Radio Versus Cellular Mobile Radio

	Conventional Mobile Radio	*Cellular Mobile Radio*
Transmitter type	Single land-based transmitter at high elevation so that received levels at the mobile are substantially above ambient noise	Multiple transmitter in a cellular system; each transmitter located at cell site; system operation limited by cochannel interference.
Transmitter power	High-power, 200–250W.	Low-power, limited to 100W per channel by the FCC
Area of coverage	20–25 miles without interference.	10–30 miles
Total area	300 to several thousand sq. miles	75–700 sq. miles
Interference	Significant beyond 20 miles if 2 transmitters are within 60–100 miles of each other	Cochannel interference can be minimized by proper frequency planning and physical layout
Number of channels	Small number of fixed-frequency channels (e.g., the 12-channel system in New York City); thus, when a system is fully loaded growth is not possible	832 divided equally between 2 non-overlapping 25-MHz bandwidths in U.S.; original allocation: 666 channels over 2 nonoverlapping 20-MHz bandwidths

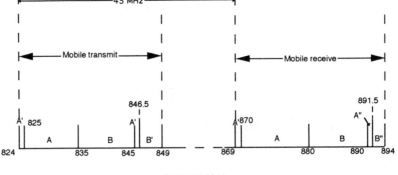

Figure 1.2 Cellular frequency allocation.

nonwireline service providers have been allocated an additional 5 MHz of spectrum, divided equally between transmit and receive bands. Thus with new spectrum allocation, mobile units transmit in the 824–849 MHz band and receive in the 869–894 MHz band, so that each full-duplex channel has a 45-MHz frequency separation. The original 40-MHz spectrum contains a total of 666 full-duplex channels to which an additional 166 full-duplex channels have been added. The total number of channels, 832, is divided equally (416 channels each) between A (RCC) and B (WCC) service providers. Here we provide an example based on the original spectrum allocation of 1974 (40 MHz). In this case, a mobile unit transmits in the 825–845 MHz band and receives in the 870–890 band. The structure of the radio frequency plan is shown in Table 1.4. The table represents 312 full-duplex voice channels, represented by numbers 1 through 312, and 21 control channels (S1 through S21). Channel 1 represents the lowest available base/mobile transmit frequency pair (870.030/825.030 MHz), and the highest number, 312, corresponds to the highest frequency pair (879.360/834.36 MHz). The base/mobile frequency pair denotes the frequency combination pair (full duplex) from cell site to mobile and from mobile to cell site. The 21 control channels are designated S1 through S21, with S1 corresponding to the lowest available base/mobile frequency pair (879.390/834.390 MHz) and S21 corresponding to the highest pair (879.990/834.990 MHz). As a standard designation, S21 is regarded as the first control channel and S1 is regarded as the last control channel. This is discussed further in Chapter 2. From Table 1.4, the voice channels are organized into 24 groups, with each group having 13 full-duplex channels. This division is used for a basic four-cell configuration. This will be further discussed in Chapters 2 and 4. Frequencies within the group are separated by 720 kHz (24 × 30 kHz). An individual cell can have a partial group assignment, a full group assignment with all thirteen channels, or multiple group assignments, based on the traffic demands. A partial group assignment is made in the beginning of a cell layout, especially if the anticipated amount of traffic over the foreseeable time period does not warrant implementation of all thirteen channels.

Unlike the United States, some European and Arab countries like Norway, Sweden, Saudi Arabia, and Oman have decided to use the 400–500 MHz frequency band for their cellular systems. Some Scandinavian countries use both the 450-MHz and 900-MHz bands. We shall cover different systems of the world in Chapter 4.

The motivation behind the choice of 900 MHz is the availability of proven technology in the UHF band and its ability to penetrate buildings. Furthermore, even in a high electrical noise area, this band is less affected than lower frequencies [3]. Figure 1.3 shows that manmade noise power is about 20 dB smaller at 900 MHz when compared to the noise power at the 40-MHz range. Long-range interference caused by ionospheric changes or by temperature follows an inverse law, with lower frequencies being severely affected and higher frequencies being generally immune to these effects. The antenna size loading effects are considerably reduced because smaller antennas are required at higher frequencies. Also, at this frequency it is possible to make mobile antennas less than a foot in length. Despite these advantages, there are disadvantages in the use of this frequency band for rural areas. For densely cultivated areas with thick vegetation, the attenuation will change considerably with the seasons. In general the direct loss will increase as the obstructions become saturated with moisture or when foliage becomes thick. Obstructions, such as mountains or rain-soaked buildings, often provide effective reflecting areas. Although they affect the direct radio path adversely, the reflected path is often improved. The effect of rain and other atmospherics at this frequency is almost insignificant.

Table 1.4
Cellular Radio Frequency Plan

Voice													Channel Set Designation												
	A1	B1	C1	D1	A2	B2	C2	D2	A3	B3	C3	D3	A4	B4	C4	D4	A5	B5	C5	D5	A6	B6	C6	D6
1	1	2	3	4	5	6	7	8	9	10	11	12	13	14	15	16	17	18	19	20	21	22	23	24
2	25	26	27	28	29	30	31	32	33	34	35	36	37	38	39	40	41	42	43	44	45	46	47	48
3	49	50	51	52	53	54	55	56	57	58	59	60	61	62	63	64	65	66	67	68	69	70	71	72
4	73	74	75	76	77	78	79	80	81	82	83	84	85	86	87	88	89	90	91	92	93	94	95	96
5	97	98	99	100	101	102	103	104	105	106	107	108	109	110	111	112	113	114	115	116	117	118	119	120
6	121	122	123	124	125	126	127	128	129	130	131	132	133	134	135	136	137	138	139	140	141	142	143	144
7	145	146	147	148	149	150	151	152	153	154	155	156	157	158	159	160	161	162	163	164	165	166	167	168
8	169	170	171	172	173	174	175	176	177	178	179	180	181	182	183	184	185	186	187	188	189	190	191	192
9	193	194	195	196	197	198	199	200	201	202	203	204	205	206	207	208	209	210	211	212	213	214	215	216
10	217	218	219	220	221	222	223	224	225	226	227	228	229	230	231	232	233	234	235	236	237	238	239	240
11	241	242	243	244	245	246	247	248	249	250	251	252	253	254	255	256	257	258	259	260	261	262	263	264
12	265	266	267	268	269	270	271	272	273	274	275	276	277	278	279	280	281	282	283	284	285	286	287	288
13	289	290	291	292	293	294	295	296	297	298	299	300	301	302	303	304	305	306	307	308	309	310	311	312
Signaling	S1	S2	S3	S4	S5	S6	S7	S8	S9	S10	S11	S12	S13	S14	S15	S16	S17	S18	S19	S20	S21	X	X	X

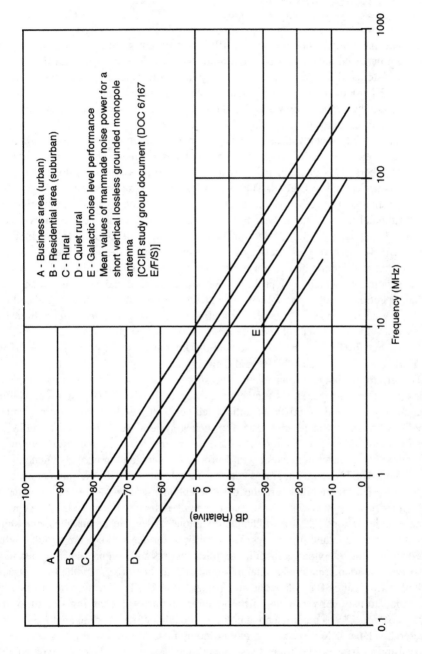

Figure 1.3 Mean value of man-made noise power in mobile surroundings. After [3].

1.6 DIGITAL CELLULAR RADIO

Cellular radio has been in operation since 1981 in different parts of the world, and the demand has increased steadily in all the countries where it is in use. With the present growth rate systems will soon reach their capacity in both Europe and in the United States. In the United Kingdom and in Scandinavian cities, planners of cellular radio are frequently forced to revise the capacity upwards. By the end of this century some cities will need 5–10 times more capacity than existing systems can provide, even when these systems have been developed to their fullest capacity. The market is growing by leaps and bounds in North America. The growth pattern indicates that soon there will be more than 10 million subscribers. This phenomenal growth is creating pressure on radio spectrum in both North America and in Europe. The exhaustion of the radio spectrum in large cities is causing problems. The result is a higher blocking rate, dropped calls, and interference that is detrimental to the cellular systems. These problems can only be solved by a new generation of cellular systems. According to the Cellular Telecommunication Industry Association (CTIA) survey, the new system should last until the year 2000. It is also envisioned that this second-generation digital technology should readily evolve towards the third generation, which will probably be based on a spread spectrum technique. Immediate spectrum relief is possible by introducing digital cellular radio. Digital cellular will also provide an opportunity to be competitive with new innovations and will enable the service providers to offer enhanced telecommunication, such as (PCS). See [5–8].

The special group on mobile communications known as group special mobile (GSM) of the European Conference of Post and Telecommunications (CEPT) was created in 1982. This group was made responsible for specifying a harmonized mobile radio communications system for Europe. This was essential in view of the diversified systems prevailing in different countries and to satisfy the anticipated needs of the 1990s and future decades. After years of work, basic requirements for such a system were specified by GSM in Docket 73 of 1985. The specification calls for the implementation of an entirely automatic mobile radio communication network with European coverage with public switched telephone network (PSTN) and integrated service digital network (ISDN) interconnection. The system will offer mobile users the basic telephone service plus a set (as wide as possible) of nonvoice services. These services will not only be provided to vehicle-mounted mobile stations but also to handheld stations, at least in urban areas, and ships in coastal waters and inland waterways. The quality of service will be comparable to that provided by existing telephone companies on toll networks and will be secure. The system will also conform to international Signaling Standard Number 7 (SS7) by the International Consultative Committee for Telephone and Telegraph (CCITT). The system will be able to accommodate national variations of the member countries in charging structures and rates. The cost aspect was recommended not to exceed that of the existing analog systems, and preferably have a lower cost for mobile equipment. While considering the above objectives and cost, a digital time division multiple access (TDMA) has evolved that became operational in a few places towards the end of 1991. In summary, the technical requirements for GSM system are:

- A base-transmit spectrum of 890–915 MHz;
- A mobile-transmit spectrum of 935–960 MHz;
- The same mobile station throughout Europe;
- The ability of subscribers to access mobile both from ISDN and from PSTN;
- The provision of voice privacy and encryption;
- International roaming capability from member countries.

In order to achieve international compatibility, standardization of the interfaces is necessary. In this context interface 3 and interface 5, shown in Figure 1.4, are important. Interface 3 is important because it partitions the function associated with networking and switching (function performed in the mobile switching center MSC) with those related to the "radio aspects" and executed in the base station system (BSS). In particular, MSC in association with the visitor location register (VLR) and home location register (HLR), executes terrestrial channel management, internetworking between MSCs, call control, and encryption of user data and signaling. The BSS is associated with the radio channel management, which includes administration of radio channel configurations, scheduling of messages on broadcast channels, power control, and so forth. Standardization of these interfaces are also necessary to account for too many variations in PSTN and *public land mobile networks* (PLMN) in different countries. These variations will, however, be reduced as interface 5 is standardized to CCITT SS7. We shall discuss details of these interfaces in Chapters 7 and 8. One may expect in the long run a defacto standard for this interface between PLMN and PSTN or ISDN. Service will be provided throughout the member countries from south of Italy to north of Norway. The member countries include Austria, Belgium, Denmark, Finland, France, Ireland, Italy, Luxembourg, the Netherlands, Norway, Portugal, Spain, Sweden, Switzerland, the United Kingdom, and Germany.

In the United States, the Telecommunication Industry Association (TIA) in association with the Electronic Industries Association (EIA) has recently come up with dual-mode mobile-station/base-station compatibility standard (IS-54, December 1989). The proposed specification calls for a dual-mode mobile station that will be allowed to access the base station whether it is an analog or a digital cellular system subscriber. Unlike the objectives of compatibility with ISDN and international roaming, the objective in the United States is to increase the system capacity by about tenfold over present analog systems. This capacity problem is dealt with in Section 6.4.2. The basic technical requirements of the American system are:

- A base-transmit spectrum of 824–849 MHz;
- A base-receiver spectrum of 869–894 MHz;
- Co-existence with present analog cellular radio;
- The provision of secure access and channel encryption;
- Seamless roaming capability, which will enable the subscribers to access different features of the system in the same way throughout the country.

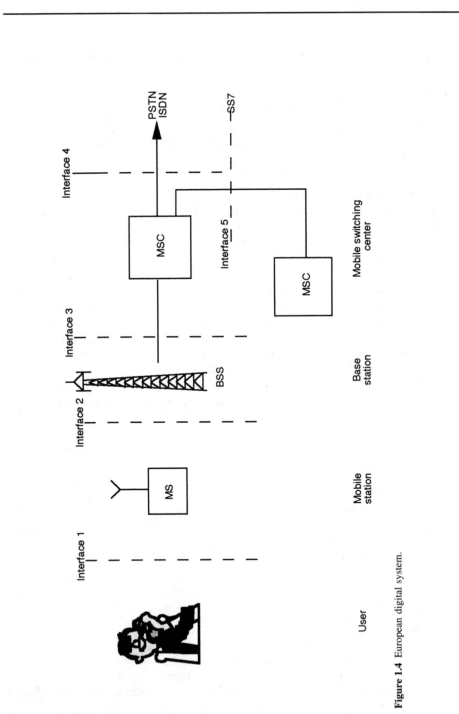

Figure 1.4 European digital system.

North America's EIA/digital system has been in operation in some part of the country since late 1991.

1.7 ORGANIZATION OF THIS BOOK

This book is based on the details found in Figures 1.5(a,b) and their modifications based on digital technologies. As seen from these figures, the essential elements of a cellular system consist of: cell sites and the number of cells for a typical geographical coverage, the number of channels and the corresponding traffic between cell sites and mobile users, the transmit and receive system at the cell site at MTSO and at the mobile user's vehicle, and the interconnection between cellular and wireline networks. Therefore, we discuss the following topics, arranged in different chapters as described below.

Chapter 2 gives the functional description of the analog cellular system and applies to both Figures 1.5(a) and 1.5(b). Applications for both metropolitan and rural areas of the country are discussed. Chapter 3 outlines the techniques for arriving at the traffic requirement of a cellular system and computing the number of channels required for different cells. This chapter also discusses some optimum channel assignment techniques. Chapter 4 discusses the different analog systems of the world and compares their characteristics. Chapter 5 outlines the problem of analog cellular signal coverage and provides solutions using low-power amplifiers, known as cell enhancers. Chapter 6 establishes the need for digital cellular radio through spectral efficiency. Chapter 7 is devoted to forthcoming digital cellular radio in Europe, the United States, and Japan. Chapter 8 deals with how the cellular radio system can be tied to a nationwide network. Chapter 9 provides practical digital microwave schemes used for transferring information between cell sites and MTSO and between MTSO and the local telephone exchange. Readers interested in analog cellular radio should focus on Chapters 2–5. Those interested in digital cellular

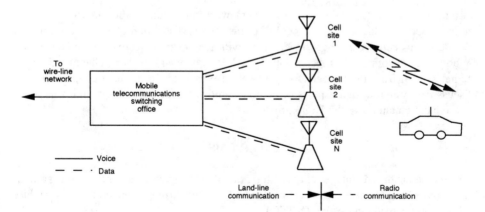

Figure 1.5(a) Elements of cellular radio.

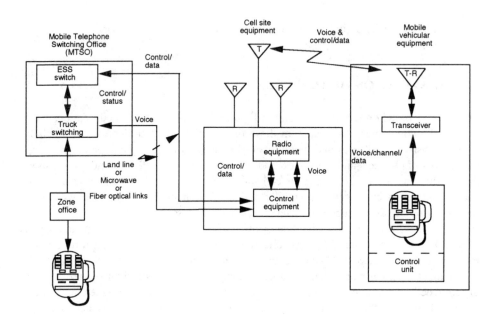

Figure 1.5(b) Typical cellular mobile radio system.

radios should cover Chapters 6–8. Chapter 9 is based on the use of microwave in transmitting digitized speech and data between MTSO and local telephone exchange and can be considered as common for both analog and digital cellular systems.

1.8 CONCLUSIONS

In this chapter we have laid down the groundwork on cellular radio on which we will build subsequent chapters of this book. To this regard, distinction between conventional mobile versus cellular has been made along with the design objectives of cellular radio. A brief introduction of European, Japanese, and American efforts in changing over to digital system from its present analog has been presented. In the next chapter we will discuss the system design aspects of cellular radio for both urban cities and rural areas of a typical country like the United States.

PROBLEMS

1.1 Enumerate the advantages of the cellular radio system over a regular telephone system.

1.2 Why is MTSO necessary in cellular radio? List the different functions of the mobile telephone switching office (MTSO)

1.3 State the major design objectives of a cellular system.

1.4 Can a car phone be regarded as a safety device?

1.5 Table 1.4 shows center frequency separation of 720 kHz between two channels assigned to the same cell. Can this separation be reduced indefinitely?

1.6 Why do you think 900 MHz is a good choice of frequency band for a cellular system? State any potential problem you see in this operating band.

REFERENCES

[1] Young, W. R. "Advanced Mobile Phone Service: Introduction, Background, and Objectives," *The Bell System Technical Journal,* Vol. 58, No. 1, 1979, pp. 1–14.

[2] Macdonald, V. H. "Advance Mobile Phone Service: The Cellular Concept," *The Bell System Technical Journal,* Vol. 58, No. 1, 1979, pp. 15–41.

[3] Pannell, W. M., and C. Eng. *Frequency Engineering in Mobile Radio Bands,* Cambridge, England, Granta Technical Editions, in association with Pye Telecommunications, Ltd., 1979.

[4] John Oetting, "Cellular Mobile Radio an emerging Technology," *IEEE Communication Magazine,* Vol. 21, No. 8, November, 1983.

[5] Chan, H., and C. Vinodrai. "The Transition to Digital Cellular," *IEEE Vehicular Technology Conference,* 1990.

[6] Hoff, Joran. "Mobile Telephony in the Next Decade," *IEEE Vehicular Technology Conference,* 1987, pp. 157–159.

[7] Uddenfeldt, J, and Bengt Persson. "A Narrowband TDMA System for a New Generation of Cellular Radio," *IEEE Vehicular Technology Conference,* 1987, pp. 286–292.

[8] *Cellular Business,* October 1991.

[9] ND-52535, *NACTS-NEC'S Advanced Cellular Telephone System Technical,* Hawthome, CA., NEC America, Inc.

[10] OKI Advanced Communications, *Technical Specification Cellular Mobile Telephone Equipment,* T-310113 issue 3, Hackensack, NJ, March 1983.

Chapter 2
Layout and Functional Description
of the Analog Cellular System

2.1 INTRODUCTION

All cellular radio telephone systems are born of the same parent: the Bell System in the United States. While still retaining the elements that make them true cellular systems, each of the systems developed in different countries of the world have their unique characteristics. The public acceptance of cellular radio to date bears out the wisdom of the FCC, which anticipated the need for a reliable, high-quality, near-non-blocking mobile radio telephone system. Until the advent of cellular radio, mobile radio telephone service in most urban areas was a disaster; there were few channels to serve many subscribers, which resulted in blockages greater than 70%. In metropolitan areas, it was not uncommon for subscribers to wait more than an hour to find an ideal channel, particularly during peak hours. The lack of available spectrum, rather than the type of system (IMTS or cellular), was responsible for the service inefficiencies of mobile radio telephone systems. The assignment by the FCC of 50 MHz of spectrum in the 800-MHz band for cellular mobile radio telephone use was the ideal solution to the problems that existed.

As valuable as cellular radio is to mobile radio, its true value in the future will undoubtedly lie in its viability as an alternative to the local loop distribution system of the PSTN, the wireline network put in place by Bell Telephone over the years. This certainly holds true for suburban and rural areas, where it is still as costly to install a local loop as it is to provide radio access to a cellular system. Some future advantages of radio (wireless) access are as follows: a right of way does not have to be acquired or maintained, no taxes are levied against a wireless path, fewer maintenance and repair personnel are needed to maintain the network, and, above all, the sending and receiving parties are free to move about. The network is less susceptible to damage from storms and other service interruptions. Also, a new subscriber can be put on the system simply by buying the equipment and registering with a system operator because the cellular system does not require a capital-intensive connecting medium. While cellular radio telephone systems have been designed to provide mobile and portable service, its use for stationary service would be more reliable and less expensive. With fixed-base service, directional antennas can be utilized from the user's premises to the cell site. Since the user is stationary, as is the cell site, fading and interference problems will almost disappear. A cellular system can serve both mobile and portable users from a fixed base simultaneously. A mature system taking advantage of conformal antenna pattern could be designed to provide service along roads and highways using RF relay stations, which are spaced as necessary. These relay stations could easily be configured to serve as "telephone booths" or as emergency telephones, thus providing a public service while extending the system coverage area.

The conservation of the radio spectrum cellular plan relies on distributing the low-power transceiver throughout the mobile service area (MSA) and efficiently reusing the allocated frequencies within this area. A given metropolitan area is divided into a number of subcoverage areas or "cells," and each cell is assigned a fraction of the total available radio channels. The channels used in one cell can be reused in spatially separated cells if they meet the specified cochannel interference criteria. The reuse of frequencies results in a high-capacity system, which requires sophisticated centralized control to coordinate the actions of the switching network, the cell sites, and the mobile units. This coordination is accomplished by data channels established between the central control (MTSO) and the base station (cell site), and between the base station (cell site) and the mobile unit. Ideally, the performance of data link between the central control and the base station, and between base station and mobile unit, should be matched in performance capacity so that neither represents a weak or over-designed part in the cellular system.

In view of the above discussion, in Section 2.2 we will deal with the initial cellular layout and various stages of cell splitting. The advantages of a regular cell structure are emphasized. The frequency reuse ratio, D/R, is computed and a scheme for the voice channel assignment is discussed. Underlying considerations for maximum and minimum cell radius are considered. In Section 2.3, fundamental concepts of the cellular system are discussed, including the mobile numbering system, RF power control, and the geographical location of cellular users. The cellular calling sequence, including mobile-originated and mobile-terminated calls, are described. Handover, which is a process of reassigning channels as the mobile crosses the boundaries of a cell, is also explained. Lastly, the call establishing and disconnecting sequences are noted. Section 2.4 provides engineering aspects of MSA and RSA. The RSA discussion focuses on the three schemes for RSA layout. Section 2.5 provides concluding remarks.

2.2 CELLULAR LAYOUT

Early on system designers of mobile radio recognized that for successful operation of a mobile telephone service, with channel availability comparable to regular landline telephone service, a substantial block of radio frequency spectrum equivalent to hundreds of voice channel would be mandatory. As stated earlier, this spectrum was provided by the FCC's allocation of a portion of the UHF television band for mobile service in Docket 18262. The system designer's concept of frequency reuse provided the roadway to a theoretically unlimited number of channels. Frequency reuse is just what the term implies: the multiple use of a specific frequency or set of frequencies within the confines of a specific system [1–12]. In practice, the 50 MHz of spectrum assigned by the FCC for cellular use is divided into two equal parts. One part is assigned to wireline operation (or WCC) in a given area, and the other part is assigned to a non-wireline (or RCC) in the same area. Thus, the bandwidth of 50 MHz has been divided into a total of 832 full-duplex channels. Of the 832 channels, the first 416 channels are assigned to nonwireline and the remaining 416 full-duplex channels are assigned to the wireline. Spectrum denoted by A, A', and A″ are assigned to RCC, and those denoted by B and B' are assigned to WCC. The detailed assignment for the first 666 channels are shown in Figure 2.1. Among these 832 channels there are 42 control channels; the remaining 790 are voice channels.

All cellular calls are initiated on control or setup channels. Each cell, or coverage area, is assigned one of the 21 setup channels (21 control channels are assigned to each

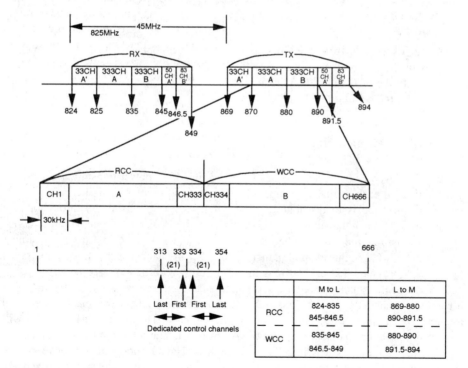

Figure 2.1 Frequency allocation plan.

RCC and WCC). Each voice channel consists of two links, called "forward" and "reverse" links. The reverse link (from mobile user to cell site) operates in the lower 800-MHz band at a carrier frequency that is exactly 45-MHz lower than that of the forward link (from cell site to mobile). The two links each require a bandwidth of 30 kHz and operate simultaneously during a telephone call. Thus, a single-duplex cellular channel requires a total bandwidth of 60 kHz. The channel number 333 is the first command channel in the RCC group and channel 313 is the last. In the WCC group, the first and last command channels are 334 and 354, respectively.

As stated in Chapter 1, initial allocation of bandwidth for cellular was only 40 MHz, which was recently increased by 10 MHz. Each service provider is seen to have received 2.5 MHz of additional spectrum in both the transmit and receive bands. The new spectrum for an RCC service provider is in two bands: one at 824–825 MHz and the second at 845–846.5 MHz. This additional 2.5 MHz goes with the former allocation of 825–835 MHz. The new receive spectrum for a WCC provider is located at 846.5–849 MHz. This is the add-on to the former allocation at 835–845 MHz. Since each service provider has a piece of the spectrum that is completely surrounded by the other service provider's band, additional spectrum may cause some filtering problems in the receiver. A common point where both the WCC and RCC networks meet is PSTN, shown in Figure 2.2(a). Here, a service area denoted by the boundary (city) is served by both the wireline common carrier (WCC) and by the radio common carrier (RCC). Figure 2.2(b) shows a typical service area served by five overlapping cells. Cells are connected to MTSO, which in turn ties to the PSTN at the local class 5 office or at the *tandem switch point* (TSP). According to the FCC rules and regulations, only 75% of the service area (CGSA) must be served by the cellular telephone system.

2.2.1 Cellular Configuration

To achieve coverage of an area, as in Figure 2.3, one high-power transmitter capable of transmitting on each available channel can be placed at point A. An alternative to this system architecture is to distribute a series of low-power transmitters throughout the service area [1]. Each transmitter would then serve a limited area or zone within a service area. If we assume the total number of available channels to be C and the total number of cells to be N (11 in this example), then the number of channels per cell is simply given by $S = C/N$ provided that the traffic is uniform throughout the coverage area.

Each of the zones in Figure 2.4 is called a cell, and the cell signifies the area that a particular transmitter serves. Cells labeled with different letters will each be assigned a unique set of channel frequencies to avoid interference. The system of Figure 2.4 will require nine sets of channels frequencies: A through I. The total number of channels is equal to the sum of channels in cells A_1 through I_1. The advantage of the above system is that through the reuse of the channels used in cells A_1 and D_1 and cells A_2 and D_2, more telephone calls can be processed in a given area at the same time. If 90 channels

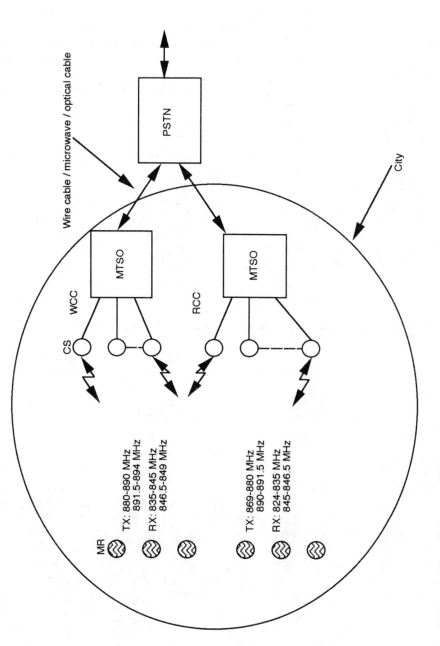

Figure 2.2(a) A typical area served by both WCC and RCC.

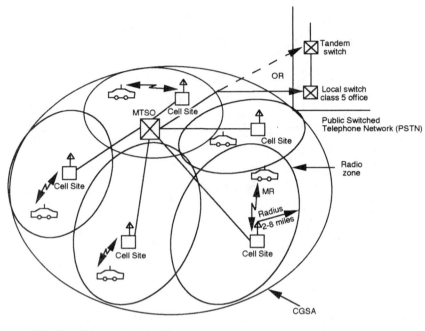

MTSO: Mobile Telephone Switching Office
MR: Mobile Radiotelephone
CGSA: Cellular Geograpical Service Area

Figure 2.2(b) Typical service area. After [13, p. 2-2].

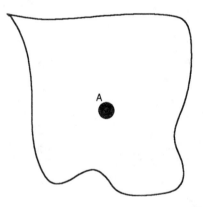

Figure 2.3 Basic coverage area.

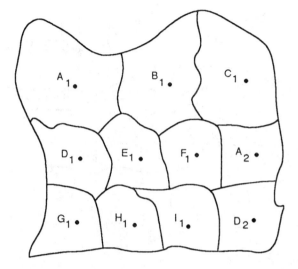

Figure 2.4 Cellular layout.

were available, the system shown in Figure 2.3 could process 90 simultaneous calls. The system shown in Figure 2.4, by reusing 20 of the 90 available channels, could process 110 simultaneous calls. The multiplier by which the system capacity for simultaneous calls exceeds the number of allocated channels depends on several factors, particularly on the total number of cells.

The reader may recall that the frequency-reuse concept, as shown for cells designated by A and D above, is not new. It is prevalent in the entertainment industry and most other radio services. It should also be noted that the radio transmitters are not equally spaced; nor are the coverage areas, shown by the boundaries of the designated area, of equal size.

2.2.2 Cell Splitting

As the traffic within a cell increases toward the point where service quality is affected, the cell can be split into smaller cells. If this is not done, "blockage" will increase. Blockage occurs when a user attempts to make a call and the system is so loaded that the call cannot be completed. A measure of telephone system performance is the amount of blockage that occurs within that system. To prevent blockage of the system from exceeding the wired telephone network, cell splitting is used. Figure 2.5(a) illustrates an early stage of cell splitting [1]. As seen in this figure, only one high-density cell is split, thus making the coexistence of smaller cells with larger cells a requirement. This also allows a lower demand area to be served by larger cells, and the higher demand areas to be served by smaller cells.

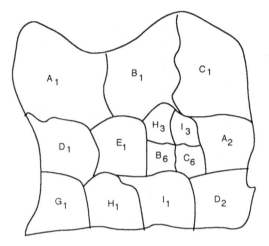

Figure 2.5(a) Cell splitting, early stage. After [1, p. 18].

In this early stage, as traffic grows within a cell a condition is reached where it is desirable to revise the cell boundaries in order to handle more traffic. So what was a single cell is now divided into a number of cells, but all within the original cell boundary. It is assumed, for example, that the cell designated as F_1 in Figure 2.4 has reached capacity. To increase traffic handling capacity within the original F_1 boundary, the cell is split into four cells, H_3, I_3, B_6, and C_6. As demand continues to grow, the original coverage area may ultimately be split into small cells, as shown in Figure 2.5(b), and the original 11 large cells will mature to 44 small cells with multiple reuse of frequency channels. This technique of frequency reuse and cell splitting makes the cellular system unique and makes it possible to meet the important objectives of serving a large number of customers in a small coverage area using a small spectrum allocation [1–10]. Additionally, cell splitting makes it possible to meet the objective of matching the density of available channels to the spatial density of demand for channels. Thus, if the bandwidth per channel is BW and if the channel is repeated k times, the bandwidth per voice path can be regarded to be BW/k (actual bandwidth per channel is BW). In another words, there are k simultaneous conversations over a single channel bandwidth of BW Hz.

2.2.3 Properties of Cellular Geometry

The main purpose of cells in a portable radio telephone system is to define an area in which either specific channels or specific cell sites will be used preferentially, if not exclusively. After designing a desired cell pattern, which includes space separation to prevent cochannel interference, the proper positioning of cell site equipment and the proper selection of equipment to service each call is necessary in order to realize the advantages of the designed cell pattern.

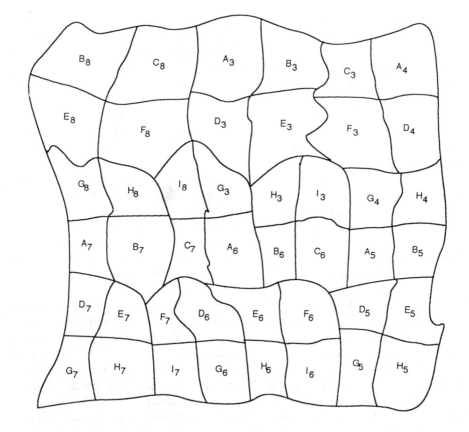

Figure 2.5(b) Cell splitting, late stage. After [1, p. 18].

To achieve this end and make orderly planning for traffic growth possible, a geometri-cally patterned structure is used rather than the irregular amorphous structure shown in Figures 2.4 and 2.5. Irregular cell structure and irregular placing of the transmitter might be acceptable in a system where the initial system configuration, including selection of transmitter sites and channels assignments, is frozen for the future, but the cellular system requires constant upgrades. As the traffic grows, new cells and channels need to be added. If an irregular cellular structure is adopted, it would lead to an inefficient use of spectrum due to an inability to reuse frequencies because of cochannel interference. It would also result in uneconomical deployment of equipment, requiring the relocation of equipment from one cell site to another. Thus a great deal of system engineering would be required to readjust the transmission, switching, and control resources each time the system went through its developmental phase. These difficulties lead to visualizing the cell as a regular structure. If, as with present mobile service, omnidirectional transmitting antennas were used, then each site's coverage contour of constant signal level would be circular, provided that the propagation did not change along different radials of the cell site. Although a circle is the recommended cell shape, theoretical transmission considerations suggest the circle as an impractical design because it provides ambiguous areas with either multiple or no coverage. To assure complete area coverage with no dead spots, a series of regular polygons can be adopted in the design of the cellular system. Since regular polygons, such as an equilateral triangle, a square, and a hexagon, remove the problems of multiple coverage and dead spots, any one of the three, shown in Figure 2.6, can be adopted for cell design. These regular polygon structures also make it easy to see where one cell ends and another begins.

For economic reasons, the hexagon has been chosen. To understand the motivation behind this choice, consider the worst case: points in a cellular grid—the point farthest from the cell site. Assuming the cell center to be the excitation point, vertex points are the worst case since they lie at a maximum distance from the center of the cell. Restricting the distance from the center to the vertex to R units for satisfactory transmission quality (to be discussed in Section 2.2.5), it can be easily shown that the hexagon has the maximum area coverage. As shown in Figure 2.7, an additional 30–100 percent of area is covered by a hexagon compared to a square and an equilateral triangle.

Consequently, a hexagonal layout requires fewer cells, and therefore fewer transmitter sites. Thus, other considerations remaining the same, a system based on a hexagonal structure costs less than triangular or square cells. It should be noted that the hexagon structure is just for analytical and theoretical design purposes. In real life, the hexagon will ideally look like a circle or a distorted coverage pattern, as shown in Figure 2.8. Figure 2.8 is the cell structure used by Nordic mobile telephone system in Stockholm. As seen from this figure, the central area of coverage is divided into six sectors to match increased traffic. Sectorization is achieved by using directional antennas at cell sites. Let us now investigate the geometrical properties of cellular structure using the hexagon as the basic building block.

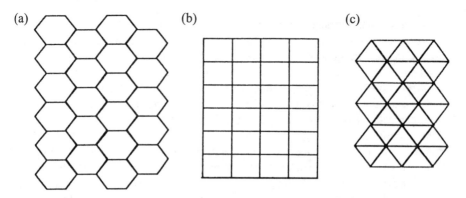

Figure 2.6 Regular polygons as cells: (a) regular hexagons; (b) squares; and (c) equilateral triangles.

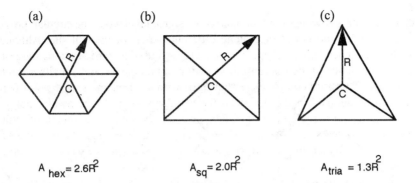

(a)   (b)   (c)

$A_{hex} = 2.6R^2$ $A_{sq} = 2.0R^2$ $A_{tria} = 1.3R^2$

Figure 2.7 Candidate regular polygons: (a) regular hexagon; (b) square; and (c) equilaterial triangle.

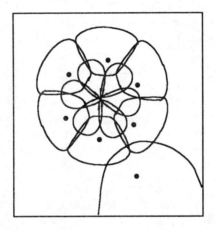

Figure 2.8 Cellular coverage area in Stockholm.

2.2.4 Hexagonal Cellular Geometry

As shown in Figure 2.9, the most convenient coordinate system for a hexagonal cellular structure are axes inclined at a 60-deg angle. If two points have the coordinates (u_2, v_2) and (u_1, v_1) then the distance between them is

$$D = \sqrt{(u_2 - u_1)^2 + (v_2 - v_1)^2 + (u_2 - u_1)(v_2 - v_1)} \qquad (2.1)$$

Assuming $(u_1, v_1) = (0, 0)$, or the origin to be the center, and restricting (u_2, v_2) to integer values (i, j), one obtains

$$D = \sqrt{(i + j)^2 - ij} \qquad (2.2)$$

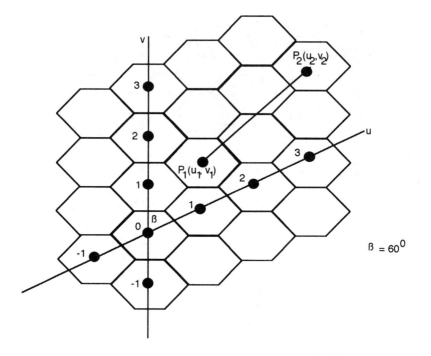

Figure 2.9 Coordinate system.

From (2.2), the normalized distance between two adjacent cell sites is unity (($i = 1$, $j = 0$), or ($i = 0, j = 1$)). Of course, the actual center-to-center distance of the adjacent hexagon is $\sqrt{3}R = (2R \cos 30°)$, where R is the center-to-vertex distance. The concept of the number of cells per cluster is important for locating the cochannel cell within the cellular structure [3]. To define this, consider the cellular structure shown in Figure 2.10. Cells designated by the letter A are the six nearest cochannel cells of the center cell A. It can be seen that these cells are located at vertices of the larger hexagonal cell of radius D.

Vectors from the center to different peripheral cells A subtend an angle of $n60°$ with respect to each other, where n assumes the values of $1, 2, \ldots, 6$. From the law of cosines, the radius of the large cell D (cochannel separation) is given by:

$$D^2 = 3R^2(i^2 + j^2 + ij) \qquad (2.3)$$

Since the area of a hexagon is proportional to the square of its radius, the area of the large hexagon is

$$A_{\text{large}} = k(3R^2)(i^2 + j^2 + ij) \qquad (2.4)$$

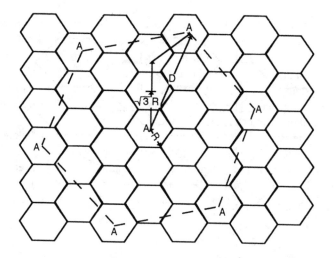

Figure 2.10 Determination of number of cells per cluster.

Similarly, the area of the small hexagon is

$$A_{small} = k(R^2) \tag{2.5}$$

Therefore,

$$A_{large}/A_{small} = 3(i^2 + j^2 + ij) \tag{2.6}$$

From the symmetry of Figure 2.10, one can see that the large hexagon encloses the center cluster of N cells ($N = 7$ in this case, center cell surrounded by six other cells) plus one-third the number of cells associated with six other peripheral hexagons. Thus, the total number of cells enclosed is equal to $3N$. Since the enclosed area is proportional to the number of cells, $A_{large} = 3N$, and $A_{small} = 1$. Thus,

$$N = (i^2 + j^2 + ij) \tag{2.7}$$

Combining (2.3) and (2.7), one obtains

$$D/R = \sqrt{3N} \tag{2.8}$$

The equation is important for estimating cochannel interference. The above D/R ratio is known as the cochannel reuse ratio. The channel reuse ratio D/R is 4.6 for $N = 7$ and 7.9 for $N = 21$.

The above equation is important as it affects two aspects of a cellular system: the traffic carrying capacity of the system and the cochannel interference. By reducing the ratio, D/R, the number of cells per cluster is reduced. Assume the total number of RF channels has a constant value c. Then the number of channels per cell is increased; thereby increasing the system traffic capacity. On the other hand, the cochannel interference is increased as D/R is reduced. The reverse is seen as the ratio D/R is increased, that is, it decreases the cochannel interference, but the traffic carrying capacity is reduced also. Often a compromise is necessary in selecting this ratio, which would balance the effects of cochannel interference and the traffic capacity of the system. The cochannel reuse ratio, D/R, for the determination of interference computation for a seven-cell cluster is shown in Figure 2.11. For equal power transmission from the center cell, A_1, and the cochannel cell, A_2, the I/S ratio at the boundary of cell A_1 (worst case S/N ratio) is approximately given by:

$$I/S \approx 10 \log \left[\frac{D - R}{R} \right]^{-n} = 10 \log[\sqrt{3N} - 1]^{-n} \tag{2.9}$$

where R is the cell radius; D is the distance to the co-located channel; and n is the propagation decay law appropriate to the environment being studied, $2 \le n \le 5$ normally.

Since there can be as many as six cochannel cells corresponding to six edges of the center cell A_1, the worst case I/S ratio (first order of approximation) is given by:

$$I/S = 10 \log \left[\frac{D - R}{R} \right]^{-n} + 10 \log(6) \tag{2.10}$$

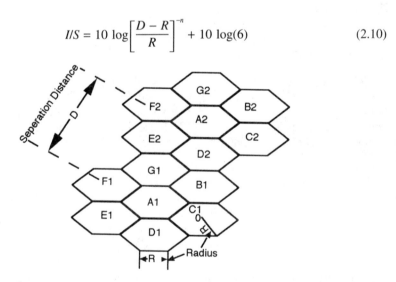

Figure 2.11 Mobile service area divided into cells. Seven cell repeat pattern ($n = 7$), $D/R = 4.6$. Each channel set is used twice with subscripts 1 and 2. For example, channel set A is used in cells A_1 and A_2.

It should be noted that in actuality the D/R ratio is $(D/R - 1)$ when using omnidirectional antennas at the cell site. This ratio is increased to $(D/R + 1)$ when using 60-deg sectored antennas at the cell site, thereby increasing the S/I ratio.

The location of a six cochannel cell can proceed as follows. Choose two integers i and j (i, j), called shift parameters, since from each hexagon, six other hexagons originate. Thus, start from the center of the cell and proceed towards any of the six edges. The cochannel cell can be located by moving i cells perpendicular to an edge before turning clockwise or counterclockwise and proceeding a distance of j. The roles of i and j can be reversed (i.e., instead of proceeding i cells first before turning, one can move j cells first and then turn clockwise or counterclockwise and go i cells). Similarly, six other cochannel cells can be located by choosing other sides of the hexagon. Counting the two choices for i and j and another two choices for clockwise or counterclockwise rotation, there are a total of four different ways one can locate the cochannel cells.

2.2.5 Reuse Partitioning in Cellular System

So far we have discussed the case where the cellular system is designed for a single D/R ratio. Here, the D/R ratio is chosen such that it meets the voice quality objective in a given propagation environment [3]. A subjective test conducted by Bell System showed that RF signal-to-noise, (S/N), ratio of 18 dB is considered adequate for good quality voice reproduction [1]. In this section we wish to show that by using two different D/R ratios the same performance objective can be met, while the system capacity can also be increased over the single frequency reuse case. The underlying principle behind reuse partitioning is to degrade the S/N ratio for mobile users who have more than adequate transmission quality, and to offer greater protection to users who require additional S/N ratio. The goal is to produce an overall S/N distribution that satisfies the system quality objectives and achieve a general increase in system capacity. It should be noted that the S/N ratio is actually the RF C/I ratio and not the baseband S/N ratio.

The reuse partitioning is implemented by dividing the total spectrum allocation into two or more groups of mutually exclusive channels. The omnidirectional case with two reuse factors $N_A = 3$ and $N_B = 7$ is shown in Figure 2.12. The channel assignment within the ith group is then determined by the reuse factor, N_i, for that group. A mobile with high S/N ratio is assigned a channel frequency from the lower reuse factor group, while those with poor S/N ratios are assigned frequencies from a group having a higher reuse factor. As the mobile travels, new channel frequencies are assigned based on the measured S/N ratio at the cell site. Typically, mobile units closer to a cell site will be served from channel groups having a low frequency reuse factor. Here the low-frequency reuse factor designates channels that have a lower geographical separation. On the other hand, channel assignment based on the high-frequency reuse factor group applies to a mobile away from the cell site. Thus one may visualize a coverage structure that resembles a pyramidal overlay of cells. The base of the structure is formed by a continuous pattern of hexagonal

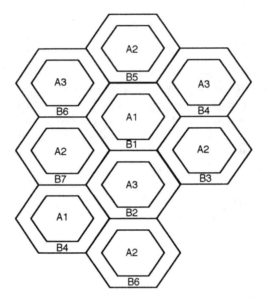

Figure 2.12 Cellular grid partitioned into two reuse groups with $N_A = 3$ and $N_B = 7$.

cells that are served from channels with a high-frequency reuse factor group. Overlaid layers are formed by a noncontiguous hexagonal group of cells that are served by channels from a lower reuse factor group. The basic concept of two reuse groups can be extended to multiple reuse groups with added complexity at cell sites.

Let the total number of channels, C, be divided into two groups; Cells with a reuse of N_A are assigned a fraction, p, of these channels, while the remaining channels are assigned to N_B. Thus, the total number of channels that can be assigned at any given cell site is

$$S = [p/N_A + (1 - p)/N_B]C \qquad (2.11)$$

Fractions of channels assigned to N_A are given by:

$$K = pN_B/[pN_B + N_A(1 - p)] \qquad (2.12)$$

While the fraction assigned to N_B is given by:

$$1 - K = N_A(1 - p)/[pN_B + N_A(1 - p)] \qquad (2.13)$$

The equivalent reuse factor, N_{eq}, for the system is obtained by dividing the total number of channels C by the number of channels per cells, or

$$N_{eq} = C/S = N_A N_B / [pN_B + N_A(1 - p)] \tag{2.14}$$
$$= KN_A + (1 - K)N_B$$

As discussed in the last section, a higher D/R ratio implies a high signal-to-interference ratio at the expense of lower capacity per cell site (less the number of channels per cell). The reuse partitioning technique provides an adjustable D/R ratio and also an adjustable capacity of the system. From (2.14), it can be shown that a higher value of N_{eq} means lower number of channels per cell (assumed uniform traffic), and thus a low capacity and high C/I ratio. On the other hand, a lower value of N_{eq} means a high number of channels per cell or a high capacity, and thus a low C/I ratio. The values of K and N_{eq} for different values of p in steps of 0.1 are shown in Table 2.1.

The reuse partitioning techniques discussed here increase the system's capacity and offer more protection towards cochannel interference to those users who need it most. Also, the scheme provides a near continuum of equivalent reuse factors providing a variable parameter, which can be adjusted during the operational phase. These advantages are achieved without much impact to the RF architecture of a cell site. As stated above, theoretical grouping of channels into different reuse groups can be carried out indefinitely. However, in practice we cannot go indefinitely as it would require frequent switching of channels to the mobile, thereby increasing the complexity of the system.

2.2.6 Cell Site Location

If we imagine a regular array of cell sites, as shown in Figure 2.13(a), then the assigned cells can be visualized in two ways. First, the cell site can be located at the center of the cell, as shown in Figure 2.13(b). In this case, frequencies allocated to the cell radiate

Table 2.1
Values of K and N_{eq} For Different Values
of the Reuse Pattern Fraction p

p	K	N_{eq}
0.0	0.0	7.0
0.1	0.205	6.18
0.2	0.368	5.52
0.3	0.5	5.0
0.4	0.61	4.6
0.5	0.70	4.2
0.6	0.80	3.88
0.7	0.844	3.62
0.8	0.903	3.4
0.9	0.95	3.2
1.0	1.0	3.0

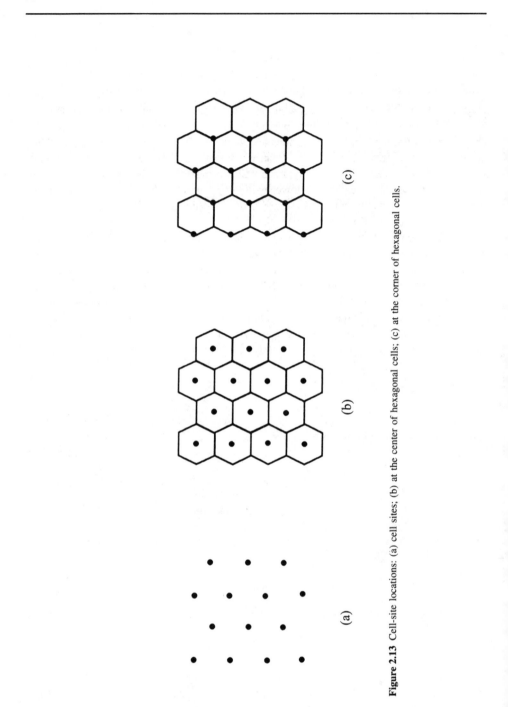

Figure 2.13 Cell-site locations: (a) cell sites; (b) at the center of hexagonal cells; (c) at the corner of hexagonal cells.

omnidirectionally. Second, the cell site can be located at alternative corners of a regular hexagon, as shown in Figure 2.13(c). In this case there are 120-deg directional antennas and the cell sites beam energy in three directions to provide spatial isolation between three cells. In order to reduce the startup cost of the system, omnidirectional antennas are usually used instead of directional antennas. As the system matures, a cost savings can be achieved by not increasing the number of cell sites; instead, omnidirectional antennas at old sites are replaced by directional antennas. There are two common types of directional antennas; either 120-deg beam antennas or 60-deg sectored antennas. For 60-deg sectored cells, the original cell-site locations are not moved. The omnidirectional antennas are just replaced by sectored antennas, as shown in Figures 2.14. Figure 2.14(a) is a 12-cell cluster using a total of 26 channels per cell (2 groups of channels from Table 1.4, assume the total number of voice channels to be 312). After cell sectorization, 13 channels per sector (1 group of channels from Table 1.4, N is assumed to be 4) can be assigned.

2.2.7 Salient System Parameter

Other than the cochannel interference ratio, D/R, three other important parameters of the cell are: cell-site position tolerance, minimum cell radius, and the maximum cell radius. We discuss these in the following sections.

2.2.7.1 Cell-Site Position Tolerance

Analysis reported in [1] allows the cell site location tolerance to be less than one-fourth of the cell radius. As also reported in [1], the cell-site position tolerance (ideal versus actual cell-site location) impacts the transmission quality adversely; for example, the received RF S/I ratio decreases gradually as the cell site position tolerance is increased from zero (ideal location) to about one-fourth of the cell's radius. Beyond this point the received signal level decreases rapidly. Therefore, the tolerance has been set to one-fourth of the cell's radius to provide latitude in the site location, which may be necessary due to practical limitations. Ideal and actual locations of the cell sites for corner cell sites are shown in Figure 2.15.

2.2.7.2 Maximum Cell Radius

Setting up the maximum radius of the cell within which satisfactory transmission quality is maintained without undue financial penalty is generally an objective during the initial layout of the cellular system. The maximum radius of a cellular cell is limited by the generated power at the cell site and at the mobile, their associated antenna gains, and the terrain contours. For the fixed antenna gain and the fixed propagation effects, the cell radius can be increased by transmitting more power. This technique can be successfully

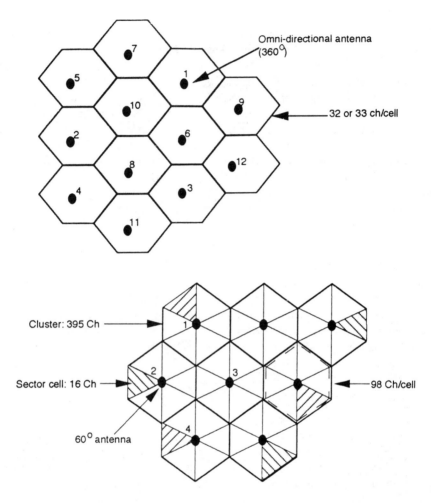

Figure 2.14 Omnidirectional antennas at different cell sites (cluster size $N = 12$); 60-deg sectored antenna at different cell sites (cluster size $N = 4$).

used to some extent and the size of the power amplifier (PA) can be increased. However, associated with increased sizing of the PA are problems of additional generated noise, cooling, and source power consumption. Obviously, in the case of mobile, dc power consumption cannot exceed a certain value. At the cell sites, as well as at the mobile station, high-power generation imposes special cooling considerations for the power amplifier (obviously, a limit on the maximum rate of cooling also exits). In addition, a high-power transmitter is not of much use once the traffic increases and the initial cells are divided into smaller cells (smaller cells require less power). On the other hand, a cell-site antenna gain increase provides the same benefit as an increase of power amplifier capacity. In a

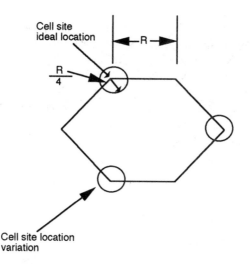

Cell site
ideal location

$\dfrac{R}{4}$

Cell site location
variation

Figure 2.15 Cell-site location tolerance.

cellular system, effective gain can be increased by increasing the antenna size or by increasing the antenna mast height. Compared to increase in PA capacity, achieving higher gain of the antenna by mounting the antenna high up on the mast is relatively cheap. The antenna height is normally limited to 100′ above the terrain of the ground, and the antenna gain is limited to about 6–10 dBi (gain in dB referred to as an isotropic gain of unity). Assuming that the power amplifier size, antenna gain, and height of the mast at the cell site are fixed, then the maximum value of the cell radius can be arrived at by the contour of desired *S/N* ratio. As stated earlier, the subjective tests have shown that a *S/N* ratio of 18 dB provides good quality voice. The AT&T AMP system requires that a *S/N* ratio of 18 dB should exceed 90 percent of the cellular coverage area. The simulation result conducted earlier by the Bell Telephone Company at Philadelphia shows that for a cell radius of about eight miles, the *S/N* ratio of 18 dB is exceeded over 90% of the covered area. Assuming that we start with the round number of an eight-mile cell radius, one can initially go through three stages of cell splitting to ultimately reach to a radius of one mile. After three stages of splitting, further division is not practical.

2.2.7.3 Minimum Cell Radius

As stated earlier, a cellular system goes through the process of cell division as the traffic demands in the cell increase. In most cases, cells carrying high traffic are split in the beginning, followed by gradual transformation of other larger cells into smaller cells. The cell division process is such that it divides the original radius of the cell into half. Thus, the area of the new cell is one-fourth of the original cell area. Since the traffic capacity is proportional to the number of new cells, each division increases the capacity by a factor of four. Thus the maximum traffic capacity of the system can be fixed by the ultimate size of the cell. Since each division increases the system complexity as well as the capacity, the cost per customer remains somewhat unchanged. Though the cell division does not impose additional cost per customer, this process cannot go on indefinitely because the smaller cell radius requires frequent handover from one cell site to another as the mobile moves around in the coverage area. This imposes additional hardware requirements in the MTSO. Combining the additional hardware requirements at MTSO with the stringent requirement of cell-site tolerance to about one-fourth its radius, the practical limit of a one-mile radius on cells has been imposed.

2.2.8 Channel Assignment

As shown in Chapter 1, a total of 416 full-duplex channels with 30-kHz channel spacing are available to both WCC and RCC. Of these 416 channels, only 395 are used for voice transmission. Using (2.7), the number of cells per cluster can be computed. Assuming an equal division of channels among cells, the number of channels per cell can be arrived at. If we choose a cluster size of twelve cells ($N = 12$), as shown in Figure 2.14(a), the cluster can have all 395 voice channels assigned and the number of channels per cell can at most be 32 (some cells can have 33 channels). The problem we now want to answer is how to assign channels to cells within a cluster.

If adjacent channels are assigned in the same cell, potential interference at both the cell site and the mobile exists. This interference can however be avoided by proper design of IF filtering stages at the receiver. This warrants using guardband, thereby reducing the total number of available channels. The problem of adjacent-channel interference for cells using omnidirectional antennas can be avoided by not assigning the adjacent channels in the same cell or in the adjacent cells. This restriction is not required for directional cell sites, since the adjacent channels can be assigned to different sectors beaming energy in different directions. One way to do this that provides maximum frequency separation between channels is to assign to the Kth cell K, $K + N$, $K + 2N$, ..., $K + nN$, where n is an integer such that $(312 - nN)/N < 1$. Thus, for our example where $N = 12$, for the first cell ($K = 1$), the assigned channels are 1, 13, 25, 37, 49, ..., 385. For ($K = 12$), assigned channels are 12, 24, 36, ... From these assignments it is seen that cell numbers 1 and 12 should never be neighbors (to avoid adjacent channel assignment). This scheme works fine for the center-exited cell. However, for corner-exited cells, transmission quality considerations calls for seven cells per cluster. It is impossible in this case to avoid having adjacent channels at the adjacent sites, because if there is a total of seven groups, then any site plus its six neighbors have all the 395 channels assigned. For assignment at the individual cell site, the assigned channel group for the cell can further be subdivided into three subgroups: n_1, n_2, and n_3. For the above example, the n_1 subgroup will contain channels 1, 37, 69, ... ; the n_2 subgroup will contain channels 13, 49, ... ; and the n_3 subgroup will contain channels 25, 61, ... An example based on omnidirectional and sectored antennas will now be discussed.

For $N = 12$, assign the channels in cells 1 through 12. We assume that the total number of voice channels available is 396, which leaves aside 20 channels for signaling. Now assign to the ith cell channels $i + 12K$, where $i = 1, 2, \ldots, 12$ and $K = 0, 1, 2, \ldots, N$ such that $i + 12K < 396$. Assuming uniform traffic in all cells; the number of channels/cell = 396/12 = 33. See Table 2.2.

The channel 1 transmit and receive frequencies from mobile are: 825.03 MHz and 870.03 MHz, respectively (channel 1 is the first channel of the old 666 channel set originally assigned by FCC). Layout of the system is shown in Figure 2.14(a). Note that adjacent cells are separated by a distance of at least a one-cell radius. As an example, the cells designated as 2 and 3 are separated by cell 8. Cells 1 and 12 are separated by a unit cell distance.

As a second example we shall discuss a channel assignment scheme based on reuse partitioning for directional antenna. Assume 120-deg directional antennas with $N_A = 3$ and $N_B = 9$. Channels assigned to group A are divided into nine subgroups A_1 through A_9; and named after the nine possible sectors of group A ($A_1, A_2, \ldots, A_9$). Channels assigned to group B are divided into 27 subgroups B_1 through B_{27}; named after the twenty-seven possible sectors of group B. All of the group A and B channels are distinct. For a given sector of A, there are three possible ways in which sector B will appear. This is shown in Figure 2.16. For example, sector A_1 appears either with B_1, B_{10}, or B_{19}. Thus, in general, the sector combination $A_k B_{k+9j}$ where $1 \leq k \leq 9$ and $j = 0, 1, 2$ will appear. Therefore, channels occurring with these combinations can only be assigned to a mobile in that sector.

2.2.9 Cell Splitting and Overlaid-Cell Concept

As pointed out earlier, cell splitting causes the radius of the new cell to be reduced by half, and thus reduces the area coverage to one-fourth of the original. The division also

Table 2.2
Cell Number and Channel
Assignment for $N = 12$

Cell Number	Channel Assignment
1	1, 13, 25, . . . , 385
2	2, 14, 26, . . . , 386
3	3, 15, 27, . . . , 387
4	4, 16, 28, . . . , 388
5	5, 17, 29, . . . , 389
6	6, 18, 30, . . . , 390
7	7, 19, 31, . . . , 391
8	8, 20, 32, . . . , 392
9	9, 21, 33, . . . , 393
10	10, 22, 34, . . . , 394
11	11, 23, 35, . . . , 395
12	12, 24, 36, . . . , 396

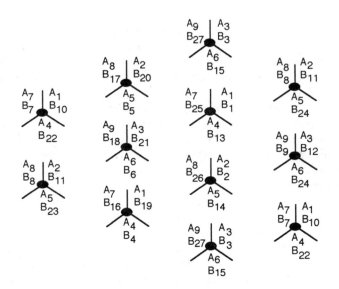

Figure 2.16 Directive channel assignment ($N_A = 3$, $N_B = 9$).

increases the cell capacity fourfold. As discussed earlier, the cell location can be anywhere within one-fourth the distance of the new cell radius.

As shown in Figure 2.17, six new sites have been established in the center. The new cell sites lie midway between two old sites, and the cochannel cell sites are little more than one (large) cell diameter away. Sites labeled 2 are circled to illustrate this relationship. Each split of the cell keeps the same geometrical relationship; however, each split causes the original cluster to rotate clockwise by 120 deg. The split requires cells of different diameters to coexist. In order to avoid cochannel interference, the *D/R* ratio has to have a certain minimum value. However, due to the presence of two different cells of different radii, this ratio is variable. For the configuration shown in Figure 2.18, the D/R ratio has a value of 4.6 for the small radius cell with respect to a small radius, and the large cell has a ratio of 4.6 with respect to a large radius.

It should be noted that the same channels are assigned to cell sites A_1 through A_5. Mobiles served by A_4 will not cause undue interference at cell site A_1. This is due to low transmit power at A_4 and A_1 being more than one radius apart from the smaller cell. On the other hand, cells served by A_1 will cause interference if the same channel frequencies are used simultaneously at cell sites A_4 and A_5. The interference can be avoided if the small radius could somehow be made applicable to site A_1. However, the assignment of a small radius for A_1 also has its negative effect, that being that the coverage of A_1 is reduced. The problem of interference can be resolved by imposing the following rule of operation: that those channels assigned to cells A_4 and A_5 be restricted to small radius, that is, the channels at A_1, A_2, and A_3 operate with reduced power and only be assigned

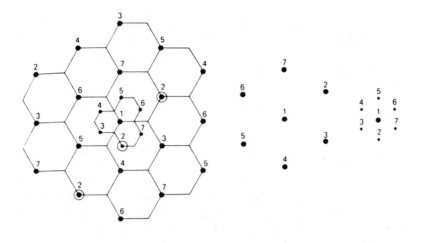

● : Previously existing or "old" cell site

●: "New" cell site installed during cell splitting

Figure 2.17 Cell splitting concept. Source: [1].

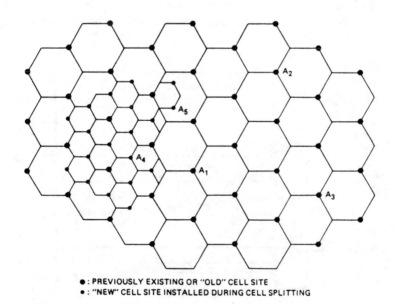

● : PREVIOUSLY EXISTING OR "OLD" CELL SITE
● : "NEW" CELL SITE INSTALLED DURING CELL SPLITTING

Figure 2.18 Overlaid-cell concept. Source: [1].

to mobiles within a small radius of these cell sites. This restriction of small radius coverage can easily be achieved by software modifications. As traffic grows, more and more channels will be assigned to a smaller group, thereby reducing the capacity of older sites to serve the large cell area.

2.3 FUNDAMENTAL CONCEPTS OF A CELLULAR SYSTEM

There are three elements of a cellular system: (i) MTSO, one per cellular system, which provides for interfacing of the mobile system to the public switched telephone network; (ii) the number of cell sites based on the area of coverage, which provides for interfacing between mobile and MTSO; and (iii) thousands of mobile users distributed throughout all the cells of a typical cellular system. Systems with single and multiple cell sites with communication paths between these elements are shown in Figure 2.19.

The mobile communicates to the nearest cell site over a radio channel assigned to that cell. The cell site is connected to the MTSO by microwave, land cable, or fiber-optic cable, which in turn interfaces with PSTN. All information exchanged over this wireline facility employs standard telephone signaling. Hence, standard switching is required within MTSO. Additionally, MTSO acts as the manager for radio channels allocated to different cells, provides coordination between moving subscribers and cell sites, and maintains the integrity of the whole system. Based on the traffic capacity of a particular cell, the number of RF channels are allocated either permanently or temporarily, based on demand. Similarly, the number of voice trunks are connected between a cell site and MTSO. In

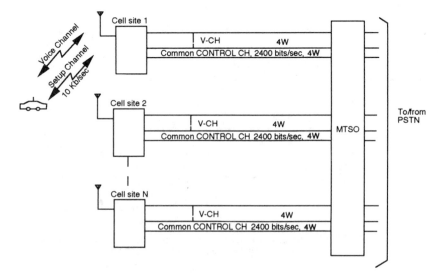

Figure 2.19 Elements of cellular system. Source: [2].

order to establish voice connection between the cell site and the mobile users, a forward setup channel (also known as a *paging channel*) is assigned at the cell site. Similarly, a reverse setup channel is assigned between the mobile and the cell site. These setup channels are for shared use within the cell. Additionally, the setup channels transmit overhead messages to ensure that the idle mobiles within the cell coverage zone are ready to communicate when a call is initiated to and from the mobile. Figure 2.20 shows a typical scenario for voice and control channel usage as the system grows. For the startup system, the functions of the paging and access channels can be combined. As the system grows and cell splitting occurs, the original paging channels are kept but a new access channel must be added for each new cell site. The reason for this is that the paging channel is a synchronous high-speed data channel and can serve a large number of subscribers, while the access channel is asynchronous and the data throughput is low due to contention. Contention occurs on the access channel due to simultaneous data transmission by more than one mobile. The data transmission speed of the forward and reverse setup channels in AMPS is 10 Kbps. A single data channel (common control channel, 4-wire) between the cell site and the MSTO carries data at the rate of 2,400 bps. The number of voice circuits assigned between the cell site and the MTSO is the same as the assigned number of channels at the cell site. Thus, the cell site does not act as a concentrator. It should be noted that the voice circuits are full duplex in nature. We discuss below the functional aspects of the mobile element, leaving aside the details of the cell site and the MTSO. This discussion will also clarify the differences between cellular radio and the regular telephone system.

2.3.1 Mobile Element

In a cellular system, a mobile unit communicates to the nearest cell site via one of the several radio channels designated to that cell. For this reason a cellular phone system is known as cellular mobile radio system (CMRS). As stated in the last section, conventional voice and data circuits link the cell sites to MTSO, which oversees the cellular system

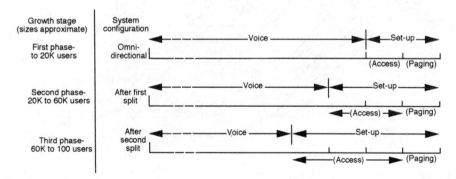

Figure 2.20 Scenario for setup channel usage as users are added.

operation and interfaces to the public telephone network. A cellular system is designed to serve customers within a given geographical area, known as mobile service area (MSA) or cellular geographic service area (CGSA). Mobile customers are expected to subscribe service in a specific MSA. When the subscriber operates within this area, the customer is referred to as a "home mobile," while outside of this area the customer is known as a roamer. MSA usually corresponds to a metropolitan area that includes a central city, its suburbs, and some portion of its rural fringe, as shown in Figure 2.21. In some cases, it could encompass a portion of a very large metropolitan area or perhaps two or more cities located close to each other, such as the Washington/Baltimore cellular system.

2.3.1.1 Numbering Scheme

One objective of cellular service is to provide dial access between home mobiles and any other telephones (landline or mobile) reached through the PSTN. A second objective is to provide access to and from roamers. In order to satisfy both these objectives, it is essential that the mobile user's telephone should have a standard ten-digit telephone number composed of a three-digit area code plus a seven-digit directory number [1]. Within the cellular mobile radio system or service (CMRS) a user is identified by a 34-bit binary mobile identification number (MIN) derived from the mobile station's ten-digit telephone number [15]. This conversion takes place as shown in below.

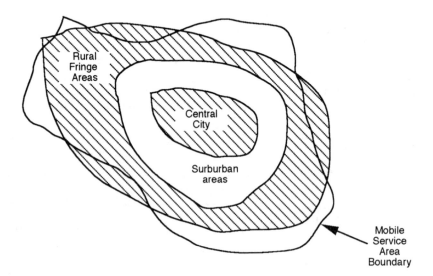

Figure 2.21 Typical mobile service area.

(1) The first three digits are mapped into 10 bits (designated as MIN2) by the following coding algorithm:
    (a) Represent the three-digit field as D_1, D_2, and D_3 with the digit 0 having the value 10;
    (b) Compute $100\,D_1 + 10\,D_2 + D_3 - 111$;
    (c) Convert the result in step (b) to a binary number.
(2) The second three digits are mapped into 10 most significant bits (MIN1) by the coding algorithm described in (1).
(3) The last four digits are mapped into 14 least significant bits (MIN1) as follows :
    (a) The thousands digit is mapped into four bits, a binary decimal (BCD) conversion, as specified in Table 2.3;
    (b) The last three digits are mapped into 10 bits by the coding algorithm described in (1).

We can demonstrate the above algorithm with an example. Identify a user directory number 321-456-7890 within the cellular system.

(1) MIN2 is computed as follows:
    (a) $D_1 = 3$; $D_2 = 2$; $D_3 = 1$;
    (b) Thus, $100\,D_1 + 10\,D_2 + D_3 - 111$;
    (c) $100(3) + 10(2) + 1 - 111 = 210_{10}$; and
    (d) The 10-bit binary equivalent of $210_{10} = 0011010010_2$.
(2) The second three-digit MIN1 follows the same rule as stated in (1):
    (a) Since $D_1 = 4$; $D_2 = 5$; $D_3 = 6$;
    (b) Thus, $100D_1 + 10D_2 + D_3 - 111 = 345_{10}$; and
    (c) The 10-bit binary equivalent of $345 = 0101011001_2$.
(3) The thousand digit $7_{10} = 0111_2$ from the BCD conversion shown in Table 2.3.

Table 2.3
Binary Code For Decimal Digits 0 Through 9

Thousands Digit	Binary Sequence
1	0001
2	0010
3	0011
4	0100
5	0101
6	0110
7	0111
8	1000
9	1001
0	1010

(4) For the last three digits, MIN1 is arrived at in a way similar to step (1).
   (a) $D_1 = 8$; $D_2 = 9$; $D_3 = 10$;
   (b) Thus, $100\ D_1 + 10\ D_2 + D_3 - 111 = 789_{10}$, and $789_{10} = 11\ 0001\ 0101_2$.
   (c) Therefore, the 34-bit binary equivalent is: 00 1101 0010 0101 0110 0101 1111 0001 0101.

This 34-bit binary number identifies the user at the cell site.

For calls originated by the mobile subscriber, the system not only needs the dialed digits but also requires the originating mobile user identification. The identification is done initially by a small integrated circuit known as the *numeric assignment module* (NAM), which is a part of the cellular telephone set. This NAM chip is programmed initially when the mobile user subscribes for service and stores the mobile telephone number and other related class-of-service information. Once the NAM chip is programmed (chip burned), the information cannot be changed without replacing the chip. The mobile initiates the call by pre-originating dialing. The called number is dialed and stored in the memory of the caller's telephone. When the cellular caller is satisfied with the accuracy of the number, the dialed number is sent in coded form along with the other call-processing information by depressing the SEND key.

Examples of dialing to and from the cellular telephone system are provided below. As stated above, each mobile has a standard ten-digit $(A_1A_2A_3N_1N_2N_3X_1X_2X_3X_4)$ assignment according to NPA registration.

1. Cellular-to-land call: $1 + C_1C_2C_3M_1M_2M_3Y_1Y_2Y_3Y_4$
   As an ordinary subscriber in the national telephone network, the cellular radio telephone user dials the office code $M_1M_2M_3$ and subscriber number $Y_1Y_2Y_3Y_4$ for a local call, along with the prefix and the associated area code $1+ C_1C_2C_3$ to make a toll call. This is designated by the number 1 in Figure 2.23.

2. Land-to-cellular call: $1 + A_1A_2A_3N_1N_2N_3X_1X_2X_3X_4$
   To reach a cellular telephone, subscribers in the national telephone network dial the cellular radio telephone number $N_1N_2N_3X_1X_2X_3X_4$ for a local call, along with the prefix and area code $1 + A_1A_2A_3$ to make a toll call. This is shown by the number 2 in Figure 2.22.

3. Cellular-to-cellular call: $1 + B_1B_2B_3L_1L_2L_3W_1W_2W_3W_4$
   To reach other cellular telephones, the originating cellular radio telephone subscriber dials the terminating radiotelephone number $B_1B_2B_3L_1L_2L_3W_1W_2W_3W_4$ with the prefix 1. In this case it should be noted that the dialing procedure remains same whether the terminating mobile is in the same area code or not. This is shown by the number 3 in Figure 2.22.

4. International call: $I_1I_2I_3C_1C_2$ + national number
   International numbers can be reached by dialing the international prefix $I_1I_2I_2$, the country code C_1C_2, and the national number. This is shown by the number 4 in Figure 2.22.

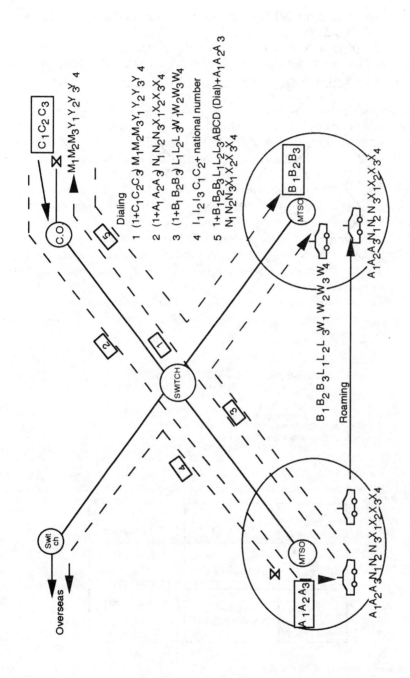

Figure 2.22 Some examples of dialing. Source: [13, p. 2-55].

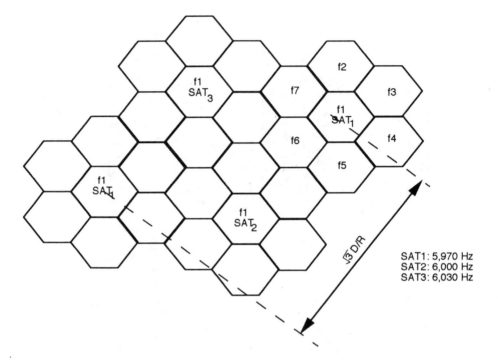

SAT1: 5,970 Hz
SAT2: 6,000 Hz
SAT3: 6,030 Hz

SAT: Supervisory Audio Tone (5970, 6000, 6030 Hz)
   - Transmitted by cell site & Transponded by MR

ST: Signalling Tone (10kHz)
   - Transmitted by MR to:
   1) Confirm Orders
   2) Signal Flash Requests
   3) Signal Release Requests

	SAT Received	SAT Not Received
ST ON	Mobile ON-Hook	Mobile in fade or Mobile TX off
ST OFF	Mobile OFF-hook	

Figure 2.23 Mobile supervision of the cell site. After [2, p. 49].

5. *Roaming calls:*
   a. If roamers want calls to be automatically transferred from their home MTSO to their roaming area, the roamer should register a temporary directory number by calling a special number at their home MTSO and use the call transfer registration method. Seamless and follow-me roaming are other techniques for roamer identification and call delivery, which will be discussed in Chapter 8.
   b. Calls to a roaming mobile telephone when the mobile user does not want the transfer to go through the home MTSO (when he or she has not registered a temporary directory number): the calling subscriber dials $1 + B_1B_2B_3L_1L_2L_3ABCD + A_1A_2A_3N_1N_2N_3X_1X_2X_3X_4$ where $1 + B_1B_2B_3L_1L_2L_3ABCD$ is the addressed MTS access number (ABCD can be ROAM) followed by the called radiotelephone original number $A_1A_2A_3N_1N_2N_3X_1X_2X_3X_4$. The total number of digits dialed is 21. This is shown by the number 5 in Figure 2.22.

2.3.1.2 Mobile Power Control

There are total of eight sizes [15] of mobile transmitters: classes I through VIII. Class I through class IV are arranged according to the decreasing nominal power output. Classes V through VIII are reserved for future definition. Class IV is available for dual-mode mobile only. The nominal maximum power level (effective radiated power-ERP) for the first four classes of mobile station transmitter are:

- Class I, 6 dBW (4.0W): automobile mounted;
- Class II, 2 dBW (1.6W): transportable;
- Class III, −2 dBW (0.6W): portable; and
- Class IV, −2 dBW (0.6W): portable.

The mobile station's maximum effective radiated power, with respect to a half-wave dipole (ERP) for all the three classes, is limited to 8 dBW (6.3W). Further, the mobile transmitter, upon command from the cell sites, is able to reduce power in steps of 4.0 dB. There are total of 10 steps for power reduction. The nominal power settings are shown in Table 2.4. Mobile units in classes I through III change their power levels in seven steps, while a wider range (ten steps) is allowed for classes IV through VIII. The power adjustment is due to the intelligence built into the cellular transceiver logic unit. If the cell site senses a higher received signal from the mobile, it sends command signals and asks the mobile transceiver to cut back its RF power. Similarly, if the signal received at the cell site decreases as it moves away from the cell, the cell site asks the transceiver to boost its power. The ability to adjust the mobile transmitted power has two purposes. First, it increases the battery life by reducing the transmitted power from the mobile, and second it adjusts the power so that the cochannel and adjacent channel interference are reduced [18].

Table 2.4
ERP of the Mobile Station

Power Level	Mobile Attenuation Code	Nominal ERP for Power Class (dBW)				
		I	*II*	*III*	*IV*	*V–VIII*
0	000	6	2	−2	−2	R
1	001	2	2	−2	−2	R
2	010	−2	−2	−2	−2	R
3	011	−6	−6	−6	−6	R
4	100	−10	−10	−10	−10	R
5	101	−14	−14	−14	−14	R
6	110	−18	−18	−18	−18	R
7	111	−22	−22	−22	−22	R
8*						−26 ± 3 dB
9*						−30 ± 6 dB
10*						−34 ± 9 dB

Note: R = reserved for definition.
*Dual mode.

2.3.1.3 Call Supervision

On the voice channel, one of the three tones that modulate the carrier at a low modulation index is used for supervision. These tones are centered at 6 kHz and are termed supervisory audio tones (SAT). The SAT is added to the voice transmission by a cell site. The three frequencies used are 5,970, 6,000, or 6,030 Hz. The function of the SAT is similar to the closing of the local loop in the land telephone system [1–3, 20–22]. A given SAT is sent from the cell site to the mobile, which in turn loops back the same SAT to the cell site. The cell-site station must make decisions to determine whether it has received the original transmitted tone, or whether the tone received is different and is the result of some interference. This decision is made through measurement in a phase-locked loop according to the data in Table 2.5. A single phase-locked loop is adequate for this measurement because the band of interest here is only 60 Hz. Phase-locked loop is a narrow band filter and can be designed to recover these frequencies accurately.

The assignment of SAT to a seven-cell configuration is shown in Figure 2.23, where a complete cluster is assigned one SAT frequency. Since an assigned supervisory carrier can be modulated by one of the three tone frequencies, a seven-cell cluster appears to be a twenty-one cell cluster from the supervision point of view, that is, the D/R ratio is $\sqrt{3}$ times the conventional D/R ratio, as shown in Figure 2.24 (see also problem 5). Thus the nearest cell, where the same carrier occurs with the same SAT frequency, is $\sqrt{3}\ D/R$.

Since these frequencies are roughly twice the maximum audio frequencies, filtering them from the voice band is relatively easy. The intermodulation products are also controllable. In addition to SAT, a signaling tone (ST) at a frequency of 10 kHz is transmitted

Table 2.5
SAT Frequencies And Their Detection

Measured Frequency (f) of Incoming Signal	Measured SAT (Hz)	Tone Frequency (Hz)
$f < f_1$	No valid measurement	
$f_1 \le f < f_2$	5,970	$f_1 = 5955 \pm 5$
$f_2 \le f < f_3$	6,000	$f_2 = 5985 \pm 5$
$f_3 \le f < f_4$	6,030	$f_3 = 6015 \pm 5$
$f \le f$	No valid measurement	$f_4 = 6045 \pm 5$
No SAT received	No valid measurement	

Note: SAT = supervisory audio tones.

to represent mobile user on-hook and off-hook conditions. In addition to the continuous signaling tone at 10 kHz, digital signals are also sent over the voice channel to the mobile user. The function represented by the combination of SAT and ST is shown in Figure 2.23. Data transmission over the forward voice channel is accomplished by a technique known as blank-and-burst. When the cell site wants to send messages to the mobile, the voice signal is blanked for about 50 ms and a burst of 10-Kbps data (AMP) is inserted in the voice channel [1–3, 22]. This signaling is used for (a) alerting the mobile user, (b) for disconnection, (c) for hold, and (d) for handoff.

2.3.1.4 Locating the Mobile

For maintaining good quality voice and data transmission service, a mobile user's S/I ratio is monitored at the cell site every few seconds. This in turn monitors the cochannel interference. When the call is initially established, the mobile locates the appropriate cell site by scanning all the control channels (paging channel) and selects the one with the highest quality (high S/N ratio). After a call is initially established, the mobile may move out of the original service area. In this condition it may become necessary to reroute the original call through the new cell site, the location of which with respect to the mobile provides a better signal quality. This process of switching the call from one cell site to another is known as handoff and is executed under the control of MTSO. This handoff process can take place several times until the mobile terminates the call. A comprehensive treatment of this is provided in Chapter 8 from the networking point of view. The cellular system has the following characteristics, which depend heavily upon the rational for the design of a mobile locating plan [12].

- A very high degree of accuracy in the location of the mobile is not required. In other words, a small percentage of inaccuracy in the coverage radius is tolerable.
- A comparative rather than an absolute measure of position or distance from the cell site is needed. Once the identity of a zone containing the mobile is established, a more precise measurement is not required.

- The measurement technique should be fast to match the dynamic mobility of the users.
- Location information is required at the MTSO so that the most desirable cell site can serve the mobile user.
- Shadowing and fading are significant factors to be accounted for in locating the mobile.

The two broad schemes that can be utilized for locating the mobile from a cell site are measurements based on direction, θ, and measurements based on the distance, (r). The scheme that is dependent on the measurement of direction, θ, can be based on a fixed antenna beam or on a rotating antenna beam. In the rotating antenna beam method, the antenna is rotated until the maximum signal is received from the mobile. The scheme dependent on distance measurement, r, is either based on measuring the received S/I ratio, or it can be based on propagation time delay between the transmitted SAT from the cell site and the received SAT from the mobile user at the cell site. The rotating beam method is in general slow and yields data in an inconvenient form for making decisions. The signal strength, (S/I), method has limited accuracy unless a large number of samples of the received signal is taken. The propagation delay method is accurate, but requires synchronous measuring techniques and therefore makes the measurement technique complex. Here, we describe the measurement techniques based on directional antennas having a fixed beam width. Figure 2.24 shows the setup for 120-deg and 60-deg sectored antennas at cell sites. As shown in these figures, the equal gain intersection points of the antennas define the hexagonal boundaries. For 120-deg sectored antennas, equal gain boundaries are represented by the firm lines, and 60-deg sectored antennas are represented by dotted boundaries of the sectors.

Initially, the mobile user finds the best zone for service by scanning the forward access channels (paging channels). Let us assume that the mobile has found the best zone for service to be zone 1 and has then radiated the RF energy on the reverse access channel of zone 1. By making the signal strength measurement of the received signal with two antennas at each cell-site location and subtracting the measured values of the two antennas, it can be determined whether the mobile is to the left or to the right of the equal gain line. Since the signal received by two antennas at each cell site is correlated, subtracting the signal levels will leave little dependence on multipath and shadows. This assures that noise has little effect on these relative measurements. In this method, by making measurements at the three different cell sites, the location of the mobile user, X, can be accurately determined. A signal from a mobile at X will be most strongly received by antenna A at cell location 1, will be received with less strength by antenna C at cell site 2, and with less strength yet by antenna B at cell site 3. When the mobile user moves to location Y, the signal is received strongly by antenna A at location 3, strongly by antenna C at location 2 and strongly by antenna A at location 1. For 60-deg sectored cells, the mobile at X will be received strongly by lobe A of antenna array 1, lobe C of antenna array 2, and lobe D of antenna array 3. For a mobile at Y, signal strength will be maximum

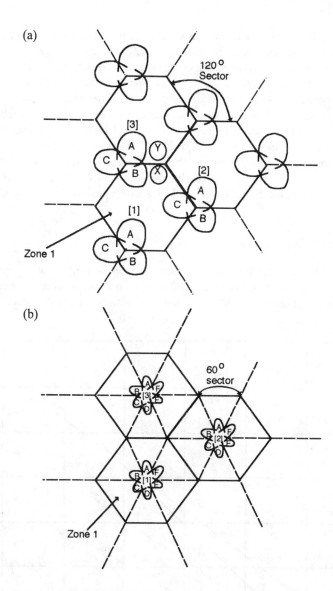

Figure 2.24 Location of a mobile in (a) 120-deg and (b) 60-deg corner-excited cells.

at lobe D of array 3, lobe A of array 1, and lobe B of array 2. From these three measurements, from instant to instant and taking the average, the MTSO can locate the user accurately and use this information to select the new cell site for the mobile. It should be noted that the weighing factors will be required for location identification. Also, calibration will be required initially when the system is put into operation. For further details on this method, the reader is referred to [12].

The theoretical and measured curve for a corner-feed 120-deg antenna is shown in Figure 2.25. The 0-deg angle is the equal antenna gain reference point that is valid on the boundary between zones. The y axis measures the difference in signals between two corner-fed antennas. The dotted curve is the experimental curve [12] measured in an urban environment at two extremes of range (0.5 mile and 7 mile). Measurements have been reported to have an accuracy of ± 2.5 dB. It was also reported that, taking all the data from the various distances and weighing the result according to distance, the probability that a source would be wrongly placed with respect to the equal gain line is about 0.02. Thus, based on this approach, the system will have a 0.98 probability of accurately locating the mobile.

2.3.2 Mobile Calling Sequence

In this section we discuss briefly the calling sequences for a mobile receiver (MR) terminated call, a mobile receiver–originated call, the handoff sequence, the call releasing sequence when originated by a MR, and the call releasing sequence when initiated by landline [2,3,13–22].

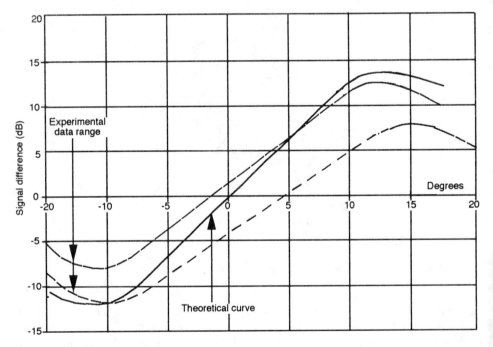

Figure 2.25 Locating tests, experimental results. After [12, p. 79].

2.3.2.1 Mobile Terminated Call

When the mobile radio telephone is switched on, it scans all 21 assigned setup channels according to the program it has in its memory and selects the strongest channel. This channel will normally be associated with the nearest cell site. When MTSO receives an incoming call through a standard wireline network, it collects the calling digits, converts them to MIN (telephone number) as discussed earlier in Section 2.3.1.1, and instructs all cell sites to page the mobile over the forward setup channels (step 1), as shown in Figure 2.26. The mobile unit, after recognizing its page, responds to the cell site over the reverse setup channel (access channel, step 2). The cell site in turn relays this information to the MTSO over its dedicated landline data link (step 3). The MTSO selects an idle voice channel and the associated landline trunk, and informs the cell site of its choice over the appropriate data link (control channel, step 4). The serving cell-site intern tells the mobile

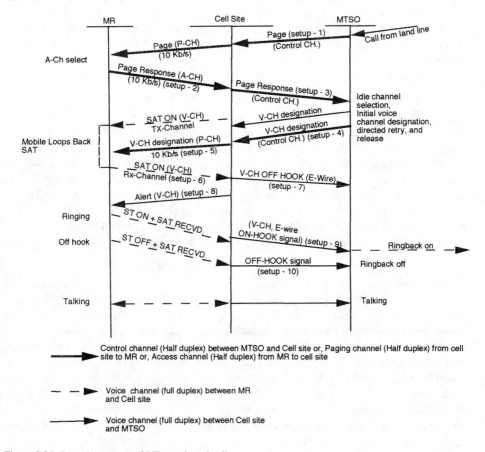

Figure 2.26 Control sequence of MR-terminated call.

of its choice of channel over the forward setup channel (step 5). The mobile in turn tunes to the designated voice channel where the SAT signal is present, which is looped back to the cell site (step 6). On recognizing the correct looped-back SAT, the cell site places the associated landline trunk in an off-hook state, which the MTSO interprets as a successful voice channel establishment (step 7).

On command from MTSO, the cell site transmits a data message over the voice channel to the alerting device at MR, which signals the MR of an incoming call (step 8). The signaling tone from the MR causes the cell site to place an on-hook signal over the previously selected landline trunk, which confirms successful alerting to the MTSO (step 9). The MTSO, in turn, provides an audible ring-back tone to the calling party. When the MR answers by going off-hook, ST is removed from the voice channel to the cell site, and in turn activates the off-hook signal on the landline trunk (step 10). The off-hook signal is detected at the MTSO, which disables the ring-back tone to the land party and establishes the talking connection. It should be noted that up to the point of SAT turnaround by mobile, all communication between the MTSO and CS is over the data link. Communication between the cell site and the MR is over the voice channel after voice channel assignment.

2.3.2.2 Mobile-Originated Call

Using the preorigination dialing procedure, the mobile user enters the dialed digit into the equipment's memory. The stored digits, along with the mobile's own identification, is transmitted to the cell site through the reverse setup (access channel) channel (step 1). The cell site receives this information and relays it to the MTSO. As shown in Figure 2.27, a voice channel is selected by the MSTO and the cell site is informed (step 2). The cell site then informs the mobile about the voice channel designation (step 3). Similar to the mobile-terminated call, the MTSO extends the connection to the PSTN after confirming the SAT of the calling mobile. The conversation can begin when the called party answers.

2.3.2.3 Handoff

Handoff, in general, is the process of switching over a call path from its old cell site to a new cell site when the voice signal drops below a certain minimum value. Here, the mobile initially located in cell A moves to cell B, and subsequently to cell site C. As the user moves from cell to cell, he or she is assigned a new channel with each move. With a deteriorating signal-to-noise ratio at the cell site, the switch-over can also take place within the same cell. This condition is shown at cell site B in Figure 2.28(a), where the mobile has been assigned a different channel within cell B (from voice channel B to voice channel C). The sequence of steps involved in the handoff process is shown in Figure 2.28(b). The location information gathered by serving the cell site and other cell sites is

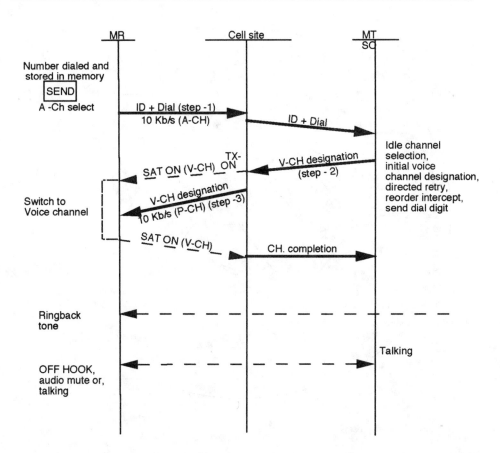

Figure 2.27 Control sequence of MR-originated call.

transmitted to the MTSO over the landline trunks. When the carrier drops below a certain level, the MTSO decides to switch over the present call from the old cell site A to the new cell site B.

The MTSO sets an idle voice channel (and an associated landline trunk) at the receiving cell site and informs the new cell site to switch on its transmitter. The message is sent to the mobile through the active serving cell site informing it of its new voice channel designation. The mobile turns off the supervisory tone from the old voice channel, which is interpreted at the MTSO as going on-hook. The mobile also retunes to the new channel and transponds the SAT found there. This is recognized by the MTSO as a successful completion of the handoff sequence. The MTSO reconfigures its switching network and connects the landline party to the mobile through the new voice channel and landline trunk. The entire handoff process takes about 0.2 seconds and is not noticed by users. Also, handoff does not degrade the quality of voice transmission.

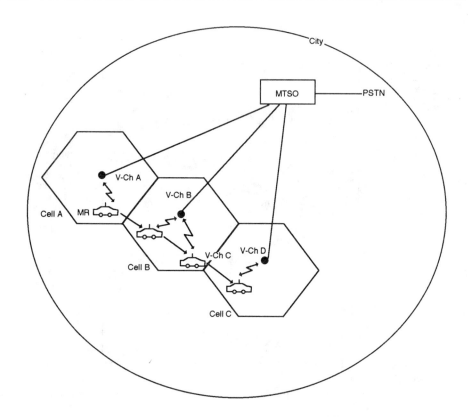

Figure 2.28(a) Handoff process.

2.3.2.4 Call Releasing Sequence

There are two types of call releasing sequence, namely, the mobile-initiated release and land subscriber–initiated release. The sequence of operations for both these follows.

The mobile initiates the releasing sequence by going on-hook (or END button is depressed). The supervisory tone on the voice channel is turned on, which is received by the cell site (step 1). As shown in Figure 2.23, the presence of ST and SAT indicates the on-hook condition of the mobile at the cell site. As a result, the cell site places an on-hook signal on the appropriate landline trunk towards the MTSO (step 2). On receipt of the on-hook signal, the MTSO idles all switching office resources and transmits any necessary disconnect signals through the wireline network. The MTSO also commands the previously serving cell site over its data link to shut down the cell-site radio transmitter (step 3). All equipment used at this time is then free to be used in a new call. This releasing sequence is shown in Figure 2.29.

In response to an on-hook signal from the landline network, the MTSO idles all the switching office resources associated with the call to be released. The MTSO sends

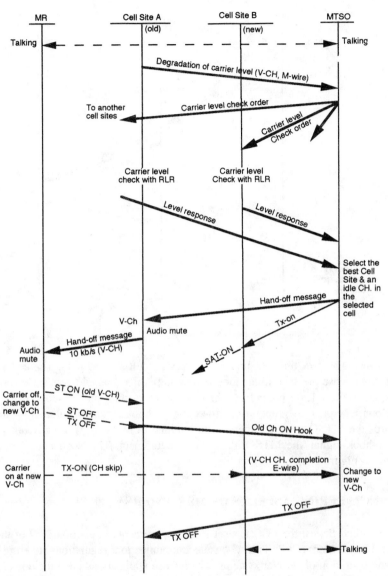

Figure 2.28(b) Control sequence of handoff.

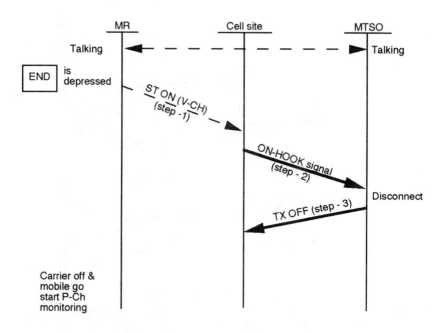

Figure 2.29 Control sequence of MR-initiated release.

a data message over the data link to the serving cell site (step 1). The cell site in turn sends the release message to the mobile over the voice channel (step 2). The mobile then responds to the release order from the cell site by turning on the supervisory tone (step 3). Upon receipt of the supervisory tone, the on-hook signal is initiated by the cell site towards the MTSO over the appropriate landline trunk (step 4). Finally, after receiving the on-hook signal, the MTSO idles all the equipment. This sequence of operations is shown in Figure 2.30.

2.4 ENGINEERING ASPECTS OF MSA AND RSA [20–22]

In the United States, the FCC has set forth requirements in Section 22.9 of the Code of Federal Regulations (CFR) Title 47 on telecommunications regulations covering the licensing and operation of cellular systems. The license applications must include engineering, legal, and financial responsibilities. In addition to technical considerations, response to public need, service proposals; and service management plans must be included. The engineering portion of the application describes what the applicant might build for both a metropolitan service area (MSA) and a rural service area (RSA). For MSAs, the FCC allows each licensee five years to expand its system to cover the entire MSA. In RSAs, engineer has the choice of designing an inexpensive one-cell system or an expensive multicell system. The technical requirements of Section 22.9 of the CFR are based on

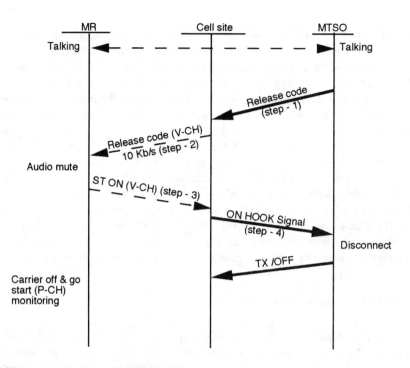

Figure 2.30 Control of land subscriber–initiated release.

the efficient use of radio spectrum, the cost minimization to subscriber, modularity, and flexibility of the system.

2.4.1 Choice of Frequency

As stated in Section 2.2 and shown in Figure 2.1, the 50 MHz of spectrum assigned for cellular use is divided into two equal parts. Spectrum assignments are made from the frequencies listed for cellular systems A and B. The details of system A and system B voice and control channel assignments, as described in Section 22.902 of the CFR, are as follows.

Cellular system A has allocation of 416 frequency pairs with 30-kHz channel spacing. The first set of mobile frequency channels will be 824.040 MHz, followed by 824.070 MHz and proceeding to 834.990 MHz. The second set of mobile frequency channels will be 845.01 MHz, followed by 845.04and proceeding to 846.48 MHz. The first set of base station frequencies will begin with 869.040 MHz, followed by 869.070 MHz and proceeding to 879.990 MHz. The second set of base station frequencies will be 890.01, followed by 890.04 and proceeding to 891.48.

A second cellular system, B, has allocation of 416 frequency pairs with 30-kHz channel spacing. The first group of mobile frequency channels will be 835.020 MHz, followed by 835.050 MHz and proceeding to 844.980 MHz. The second group of mobile channels will start at 846.51 MHz, followed by 846.54 MHz and proceeding to 848.97. The first group of base station frequencies will begin with 880.020 MHz, followed by 880.050 MHz and proceeding to 889.980 MHz. The second group of base station frequencies will begin at 891.51 MHz, followed by 891.54 MHz and proceeding to 893.97 MHz.

Twenty-one control channel pairs will be assigned in each cellular system. For systems operating on the frequencies specified for cellular system A, the 21 channel pairs are 834.390 MHz through 834.990 MHz and 879.390 MHz through 879.990 MHz. For systems operating on the frequencies specified for cellular system B, the 21 channel pairs are 835.020 MHz through 835.620 MHz and 880.020 MHz through 880.620 MHz.

In addition, the frequency assignments to the applicants shall be made in coordination with proposed frequency usage with existing users in cellular geographic areas within 75 miles of all base stations affected, and with other applicants who have previously filed applications and whose facilities could affect or be affected by the new proposal in terms of intersystem frequency interference or restrictions on ultimate system capacity.

2.4.2 Cellular System Service Area

The cellular geographic service area (CGSA) of a cellular system is defined by the applicant as the area to be served within the SMSA. CGSA includes areas within an MSA, RSA, or New England county metropolitan area (NECMA). These service areas may extend beyond the boundaries of the MSA, RSA, or NECMA, except where such extensions are minimal and do not include areas within another central MSA, RSA, or NECMA. Further, the CGSA must be drawn on one or more U.S. Geological Survey map(s) with a scale of 1:250,000. With the CGSA, the applicant must depict each cell site and its respective 39-dBu contours. The 39-dBu (dB refers to 1 μv) contours of all cell sites must cover at least 75% of the total CGSA. The applicant will apply for the change of CGSA in case the proposed 39-dBu contours extend beyond the authorized CGSA. The exhibit generated as a result of engineering considerations will detail the processes used to choose a proposed CGSA, select cell sites, and calculate coverage contours. The geography, general shape, and population distribution of the standard metropolitan statistical area (SMSA) present unique problems in designing a cellular system to comply with the requirements of an optimum configuration. In addition to 75% area coverage, the CGSA must include major population and transportation arteries that may add a substantial number of roamers to the system, particularly in the summer when traffic in recreational and resort areas is significant. In Figure 2.31, the CGSA 39-dBu contours were chosen to extend coverage to the major portion of the permanent population plus major highways.

The choice of the cell-site locations is paramount to provide proper coverage within a CGSA. Cell-site locations should be chosen so that the coverage patterns of each cell are continuous to its neighbors, with only enough overlap to allow handoff from cell to cell as a subscriber travels between cells. If the cells do not overlap, especially along

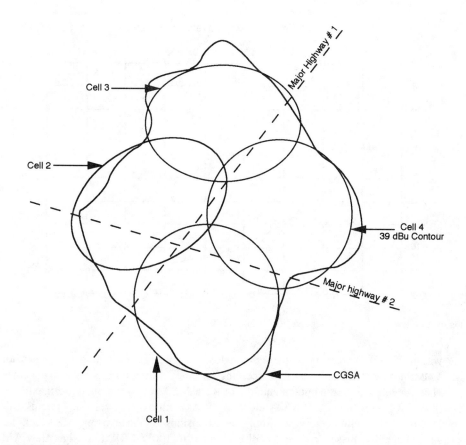

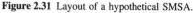

Figure 2.31 Layout of a hypothetical SMSA.

major roadways, too many calls can potentially be lost as the subscriber leaves one cell and is no longer in the covered area. Care must also be exercised to prevent coverage from extending to the neighboring MSAs. To keep the construction costs down, the sites should be chosen such that the cell antenna can be mounted on an existing tower or on the roof top of a existing high-rise building.

2.4.3 Power Limitations

For RSA cell sites that are at least 24 miles from MSAs or NECMAs, the FCC has authorized an effective radiated power as high as 500W. The effective radiated power of base stations with transmitting antennas in excess of 500 feet above average terrain (AAT) must be reduced below 500W by not less than the amount shown in Figure 2.32. The only exception to this is when agreements are reached with all neighboring carriers that

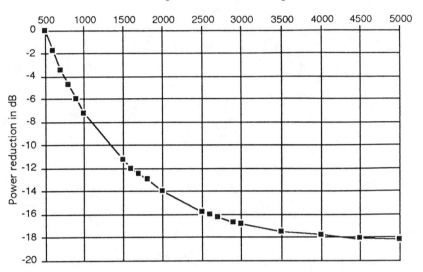

Figure 2.32 Power reduction versus antenna height AAT.

are within 75 miles. Carriers in the case of RSAs may share switching facilities with MSAs but not radio transmitting and receiving facilities. Based on this, several schemes have been developed for the implementation of RSA. MSA cells are limited to a maximum of 100W ERP. It is likely that RSAs close to MSAs with existing cellular systems will be developed for cellular service simply as extensions of the existing MSA cellular systems. Generally, these RSA cells will be limited to 100W ERP. In design, they will resemble MSA cells.

2.4.4 Rural Service Area Configurations

Rural service area (RSA) applications to provide cellular telephone systems are now being developed in the United States. Decisions on how to build these systems are being made in many areas of the country. The technical requirements for RSAs are covered in Section 14.1 of the CRF. Here we choose to discuss the several system layouts that are, in general, different than the layout of the systems in MSA. Three approaches have emerged for the system layout: stand-alone, backhaul, and satellite. We discuss below each of these approaches for the cellular layout in RSAs and bring out some of the reasons for their selection. Proper system selection is important, as the RSA can be required to provide service to areas with different population concentrations. The RSA market ranges from populations as low as 20,000 to even greater than 500,000.

2.4.4.1 Stand-alone System

The stand-alone cellular system shown in Figure 2.33 is similar to systems installed in many MSAs. This configuration is suitable for areas where the population is high. The most costly part of the system is the central switch with its associated controls and memory. Traffic recording equipment must also be included for day-to-day peak hour analysis and future planning of the system. The switch must be connected to roamer validation, call forwarding, and the billing databases of different companies using terrestrial telephone circuits.

2.4.4.2 Backhaul System

Owners of RSAs that are adjacent to stand-alone MSAs may choose to use an adjacent switch and connect them in the form shown in Figure 2.34. This will reduce the system operator, thereby reducing its cost. In this case, both voice and control circuits must be backhauled to the adjacent MSA host switch. Additionally, trunks from the local telephone company must also be terminated at the switch. Apparently, though, the cost of backhauling can be high. Reliability also becomes a problem as the loss of a link between the RSA and the MSA can knock the entire system off the air. An additional problem of concern arises due to complete dependence of the RSA system on the host MSA operators for the control and maintenance of the backhaul host switch. The question also arises as to which system (whether its own MSA or RSA backhaul system) gets priority. Along with these disadvantages, there are several advantages to this configuration, namely, combining these two systems blends the sales and service under one agreement and thus cost can be reduced.

2.4.4.3 Satellite System

As RSAs have become more popular, several important factors have come to limelight. It is apparent that the rural areas in America have lower populations, larger service areas, and higher networking costs. This is precisely the situation where the satellite-based system can provide the maximum advantage. Here, the networking part is performed by the satellite subsystem while the voice path to the mobile is provided by the ground-based cellular system. It should be realized that the biggest revenue to the RSA can be due to roamer, where the networking plays a very important role. Satellite data networks in this case will provide positive customer verification, call forwarding, and customer billing. Stated differently, if a cellular subscriber wants to make or receive a call from virtually any RSA or MSA within the country, satellite data networking is the answer. Call validation, call forwarding, and system-to-system handoffs can all be supported over the satellite data network. Figure 2.35 depicts one answer to future seamless cellular networking in the United States.

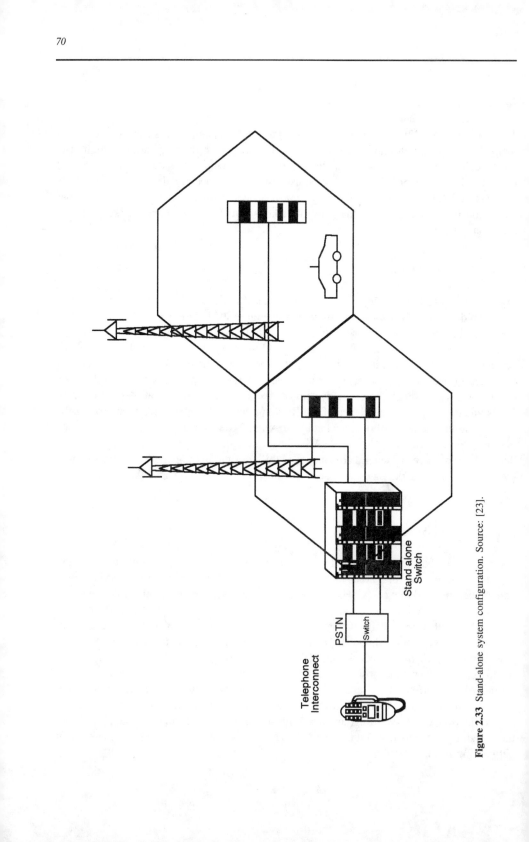

Figure 2.33 Stand-alone system configuration. Source: [23].

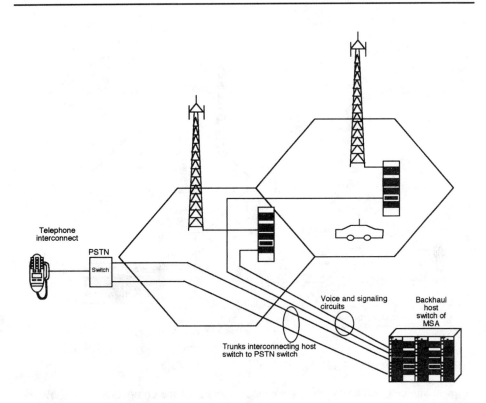

Figure 2.34 Backhaul system configuration. Source: [23].

2.4.5 RSA Highway Cell Design

For interstate highway coverage, a cell's service area must be as large as possible. In these cases, cells can serve an elongated area to match the course of the highway as the nonhighway areas need not be included in the coverage. Figure 2.36 shows a special highway cell. By creating an elongated service area that stretches along the highway as far as possible in both directions, the number of cells is minimized. At the same time, trunking efficiency is maximized because all channels are available to serve subscriber traffic in both directions. As a result, the number of transceivers required is minimized, reducing the overall cost of the cell. If the RSA cell site is at least 24 miles from the nearest MSAs, the FCC permits an effective radiated power (ERP) of 500W. If this high-powered cell is not designed properly, the uplink and the down link can become unbalanced. As a result, the uplink path often will become too noisy to be used while the downlink path remains clear. Below we provide an example of a cell design where the two links are balanced out for proper operation. This balanced design ensures that the downlink and uplink portions of cellular conversation will be equally strong so both

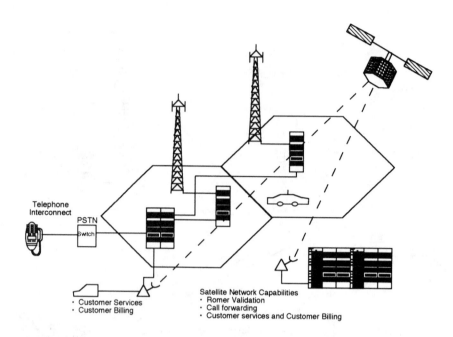

Figure 2.35 Satellite system configuration. Source: [23].

parties to any cell will enjoy quality reception. Details of the computation are shown in Table 2.6.

A key design element is the group of four-sided mounted antenna systems. Two transmit and two receive antennas pointing on two sides of the highway are used. Thus, the downlink signals are radiated in both directions along the highway from the cell. Also, the two receive antennas are pointed in a similar manner. Each antenna is connected to a multicoupler consisting of a preselector filter, an amplifier, and a splitter. Both receiver multicouplers (one for each direction) are connected to one access channel and all voice channels. Using the 12-foot receive and 6-foot transmit parabolic antennas, achievable gains are 26.9 dBi and 21.3 dBi, respectively, at minimum transmit and receive frequencies of 869 MHz and 824 MHz. The computation assumes 50% antenna efficiency. Limiting the cell transmit power to 500W, the downlink budget provides a net gain of 171.1 dB. For 3W mobile ERP net link gain once again becomes 171.1 dB.

2.5 SUMMARY AND CONCLUSIONS

This chapter was divided into three parts: cellular layout, the operational concept of a cellular system, and the engineering aspects of MSAs and RSAs. The section on cellular layout included the hexagonal structure choice for a cell, considerations for the location of cell sites, allowed tolerance in the cell-site location, the basis on which the minimum

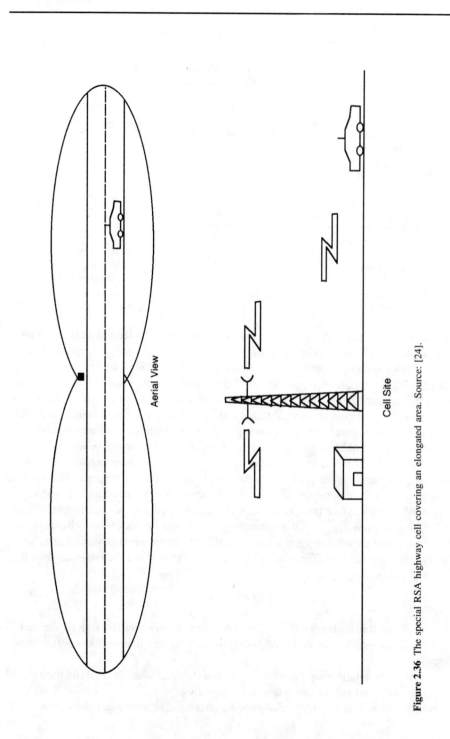

Figure 2.36 The special RSA highway cell covering an elongated area. Source: [24].

Table 2.6
Highway Cell Design with 500W Transmit Power

	Downlink Design (Cell Site to Mobile)	Uplink Design (Mobile to Cell Site)
Transmitter output	16.4 dBW (44W)	
16-channel combiner loss	−4.0 dB	
Transmission line loss 550′		
Air-diletric helix @ 0.28 dB/100′	−1.5	
Coupling loss	−3.0 dB	
Cell-site antenna gain	19.2 dBd	24.8 dBd
Cell-site transmit power (ERP)	27.1 dBW (513W)	
Mobile receive antenna gain	0.0 dBd	
Transmission line loss	−2.0 dB	−4.7 dB*
Receiver sensitivity	−146.0 dBW	−146.0 dBW
Allowable path loss	171.1 dB	171.1 dB
Mobile ERP		5.0 dBW (3W)

*Includes multicoupler loss.

choice of cell radius depends, and the overlaid cell concepts. The second section primarily discussed the operational concept of cellular radio through the mobile element. Here we started our discussion by describing the relationship of the mobile with respect to cell site and MTSO. The important aspects of the numbering scheme were discussed. Techniques of locating the mobile are essential because of the handoff mechanism, which the mobile goes through during a typical conversation. Lastly, the different types of calling sequences were extensively discussed. The third section dealt with the engineering requirements and the system layout of MSAs and RSAs. Three different layouts for RSAs were discussed. In some sense this chapter is basic and the reader is advised to conceptualize the working of the cellular system before going on to the details in subsequent chapters. This chapter is a must for both the practicing engineer and for students who are being exposed to this type of material for the first time. In the next chapter we shall take up an important aspect of cellular radio known as traffic engineering. The initial layout of the cellular system and its subsequent growth is mainly a function of traffic estimates at the initial installation of the system. Automatic traffic monitors are used as a part of the system to achieve this.

PROBLEMS

2.1 For a light traffic density (i.e., startup system), justify that the size of the cell is determined by the required signal-to-noise ratio rather than signal-to-interference ratio.

2.2 Establish the relationship $P_{T2} = P_{T1}/16$, where P_{T2} and P_{T1} are the cell site transmitted power after and before one stage of cell splitting.

2.3 Justify that within a cluster of N cells no channels are used more than once.

2.4 Compute the ratio D/R for the three cases in Figure 2.37.

2.5 For $i = 3, j = 1$ and for $i = 3, j = 2$ find the number of RF channels/cell, the channel frequency assignment, and show the layout of the cellular system. Assume uniform traffic in the system.

2.6 Enumerate advantages of cell splitting down to a one-mile radius. What happens if you split the cells down further? How will this effect the location accuracy of the mobile?

2.7 For $i = 3, j = 1$, draw the layout of the cellular structure and provide the audio channel assignment for a total bandwidth of 20 MHz with FM modulation having channel separation of 25 kHz.

2.8 For a frequency reuse partition case with $N_A = 3$ and $N_B = 9$ compute k, and N_{eq} for $p = 0.0, 0.5, 1.0$.

2.9 Is cochannel interference a function of transmitted power from the cell site?

2.10 Find the encoded binary number for the mobile ten-digit directory number 301-520-9209.

2.11 Prove that the cell separation is $\sqrt{3}D/R$, for SAT frequency assignment where D is the cochannel cell separation distance.

2.12 For the 60-deg sectored antennas in Figure 2.38, develop a mobile locating strategy similar to the one described in Section 2.3.1.4.

2.13 Discuss the handoff phenomenon in detail, giving the sequence of operations among mobile, cell site, and the MTSO. Can you think of one situation where the handoff is necessary but cannot be achieved in the system? Is handoff transparent to both the calling and the called party? Assuming 0.05 sec to 0.2 sec disruption during handoff, can 10-Kbps data be supported?

2.14 What is blank-and-burst phenomena and where does it take place? What is the nominal duration of blank-and-burst in the cellular system?

REFERENCES

[1] Macdonald, V. H. "The Cellular Concept," *The Bell System Technical Journal,* Vol. 58, No. 1, January 1979, pp. 15–41.

[2] Fluhr, J. C., and P. T. Porter. "Control Architecture," *The Bell System Technical Journal,* Vol. 58, No. 1, January 1979, pp. 43–69.

[3] Halpern, S. W. "Reuse Partitioning in Cellular Systems," *IEEE Vehicular Technology Conference,* 1983, pp. 322–327.

[4] Hanson, B. L., and C. E. Bronell. "Human Factors Evaluation of Calling Procedures for the Advance Mobile Phone System (AMPS)," *IEEE Transactions on Vehicular Technology,* Vol. VT-28, No. 2, May, 1971, pp. 126–131.

[5] Hachenbury, V., et al., "Data Signaling Functions for a Cellular Mobile telephone System," *IEEE Transactions on Vehicular Technology,* Vol. VT-26, No. 1, February, 1977.

[6] Fluhr, Z. C., and E. Nussaum. "Switching Plan for a Cellular Mobile Telephone System," *IEEE Transactions on Communication,* Vol. com -21, No. 11, November, 1973, pp. 1281–1286.

[7] Mikulski, J. J. "A System Plan for 900 MHz Portable radio Telephone," *IEEE Transactions on Vehicular Technology,* Vol. Vt-26, No.1, February, 1977, pp. 76–81.

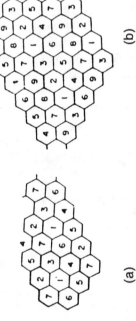

(a)　　　　(b)　　　　(c)

Figure 2.37 Cochannel cell locations: (a) $i = 2, j = 1$; (b) $i = 3, j = 0$; (c) $i = 2, j = 2$.

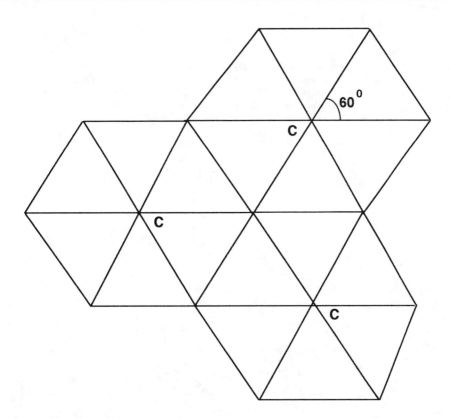

Figure 2.38 60-deg sectored antennas.

[8] Blecher, F. H. "Advanced Mobile Phone Service," *IEEE Transactions on Vehicular Technology,* Vol. VT-29, May, 1980, pp. 238–244.

[9] Rypinski, C. A. "Economic Design of Interference Limited Radiotelephone System," *IEEE Vehicular Technology Conference,* 1983, pp. 332–335.

[10] Frenkiel, R. H. "A High-Capacity Mobile Radiotelephone System Model Using a Co-ordinate Small-Zone Approach," *IEEE Transactions on Vehicular Technology,* Vol. VT-19, No.2, May, 1970, pp. 173–177.

[11] Staras, H., and L. Schiff. "A Dynamic Space Division Multiplex Mobile Radio System," *IEEE Transactions on Vehicular Technology,* Vol. VT-19, No. 2, May, 1970, pp. 206–213.

[12] Porter, P. T. "Supervision and Control Features of a Small-Zone Radiotelephone System," *IEEE Transactions on Vehicular Technology,* Vol. 20, No. 3, August, 1971, pp. 75–79.

[13] ND-52535, *NACTS—NEC's Advanced Cellular Telephone System Technical Information.*

[14] *Advanced Mobile Phone Service System Description,* AT&T, October 12, 1982.

[15] CIS-3, *EIA Interim Standard on Cellular Mobile Station-Land Station Compatibility Specification,* Electronic Industries Association, July 1981.

[16] OKI Advanced Communications, *Technical Specification Cellular Mobile Telephone Equipment,* T-310113, Issue 3, March 1983.

[17] Jakes, W. C. *Microwave Mobile Communications,* John Wiley & Sons: New York, NY, 1974.

[18] IS-54. *Dual-Mode Mobile Station-Base Station Compatibility Standard,* EIA/TIA Project Number 2215, December, 1989.

[19] Johnson, A. K., and S. Wilkus. "Front End Filter Design for the Newly Acquired Spectrum from the FCC," *IEEE Vehicular Technology Conference,* pp. 469–470.

[20] *Code of Federal Regulations,* Telecommunication §22.901, Domestic Public Cellular Radio Telecommunication Service, 10-1-89 Edition.

[21] Carey, R. D., *Technical Factors Affecting the Assignment of Facilities in the Domestic Public Land Mobile Radio Service,* Federal Communications Commission Report No. R-6406, June 1964.

[22] *Cellular System Mobile Station - Land Station Compatibility Specification,* Communication Interim Standard #3, Electronic Industries Association, July, 1984.

[23] Petranek, J. A., "Three Ways to Build Your RSA," *Cellular Business,* Sept. 1990, p. 56.

[24] Adcock, T. G., "Special RSA Highway Cells," *Cellular Business,* Sept. 1988, p. 36.

Chapter 3
Cellular Traffic

3.1 INTRODUCTION

The basic consideration in the design of a cellular system is the sizing of the system. Sizing has two components: coverage area and the traffic handling capability. The coverage area in a primary market is dependent upon SMSAs or NECMA. In the case of markets other than primary markets (as defined by the FCC), the coverage area is dependent upon the population distribution of the area of radio coverage. The second factor, traffic handling capacity, is based on the analysis of a complex group of data gathered from demographic and market analysis. After the system is sized, channels are assigned to cells, which is a function of user density in the cell, the frequency reuse distance, D, and the available spectrum. The channel reuse distance, D, as described in Section 2.2.4, plays an important part in reducing the cochannel interference to an acceptable value. After a cell has been assigned to serve a mobile radio, various channel assignment schemes can be followed to determine if a channel is available for use or not. There are several assignment schemes. Among the prominent schemes are: fixed channels assignment, dynamic channel assignment, and hybrid channel assignment [1–10]. Some heuristic techniques for channel assignment based on considerations of cochannel, adjacent channel, and spurious and intermodulation interference has also been reported [11].

In the fixed-channel assignment scheme, a subset of the total number of (416 channels in AMPS) voice and signaling channels can be permanently assigned to a cell. The same subset can, however, be used again on cells separated by an acceptable reuse distance. Only channels assigned to the group can be used within the cell. If all channels in a cell are busy, calls arriving or originating in the cell are rejected even though there may be channels available in the adjacent cells. In this scheme, the number of channels assigned to a cell provides an estimate of the amount of equipment provided at the cell site. If the quality of service is met, the amount of equipment provided at the cell site is adequate. These adequacies are judged by the number of subscribers each channel can support during the mean busy hour while providing a desirable quality of service. It is known empirically that the mobile telephone system will carry about 25 to 40 subscribers per channel with acceptable traffic congestion or blockage. The number of channels assigned in a fixed-channel assignment scheme is based on traffic calculations for the busiest hour of the day on a typical business day. An unusual or special day because of catastrophe, holiday, or planned entertainment events is not considered. Unlike the traffic generated due to special events, low-traffic subscribers are counted in the total traffic though they contribute little load on the system.

In the dynamic channel assignment scheme, all channels are kept in a central pool and every channel can be used in every cell. In order to avoid cochannel interference, an up-to-date status of the channels used at different cells must be known at the MTSO. As noted above, the same channels can be assigned to different cells when they are separated by a minimum reuse distance of D. Before the MTSO assigns a channel to a cell site, a computer search is required to determine the validity of the channel assignment. If the chosen channel passes the threshold criteria for cochannel interference, the channel is assigned to the cell demanding service. Thus, control of the dynamic channel assignment requires an access to and processing of a large amount of stored data. Therefore, a fast digital computer or a microprocessor-based system is required at the MTSO. From the individual cell point of view, the dynamic channel assignment forms a larger trunk group and thus serves the system more efficiently, as discussed in Section 3.5.

For the hybrid channel assignment, the total number of channels are divided into two groups. The first group contains channels assigned to cells on a fixed basis and the second group contains channels assigned on per-user-call basis. It has been reported that the division of channels among two groups (dynamic and fixed) should be based on the future growth of traffic density [4].

Since market survey is a very important phase before the design of a new system starts, we discuss the methodology of market survey in Section 3.2. This is considered in two phases: need survey and demographic analysis. Section 3.3 covers fundamentals where we discuss the terminology used in traffic theory. Experimental findings on the three different systems based on traffic statistics of cellular radio are recommended and extensively discussed in Section 3.4. Section 3.5 outlines Erlang B, Erlang C, and Poisson disciplines for a fixed channel assignment in the cells. Future monitoring of the system statistics is an important aspect of the cellular system, and is a part of Section 3.5. Different accessing schemes for the mobile system and why a dedicated control channel scheme has been chosen for cellular radio is discussed in Section 3.6. Sections 3.7 and 3.8 provide details for the hybrid channel assignment, and the improvement one obtains by reassignment of channels. Lastly, Section 3.9 provides a summary and conclusions.

3.2 TRAFFIC CONSIDERATIONS

Cellular radio was developed to provide telephone quality service to mobile radio subscribers. Inherent in this concept is the necessity to provide a grade of service (GOS) equivalent to the landline telephone system. Once the system is in place, traffic statistics can be gathered and facilities to provide service can be added, or existing facilities can be shifted to where they are needed. But what about the initial design of a system where traffic data is not available? The following sections describe a process generally followed by cellular radio applicants for arriving at an initial estimate of traffic.

The FCC requires cellular license applicants to include a market survey to justify their system size. The rationale for this is twofold. First, it provides a reasonable assurance that adequate facilities providing a good grade of service will be installed. Second, the subscriber cost for service will, to some extent, depend on the cost of building the system. A balance is required to prevent overbuilding resulting in an increased cost and burden to the subscriber. This would prevent applicants from overbuilding just to make an application look better. The specific criteria established by the FCC are:

- *Demand projection* is the determination of the demand for service. This includes a forecast of both the current and the projected demand, and the distribution of total demand among WCC and RCC.
- *Demand per cell* determines the geographical demand within the proposed CGSA.
- *Demand handling* involves the applicant's ability in terms of handling the projected demand on the system, and accounting for such factors as the temporal distribution of demand.
- The efficiency with which the proposed system will fulfill *expansion needs* created by future demand.
- The degree to which the applicant's rate structure promotes the *efficient use of a cellular system.*

The experts who performed the market and need surveys for each initial application in the first cellular markets (markets 1 through 90 in the U.S.) all claimed to have used the same basic methodology. However, a closer examination of the survey results indicates variations on the order of 100%. Despite of the wide diversity of results reported by each practitioner, we will present the elements of a market survey that all applicants seem to agree are necessary.

3.2.1 Methodology of Market Survey

Cellular telephone marketing assessment is generally based upon the surveys of two major market segments: commercial establishments and households. The reason for separate market surveys of these segments lies in the possibility that the volume, frequency, and character of demand will differ somewhat. A set of separate questions is framed for business as well as for the household. These questions are about the same in nature, and usually the point of departure pertains to the differences between household and business. We present below a brief summary of the process as followed by the applicants to arrive at a traffic assessment of the market. Since the steps and the techniques are identical for both, we simply present here the details of a common market assessment without differentiating these two segments of the survey. However, before we take up the discussion of a market survey, a few items that differentiate these two segments should be noted. It is expected, especially at the infancy of the service, that market penetration will be greater for the business population than for the residential population. A common result of almost all market studies to date is that the price sensitivity threshold for business is significantly higher than for residential or private users. A commercial user can write off the cost of the cellular radio service and at the same time use it to conduct business more efficiently.

As a first step in this process, a survey of the full-time wage earners is conducted to determine the potential level of interest in the proposed cellular mobile service area. The questionnaire seeks information concerning the relation between price and the individual level of interest for the proposed service, the intended use of service (personal, business, or both), the frequency of use, the geographical area of use, interest in special features, and other related factors. Additional information is sought regarding the socioeconomic conditions of the respondents, including age, income level, sex, occupation, type of industry they are employed in, and whether or not they have any physical impairments to their mobility. The result of a typical response for a person "very interested" in subscribing for the cellular service in the Kansas city SMSA is shown below in Table 3.1 [12]. The survey clearly indicates an increase in the level of interest as the monthly cost reduces.

In addition to the level of interest, additional factors are considered to arrive at the actual demand for cellular mobile services. They include the proposed rate structure, consumer awareness of the availability and the capability of the proposed service, consumer perceptions of the usefulness and desirability of cellular radio, and the effectiveness of marketing activities (nonwireline) compared to competitive wireline supplier.

The actual demand in terms of the number of subscribers is determined from the information similar to that in Table 3.2. This is based on the projected full-time employment in the year of service and growth. In this example, we will consider the growth during the fifth year. Assuming the initial service cost to be $140/month, which is reduced to $120/month within a five-year period (assumed), the projected total potential interest can be determined. The next step in the process is to convert the projected total potential interest into the actual number of expected subscribers for the service. The estimates are based on three key factors: market exposure, sales conversion rate, and market allocation between nonwireline and wireline carriers. The level of market exposure (i.e., consumer

Table 3.1
Typical Survey Responses

Monthly Cost ($)*	Respondents "Very Interested" in Cellular Service (%)
200	2.6
140	5.8
75	13.0

*Excluding long-distance charges.

Table 3.2
Projected Potential Interest for Cellular Service

	1984	1989
Full-time employment	551,900	583,500
Monthly cost ($)	140	120
Penetration potential (%)	5.8	7.0
Total potential interest	32,000	40,800

awareness of the proposed service) is a function of marketing efforts and the length of time the service has been available. Consumer awareness of the service will increase over time due to greater exposure to the service as well as marketing efforts and verbal discussions. Sales conversion rate is a function of the quality and characteristics of the proposed service as well as the effectiveness of the marketing program. Market allocation between RCC and WCC can be assumed to be in the range of 40–60%.

Considering the above three criteria, the figures in Table 3.2 can be converted to those Table 3.3. It should be noted that all these factors assume stable economic conditions. After arriving at the expected number of subscribers, these numbers are generally correlated into demographic patterns that include population, employment, and traffic concentrations.

3.2.2 Demographic Analysis

Demographic analysis is the second most important effort undertaken for arriving at the sizing of a system. A combination of a census tract map provided by the U. S. Department of Commerce, and of Census Bureau data from another source, such as the county and city data book, can be used to determine the population distribution throughout the proposed CGSA. The census tract maps have proven to be particularly accurate for this application. One such example of population and land distribution for the Memphis SMSA is shown in Figure 3.1 [13]. The Memphis SMSA covers a land area of about 2,305

Table 3.3
Expected Number of Cellular Subscribers

	1984	1989
Total potential interest	32,000	40,800
Market exposure (share) (%)	50	100
Number of interested consumers	16,000	40,800
Anticipated sales conversion rate (%)	50	70
Total expected number of subscribers	8,000	28,600
Market share (nonwireline) (%)	40	40
Actual expected number of subscribers	3,200	11,400

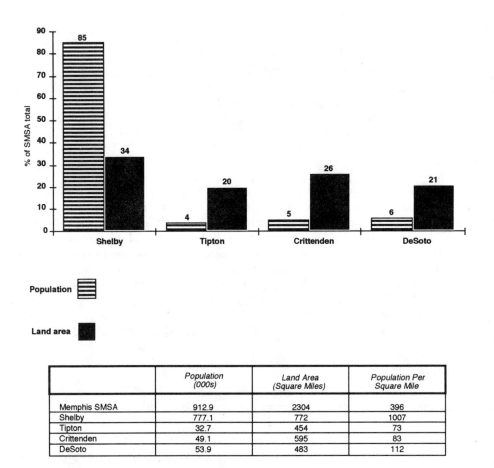

	Population (000s)	Land Area (Square Miles)	Population Per Square Mile
Memphis SMSA	912.9	2304	396
Shelby	777.1	772	1007
Tipton	32.7	454	73
Crittenden	49.1	595	83
DeSoto	53.9	483	112

Figure 3.1 Memphis SMSA population and land characteristics by county in 1980. Note that columns may not be added to totals or subtotals due to rounding. Source: U.S. Department of Commerce, Bureau of the Census. *Land area and Population: 1930–1980,* Washington D.C.: Government Printing Office, 1980.

square miles and includes Shelby and Tipton Counties in Tennessee, Crittenden County in Arkansas, and DeSoto County in Mississippi. According to the Bureau of the Census, the SMSA contained some 912,900 people in 1980. Shelby County had 777,100 people and accounted for some 85% of the SMAS's total population in 1980. Tipton, Crittenden, and DeSoto Counties accounted for 4%, 5%, and 6% of the SMSA's population in 1980, respectively. The population-density figures show an average of 396 people per square mile in the SMSA, while in Shelby County 85% of the population lived on 34% of the SMSA land area. The other counties had relatively low population density figures, and agriculture was their prime business. In addition to the population density for demographic analysis, population trends over the last decade need to be understood so that the future population growth can be projected.

Knowing the expected number of subscribers for a cellular service area from the market survey and knowing the population distribution in the SMSA, the probable number of users in different cells associated with different geographical areas can be calculated. After arriving at the number of potential users for an individual cell, a sufficient number of radio channels is assigned in order to provide an adequate grade of service. We describe this process below.

3.2.3 Channelization of a Cell

Once the geographic coverage and demand information is acquired, the system designer must determine how many channels will need to be assigned to each cell in the system. The procedure used is relatively straightforward. First, a GOS for the system is chosen. Since cellular systems have to provide service quality similar to the landline telephone system, the usual GOS chosen by the system designers has to be 0.02 or better. This means that a maximum of two calls, on average, out of every one hundred attempted may be blocked within the system during the peak hour of service.

Next the demand of traffic per cell, usually measured in erlangs or traffic units (TU), is calculated for the peak busy hour. This is done by multiplying the number of users in a cell by the anticipated number of calls per user in the peak busy one-hour period and by the projected duration of a call during the same period. The number of channels can then be arrived at by using either the Poisson, Erlang B, or Erlang C formulas. Each of these formulas is based on different assumptions, and the reader must understand these clearly before selecting an individual formula. Therefore, the next section is fully devoted to assumptions and terminology used in traffic engineering. Once the system is built and operational, the traffic statistics are gathered and analyzed. Any modifications necessary to retain service quality at the desired level will become evident from traffic analysis.

This process of channelization of a cell is known as a fixed channel assignment. The sequence of steps necessary to arrive at the number of channels per cell, and thus for the complete system, is shown in Figure 3.2

3.3 TERMINOLOGY IN TRAFFIC THEORY

Before we discuss the various formulas used in arriving at the number of channels needed for a specified or predicted traffic and the blocking probability, it is necessary to briefly define commonly used terms in traffic theory. Knowledge of the terms is mandatory for a proper understanding of cellular traffic: traffic flow or traffic intensity, the grade of service, the units used in traffic theory, call-holding time distribution, busy hour or peak busy hour traffic, interarrival time and arrival time distribution.

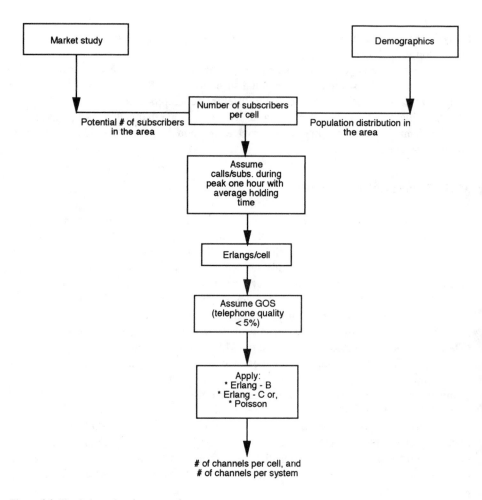

Figure 3.2 Fixed-channel assignment scheme.

3.3.1 Traffic Flow or Traffic Intensity

Cellular traffic, referred to here as traffic, is defined as the aggregate of mobile telephone calls over a group of channels with regard to the duration and the number of calls. Traffic flow or intensity through a cell site is defined as the product of the number of calls during a specific period of time, and the average duration known as call holding time, t_h. In traffic theory, the unit of time generally considered is a period of one hour. Thus the number of calls can be expressed in terms of the arrival rate (number of calls per unit time), and the average duration is expressed in terms of unit time per call. The traffic intensity E is given by:

$$E = \lambda t_h \text{ erlang} \tag{3.1}$$

where λ is the arrival rate expressed as calls/hour and t_h is the average holding time expressed as hour per call. It should be noted that the resulting quantity E is dimensionless and usually expressed in erlangs in honor of the Danish pioneer traffic theorist A. K. Erlang.

The following examples illustrate the computational process in two steps: First we compute the traffic intensity in erlang per call assuming that one call is generated by each subscriber during a busy hour period, and second, we compute the number of calls generated by each subscriber during the same busy period of one hour. These two together can provide the estimate of the number of erlangs per cell and the total traffic in erlangs per system.

Example 1

The results of the market survey represent the peak hour call length in minutes versus the percentage of subscribers in the individual groups. Compute the traffic intensity in erlangs per subscriber.

- 0–1 call; 50% of likely subscribers
- 1–2 call; 30% of likely subscribers
- 2–3 call; 15% of likely subscribers
- 3–10 call; 5% of likely subscribers

Normalizing to 100 subscribers (i.e., 100 subscribers = 100%) and assuming an average call length for each group, the total call-time duration can be computed as follows:

$$
\begin{array}{rl}
0.5 \times 50 = & 25 \\
1.5 \times 30 = & 45 \\
2.5 \times 15 = & 37.5 \\
6.5 \times 5 = & \underline{32.5} \\
& 140.0 \text{ minutes}
\end{array}
$$

Thus, 100 subscribers use a total 140.0 minutes of airtime during the peak busy hour. This is equal to an average call duration of 1.4 minutes equals 0.023 hours or hours per call (during rush one-hour period). Assuming an arrival rate of one call per hour ($\lambda = 1$), we have total traffic = 0.023 erlangs per subscriber.

Example 2

The market study of a hypothetical cellular system shows the distribution of the number of calls during the busy one-hour period versus the percentage of subscribers making those calls. Find the number of calls/subscriber.

- 0–1 calls during a busy one-hour period; 60% of likely subscribers
- 1–2 calls during a busy one-hour period; 30% of likely subscribers
- 2–10 calls during a busy one-hour period; 8% of likely subscribers
- 10+ calls during a busy one-hour period; 2% of likely subscribers

Normalizing 100% to 100 subscribers and using the average number of calls within each group, and assuming 10+ number of calls to be 10 calls for computational purposes, we obtain an average arrival rate of

$$\frac{0.5 \times 60 + 1.5 \times 30 + 6.0 \times 8 + 10 \times 2}{100} = 1.43 \text{ calls/hr}$$

Therefore, if the average arrival rate is 1.43 calls per hour during the peak busy hour (calls/hr), then the total traffic generated from Examples 3.1 and 3.2 is 1.43 calls/hr. × 0.023 hrs/call = 0.033 erlang per subscriber. If we now assume the number of subscribers per cell to be N, then the total traffic during the peak busy one hour period will be $0.033N$ erlang per cell.

Besides the erlang unit, occasional use is made of other units in traffic, which are discussed subsequently in this section. In connection with traffic intensity or traffic handled by a group of trunks, it is customary to use terms like "the carried traffic" and "the offered traffic." The carried traffic is the actual traffic load carried by a group of trunks, while the offered traffic is the actual demand of the traffic by mobile users. The difference between offered traffic and carried traffic is a function of GOS, which is the subject of our next discussion.

3.3.2 Grade of Service

If calls are to be handled without delay or loss, it is necessary to provide as many full-duplex radio channels as there are mobile subscribers. However, for economic reasons it is unrealistic to have as many radio channels as there are subscribers. Thus, to limit the number of RF channels to a reasonable number, a controlled amount of blocking is introduced into the system. Subscribers have to realize that some of the calls may not be completed when all channels are being used. In these cases, subscribers have to wait and try calling the desired party again. The term "grade of service" is defined as the number of unsuccessful calls relative to the total number of calls attempted. Thus the grade of service is defined as the measure of insufficiency of the number of channels available to mobile users. In practice, it is expressed as the portion of calls allowed to fail during the peak busy hour due to a limited number of RF channels. In cellular radio service, the system design is usually based on a grade of service of 0.02 or better. A 0.02 GOS means that a subscriber will find on average an available channel 98% of the time during the peak busy hour loading of the system. During nonpeak hour loading, the GOS will improve, and in fact most systems will appear to be unblocked.

3.3.3 Traffic Unit

There are two commonly used units of traffic; namely the traffic in terms of erlangs and the traffic expressed as unit calls (UC) or the synonymous term hundred calls seconds (HCS) per hour, abbreviated as CCS. The erlang, as a unit of traffic, represents a radio channel being occupied continuously for a duration of one hour. Thus, we can define one erlang as a single call occupying a channel for one hour. The relationship between erlangs and CCS units can be derived by observing that there are 3,600 seconds in one hour. The other less popular units of traffic are: traffic unit (TU) and the equated busy hour call (EBHC). The intensity when expressed in CCS, HCS, and UC represent the average number of calls per hour on the basis of a mean holding time of 100 seconds. The equated busy hour call assumes a mean holding time of two minutes. The relationships between CCS, EBHC, and erlang are:

- 1 CCS = one call for 100 seconds
      = one call for 1/36 hour
      = 1/36 erlang;
- 1 EBHC = one call for 120 seconds
      = one call for 1/30 hour
      = 1/30 erlang.

A complete summary of their relationships is provided in Table 3.4.

3.3.4 Call Holding Time

One of the most important characteristics of traffic is the time required for the facility to satisfy a service demand. Normally we are interested in the distribution of demand in terms of the mean and variance. From the histograms of the measured holding time, the probability density can be deduced. The two most used holding time (t_h) distributions are exponential and uniform. A worthy property to note for the exponential distribution is that the variance of the distribution is equal to its mean. In cellular systems, a uniform holding time of 140 seconds has been recommended by the vast majority of companies, including the 1971 Bell Telephone Company report on cellular telephone systems. The 140 seconds is derived from improved mobile telephone service (IMTS) stations. While

Table 3.4
Different Traffic Units and Their Relationships

Traffic Units	Erlang	CCS	EBHC
1 erlang, 1 TU	1	36	30
1 CCS, 1 HCS, 1 UC	1/36	1	5/6
EBHC	1/30	6/5	1

140 seconds is generally accepted, it is recognized that it is probably incorrect. As cellular service starts to approach the service provided by the landline system, the average call holding time is expected to increase.

3.3.5 Busy Hours or Peak Busy Hour Service

Since a cellular system is designed such that even during the busiest time traffic can be handled smoothly and to the satisfaction of subscribers, that is, with a prescribed grade of service, all system designs are based on the amount of mobile traffic anticipated during the busy hour of a normal weekday at the busy time of the year. According to CCITT definition for regular telephone services, the busy hour is defined as a period of 60 minutes during the day when the traffic intensity of the trunk group under consideration, averaged over several weekdays, is at its greatest. In practice, the busy hour is usually fixed to the nearest quarter of an hour. For cellular mobile systems, peaks usually occur between 10:00 A.M. and 12:00 A.M. with a second peak between 1:00 P.M. and 3:00 P.M. due to commercial and business users. Morning and evening commuters do not generate enough load to cause any shift in traffic peaks. This is contrary to what we may think is normal. In arriving at the peak hour traffic, various types of users must be considered: business and commercial; dispatch, including police, taxi, trucking, and repair services; entertainment; personal security; and vanity.

As stated earlier, business and commercial users make up the bulk of the peak hour traffic. The transportation industry includes operators of taxi fleets, limousines, tour buses, and a few other types of services. Cellular service is likely to attract dispatch users because they are not members of organizations large enough to be able to support a private dispatch system of their own. These users contribute substantially to traffic, but they generate short-duration calls and have less usage per hour per vehicle than business and commercial users. Users from the entertainment industry will generate the bulk of their traffic during off-traffic hours. An example of these users are customers of Atlantic City and Las Vegas casinos. Personal security users are those who install cellular radio in their vehicles for personal safety reasons. There is also a class of user who installs the cellular radio for reasons of prestige. These users are generally a small segment of the cellular community and are not major contributors to peak hour traffic.

Lastly, in remote locations in the United States and Third World countries where landline telephone is not fully developed, automobile telephones may be used in lieu of the wire telephone service. The special probability distribution of these users must be known before the final traffic figures can be arrived for the cellular radio system design serving these special areas.

The traffic characteristics of police, taxi, trucking, repair, and maintenance services have been measured in several large cities of Canada and have been reported in [14]. The characteristics of these users have been expressed in terms of ''monologue'' or ''dialogue'' types of behavior. If B and M_0 represent the total base and mobile transmission duration within a message, and if M represents the message duration, then by knowing the ratio B/M and $(B + M_0)/M$, the total base station occupancy percentage per message and the combined base and mobile occupancy percentage per message can be known. When the ratio B/M is approximately equal to half of $(B + M_0)/M$, this indicates a dialogue type of behavior, which means that both the base station and the mobile occupy half the message duration. When $(B + M_0)/M$ and B/M are approximately the same, the behavior is that of a monologue type. Based on the previously defined ratios, the traffic parameters for each class of dispatch users can be derived on a per message basis. The ratio between the number N_B of transmissions due to base station and the number N_{M_0} of transmissions due to mobile is

$$N_{M_0} = M_0/\mu_{M_0} \qquad N_B = B/\mu_B \tag{3.2a}$$

where μ_{M_0} and μ_B are the average durations of the base and mobile transmissions. Therefore,

$$\frac{N_{M_0}}{N_B} = \frac{\mu_B}{\mu_{M_0}} \frac{M_0/M}{B/M} \tag{3.2b}$$

Since $M_0/M = [(B + M_0)/M] - (B/M)$, we can alternately express (3.2b) as:

$$\frac{N_{M_0}}{N_B} = \frac{\mu_B}{\mu_{M_0}} \frac{[(B + M_0)/M] - (B/M)}{B/M} \tag{3.2c}$$

For a given type of user, if N_{M0}/N_B is very small, there are more transmissions from the base than from the mobile. If the ratio is close to unity, it represents an approximately equal number of transmissions from both the base and the mobile (dialogue behavior). On the other hand, if the ratio is greater than unity it signifies that the mobile is taking a predominant role in the conversation. The message probability (number of messages versus probability of their occurrence) for these four classes of users are shown in Figure 3.3. The measured values of B/M, $(B + M_0)/M$ and N_{M0}/N_B for police, taxi, trucking, and maintenance repair services are shown in Table 3.5. From the message probabilities of this class of users, the average holding time can be arrived at. This holding time is also related to μ_{M0} and μ_B, the average holding times for the mobile and base transmissions. Once we know the average holding time for all classes of users, the average holding time for cellular users can be estimated.

3.3.6 Arrival and Interarrival Time Distribution

Assume the call arrival to be independent with an average arrival rate of one call per second, we can compute the probability of K arrivals in an interval of t by using the Poisson distribution or $P(K$ arrival in the time interval $t)$:

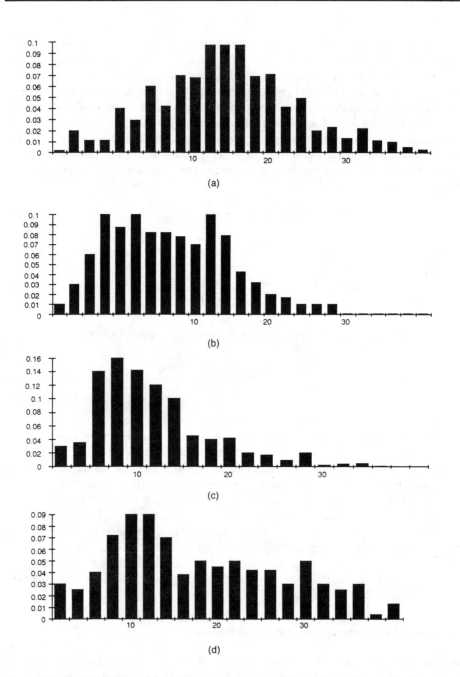

Figure 3.3 Probability density function of the message length for different classes of users: (a) police; (b) taxi; (c) trucking; (d) maintenance and repair.

Table 3.5
Measured Behavior of Dispatch Users

	Ratio		
	B/M	$(B + M_0)/M$	N_{M0}/N_B
Police	0.45	0.77	1.17
Taxi	0.53	0.65	0.35
Maintenance and repair	0.36	0.68	2.31
Trucking	0.38	0.68	0.58

$$P_K(\lambda t) = \frac{(\lambda t)^K}{K!} e^{-\lambda t} \tag{3.3}$$

The underlying assumption of arrivals from a large number of independent sources is implied in (3.3). From the above equation, the often asked question of K or more arrivals in the time interval of t seconds can be answered.

$$P_{\geq K}(\lambda t) = \sum_{i=K}^{\infty} P_i(\lambda t) = 1 - \sum_{i=0}^{K-1} P_i(\lambda t)$$

$$= 1 - P_{<K}(\lambda t) \tag{3.4}$$

Similarly, the interval time distribution or the probability of no call being received in the time interval of t seconds can be arrived by evaluating (3.3) for $K = 0$ or

$$P_0(\lambda t) = e^{-\lambda t} \tag{3.5}$$

From this discussion, P (K arrival in time interval t) can be interpreted as the probability of no arrival in the time interval t, $P_o(\lambda t)$, multiplied by the coefficient $(\lambda t)^k/K!$ If the coefficient is greater than 1, the probability $P_k(\lambda t)$ increases, if the coefficient is less than 1, then $P_k(\lambda t)$ decreases. Being a Poisson distribution, it is assumed that the probability of an arrival in a sufficiently small interval of time, Δt, is proportional to the length of the interval.

3.4 TRAFFIC CHARACTERISTICS OF THE CELLULAR SYSTEM

There are three sources of traffic data from which the traffic statistics of the cellular mobile radio telephone system can be anticipated: Bell Laboratories' report on high-capacity mobile telephone systems, published in 1971 [1]; measurements made on the mobile telephone system at Honolulu in 1975; and the measurements made on 120 UHF channels in the range of 405–420 MHz at Vancouver, British Columbia [15]. We now summarize the traffic statistics reported in these three studies.

Traffic characteristics of the mobile telephone system was reported by Bell Telephone in 1971. Since the objective of the cellular system is to provide a grade of service equal to that of the PSTN, a GOS of 0.02 is recommended. This is a marked improvement over the existing grade of service of 0.65 as a nationwide average of a large city dial mobile system. From this study, it was estimated that the cellular system will carry an average of 0.03 erlangs of busy hour traffic for each subscriber, and that the average call duration will be 140 seconds. Traffic was found to spread throughout the day with approximately one-seventh of the total business day traffic occurring in the busy hour period. Of the completed calls, 60% were mobile-originated and the 40% were land-originated. Among the total numbers of calls originated by the mobile only 60% were completed, (i.e., the actual conversation took place). This ratio was somewhat lower for the land-originated calls, where about 50% of the calls never terminated because the mobile unit were either turned off or outside the coverage area (proposed seamless coverage in the U.S. will change this). Calls originated by the mobile user's radio channel during DDD connection and ringing were partially counted in the traffic computation as the mobile uses the radio channel. However, if the called party did not answer, the attempt was counted as a completed call of somewhat a shorter average duration. Land-originated calls caused the mobile telephone number to be transmitted over the paging channels on all the cells. If the mobile did not answer the page, a voice channel was never assigned. Thus if the mobile user was not in the car or if the radio was turned off, no traffic was generated on the voice channel and the attempt was not counted as a complete call, although both the switching and the paging facilities have been used. This is explained further in Section 2.3.2.1. Thus these calls were considered as a load to those subsystems (paging and switching facilities) which come into action before the voice channel is assigned. A summary of the relevant traffic statistics, as recommended in the report for both the mobile and dispatch systems (ADS), is shown in Table 3.6. It should be noted that dispatch users may very well be served by the cellular telephone system as discussed in Section 3.3.

Table 3.6
Traffic Characteristics of Cellular System

	MTS	ADS
Busy hour traffic/mobile unit	0.03 erlang	0.004 erlang
Ratio of total daily traffic to busy hour traffic	7:1	7:1
Average call duration (seconds)	140	25
Completed calls (mobile-originated)	60%	N/A
Completed calls (land-originated)	50%	N/A

Note: N/A = not available   Source: [1].

Based on the measurements by Rydax automatic mobile telephone system at Honolulu, the measured traffic was 0.01 Erlangs per listed subscriber during the busy hour. The number of calls per listed subscriber was reported to be 0.33 per hour. Thus, the computed value for the average call duration was 109 seconds.

The measurements at Vancouver, British Columbia were also made by the Rydax automatic radiotelephone terminal on a UHF telephone system known as AUTOTEL, which provided coverage throughout the metropolitan free calling area. The main purpose of the AUTOTEL system study was to come up with the best approach for future mobile systems development and to arrive at schemes which would provide temporary relief to the system at Vancouver. Customer calling patterns and peak traffic statistics during the busy hour and for a busy day were reported. They are summarized in Table 3.7. There, one can see that the customers who are active during the busy hour make a mean average of about three calls during the busy hour. The AUTOTEL pilot system had a reasonable mixture of low-volume "prestige" users with medium and heavy users, and thus was a fair representation of the traffic of a typical cellular radio telephone system. The low channel loading levels involved enabled the peak offered traffic to be studied without the usual clipping effects caused by system overload, which is found on most operating systems.

The AUTOTEL customer job classification is shown in Figure 3.4. From the figure, service, construction, and the private sector are the biggest groups of users for the system. Hourly normalized traffic distributions are shown in Figure 3.5. As seen from this Figure and as discussed in Section 3.3.5 there are two peaks. The first traffic peak occurs between 11 A.M. and noon, and the second traffic peaks occur in the afternoon between 2 and 3

Table 3.7
Mobile Statistics Measured at Vancouver

	Mean	Standard Development
Busy-hour data:		
Customers active during the busiest hour of the month (%)	25.3	6.5
Busy-hour calls per active busy-hour customer (erlangs/subscriber)	2.97	0.91
Busy-hour traffic per active monthly customer (erlangs/subscriber)	0.023	0.006
Busy-hour traffic, busiest-day traffic (%)	15.4	4.0
Busy-day data:		
Customers active during the busiest day (%)	67.6	8.3
Busy-day calls per active busy-day customer	8.1	2.5
Busy-day completed traffic per active monthly customer (erlangs/subscriber)	0.141	0.043
Busiest-day traffic, total monthly traffic (%)	7.5	2.5
Monthly completed traffic per active customer billed (minutes)	156.5	29.7

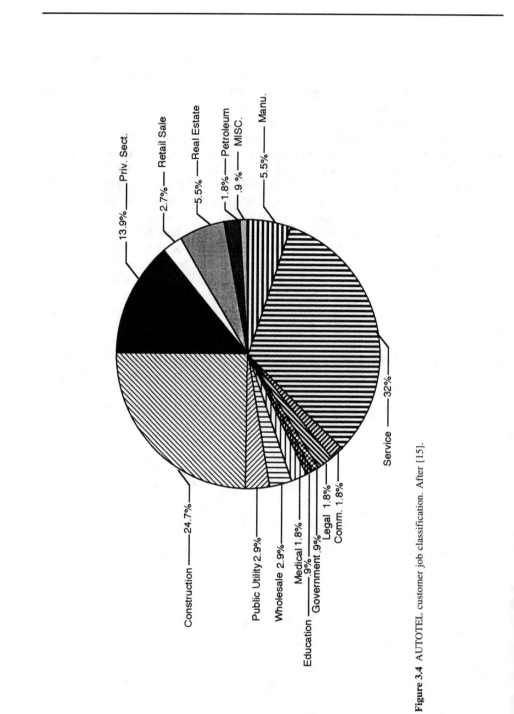

Figure 3.4 AUTOTEL customer job classification. After [15].

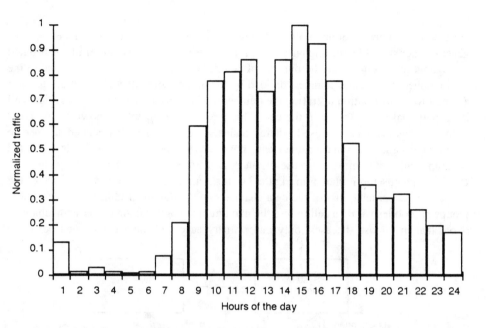

Figure 3.5 Normalized hourly traffic for AUTOTEL system.

P.M. The usual rush hour morning and afternoon traffic does not contribute all that much to traffic.

As seen from Table 3.7, the busy hour traffic per active monthly subscriber was reported to be 0.023 erlangs. Some of these users may not be active during the peak hour of the day. During the busy hour, 25.3% of the monthly customers were active. Thus the average number of busy hour calls = $2.97 \times 0.253 = 0.75$. Therefore, the average holding time, t_h, for a call = $0.023/0.75 = 0.03067$ hr = 110.2 sec.

Based on these measurements, the following range of traffic numbers can be recommended for the cellular mobile radio system.

- Traffic per subscriber during busy hour of the busy month: 0.02–0.04 erlangs;
- Average call duration: 100–140 sec;
- Blocking probability (GOS): 0.02–0.05.

The traffic per subscriber during the busy hour will decrease as the system matures, and as the system users increase the use of the system during off-peak hour.

3.5 FIXED-CHANNEL ASSIGNMENT SCHEME

As we will further discuss in Section 5.1, the number of potential subscribers desiring cellular services can be projected with the help of demographics and market surveys. In order to arrive at the total number of channels required for the service, two other elements are taken into consideration: average holding time of the call and the number of calls generated per subscriber during busy hour. If traffic is assumed to be uniformly distributed throughout the system, then the number of channels assigned to each cell is equal to the total number of available channels divided by the number of cells within the cluster. For a nonuniform distribution of traffic, the number of channels assigned to an individual cell is proportional to the traffic in the cell. This type of assignment is known as the fixed channel assignment and is used where channels are permanently assigned for use in a particular cell. The same channel can, however, be allotted to other cells separated geographically by a minimum reuse distance, D, as discussed in Chapter 2.

A typical setup is shown in Figure 3.6. It should be noted that since the role of the data channel is crucial, loss of the data channel due to equipment failure can cause large groups of mobiles within a cell to lose the maximum strength paging or the setup channel. (Note that in almost all cases, only one primary access channel is assigned per cell.)

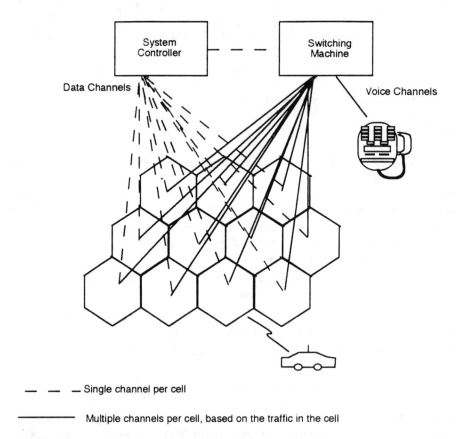

— — — Single channel per cell

————— Multiple channels per cell, based on the traffic in the cell

Figure 3.6 Cellular voice and data channel interaction.

However, this problem can be avoided if backup channels are assigned to the cell upon losing the primary setup channel. It should be noted that the dedication of a channel for signaling in the individual cell (21 signaling channels total) does reduce the total number of simultaneous voice communications.

In a cellular system requiring all 416 channels, the trunking efficiency is reduced due to the 21 dedicated signaling channels. At the same time, due to the continuous presence of synchronizing data on the setup channel, a unit just switched on would be able to receive or originate a cell after a brief period of synchronization. Since the data transmission rate is high (10 Kbps in AMPS) between the mobile and the cell site, elaborate provisions must be made to avoid and correct data transmission error due to the multipath fading and noise. These provisions include repeating the messages several times along with error correction coding. Other systems of the world transmit data at different rates. The data rate for different systems of the world are discussed in the next chapter.

Since voice channels are shared among users in the cell, they are regarded as "trunks." A basic fact of a trunked system is that the larger the trunk group, the more efficient the communication system. The efficiency is based on the following assumptions: the probability that many subscribers will communicate simultaneously is small and the average holding time of the call is small.

It is simple to see that if m subscribers are assigned to n channels at a cell site, then the probability that all n channels are busy is p^n, where p is the probability of a single channel being busy. Thus for a fixed p, p^n is a small number. In other words, if a subscriber has access to a single channel, the probability that he or she will be blocked at a given instance is simply p, but with access to n channels, the probability of blockage is reduced to p^n. Thus, chances of accessing the trunks are greatly improved by having more voice channels at a cell site and access to all. Alternately, for a fixed p, the number of mobiles served, m, can be significantly increased for the same GOS when more trunks are within one group than if the mobile would have access to just one or two voice trunks. In the extreme case (one cell system), mobile users can have access to all 411 (single signaling channels) trunks, thereby making the efficiency high, as illustrated below in example 3. As discussed above, there are two type of channels involved from the cell site to the cellular subscribers: the voice channels and a setup channel. We discuss below the traffic requirements on these channels.

Example 3

Assume that a single channel providing service to mobile users has a busy probability of 20% during peak hours of the day. If the number of channels is increased to six, find the additional traffic per channel that the system will carry with the same blocking probability. Use Erlang B discipline for the computation; given P (blocking) = 0.2.

- For $P_B = 0.02$, a single channel system will carry = 0.25 erlangs (Table B.1).
- For the same blockage, a six-channel system will carry 5.11 erlangs.

Thus, the traffic on a per-channel basis is 0.852 erlangs. Therefore, additional traffic carried per channel = 0.652 erlangs.

3.5.1 Voice Channel Traffic

From the market survey, demographics, and the assumed statistics of call arrival rate, the number of trunks per cell can be arrived at for a specified GOS. As stated in Section 3.4, the recommended grade of service for cellular systems is 0.02 or less. There are three basic formulas in traffic engineering: Erlang B, Poisson, and Erlang C. The underlying assumptions in all three formulas are: calls occur at random intervals, the number of users per channel is very large, all users offer the same traffic, the system is in statistical equilibrium and, the traffic offered is either known or accurately predictable.

The assumption with respect to the handling of a call when it arrives to the system, however, is different for each case. A system designer must understand this assumption before using one of these formulas. The objective of this section is not to derive these equations, but rather to make use of the existing theory for an accurate design of the system. An extensive treatment of traffic theory in terms of curves and equations has been documented in [5]. Table B.1 provides the traffic in erlangs (Erlang B) with different blocking probabilities for the number of channels $S = 1$–200 [19]. Figures B.1–B.4 provide traffic in erlangs versus probability of block for Poisson and Erlang C loads.

In the Erlang B case, if all channels are in use when the system receives a new call, the call will not be serviced. For Poisson, blocked calls wait in the system proportional to the average holding time. If a channel becomes available before the holding time expires, service will be provided and the caller will be allowed to use the channel for the rest of the holding time. For Erlang C, blocked calls wait in the system indefinitely. For new system like cellular, it is difficult to choose between the three formulas. However, each provides close estimates of the number of channels for a specified blocking, especially for a designed GOS of 0.02 or less (usual in most systems). A comparison of the load provided by three formulas for a 10-trunk system is shown in Figure 3.7. The divergence of the formulas is small when blocking is < 1% and appreciable when > 10%. Below 0.02, Erlang C and Poisson provide almost identical results, while Erlang B will carry approximately 0.06 erlangs more traffic at 0.02 blocking.

Most cellular systems have no capacity for an attempted call to be held, waiting (Poisson case) until an idle channel is available. Therefore, Erlang B should be the best choice. However, if the caller tries to redial the number again and again, the behavior of the system is different than what one obtains by applying Erlang B. The Erlang B formula in this case is the optimistic choice and provides a lower probability of blocking than a real-life traffic situation. Contrary to Erlang B, the Erlang C formula forces a user to wait indefinitely until an idle trunk is available. In cellular surroundings, the user will often be in an automobile, on a train, on a street, or perhaps in a hotel or restaurant. Since the cellular system allows the called number to be stored before one can press the call

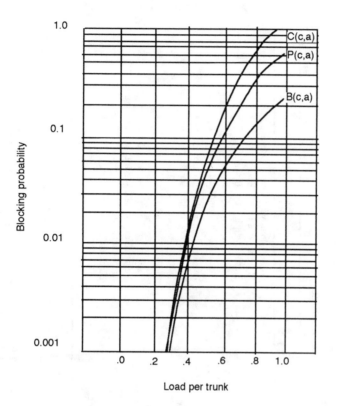

Figure 3.7 Comparison of loads as specified by Erlang C (C), Poisson (P), and Erlang B (B) blocking formulas.

origination key, the user can cause this feature to redial again and again. Thus, application of the Erlang C equation should be more appropriate. The Erlang C formula is more pessimistic than the Erlang B for cases where calls are not held in the queue. The Poisson formula is a compromise between Erlang B and Erlang C and may be applicable to cellular systems because of the ignorance about exact traffic patterns. It should also be recognized that Poisson formulas will actually carry more traffic and are better utilized than Erlang B, where callers are outright rejected. The most efficient way to keep radio channels occupied is to have a queue of waiting subscribers, just like the lines of a typical airline reservation system or bank tellers for service. Systems that do not permit this may in reality perform poorly. In practice, a cellular user may accept a 10% or even 20% blocking probability if the wait time for the channel to become available is only 10 or 20 seconds. Since blocking only occurs during peak traffic conditions, subscribers may gladly accept the service. Table 3.8 provides the different traffic formulas and their assumptions. It should be noted that the assumptions used in these equations may not hold exactly for cellular traffic, and it may be necessary to refine these equations based on measured

Table 3.8

Traffic Formulas and Their Assumptions

Formula	Blocking Probability*	Assumptions
Erlang B	$P_b = B(S, A) = (A^S/S!)/(\sum_{k=0}^{S} A^K/K!)^\dagger$	Poisson input with mean of λ arrivals/sec Mean service time $= 1/\mu$ Traffic intensity $A = \lambda t = \lambda/\mu$ Number of serving trunks S Blocked calls abandoned.
Poisson	$P_b = P(S, A) = \sum_{k=S}^{\infty} A^K/K! \, e^{-A}$, where P_b = probability that all trunks are busy‡	Poisson input with mean arrival rate λ Negative exponential service time with mean $= 1/\mu$ Traffic intensity $= A = \lambda/\mu$ Blocked calls held.
Erlang C	Probability of delay $= C(S, A) = P[\tau_D > 0]$ where $C(S, A) = \dfrac{(A^S/S!)[S/(S - A)]}{\sum_{i=0}^{S-1}(A^i/i!) + \{(A^S/S!)[S/(S - A)]\}}$ Probability of delay greater than $t = P_r[\tau_D > t] = C(S, A)e^{-(1-A)S\mu t}$ Average delay: $E[\tau_D] = C(S, A)/(1 - A)S\mu^\diamond$	Poisson input with mean arrival rate λ Negative exponential service time with mean $= 1/\mu$ Traffic intensity A Block calls held until served.

*Probability that all S servers are busy.
†See equation (B.13).
‡See equation (B.9).
◇See equation (B.18).

traffic characteristics. However, in order to mathematically track and come up with closed form solutions, these assumptions are accounted for here.

In the fixed-frequency allocation method, the total number of 395 voice channels (U.S.) is divided into m separate frequency groups or cells. Each of these cells contain j channels. A cell is assigned to one of the m groups. Each cell site can communicate with as many as j frequency channels simultaneously. The number of groups, m, is closer to the smallest number that satisfies the interference-buffering requirements. For 2-belt or 19-zone buffering, it is easy to see that the total number of channels may never be divided into more than seven groups. Each group contains a total of j frequencies. For example, group 1 may contain voice channels 1–56, group 2 may contain voice channels from 57–113, and so forth. These frequencies cannot be reused within the area shown by the hatched lines in Figure 3.8. Since each cell site has only one group of j frequencies, the number of transceivers at the cell site is also j. The mobile, when communicating within the cell, must be able to tune to one of these j frequencies also.

In an application of the above traffic formulas, it is assumed that the central processor knows the mobile location. Another important consideration is the treatment of calls in

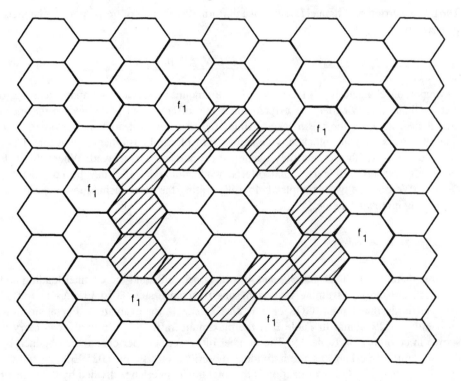

Figure 3.8 Frequency assignment of a typical seven-cell group. Shaded area is for zones on which reuse of frequency, f, used in the central seven cells are forbidden.

which the mobile crosses a cell boundary during a conversation. For the purpose of applying these equations, it is assumed that the cell size is large enough and the average message length short enough so that the handover never happens (i.e., the probability of this occurrence is zero).

Let us consider the case of N cells with a uniform distribution of traffic. Then the values of $S = C/N$ and $a = A/N$, where A is the total traffic per system and C is the total number of channels. Here, S is the total number of channels per cell, and a is the traffic in Erlangs per cell. In this case the blocking probability, P_b, for Erlang B can be computed by modifying the equation as follows:

$$P_b = B(S, a) = B(C/N, A/N) \qquad (3.6a)$$

The traffic carried in the cell is given by:

$$a' = a(1 - P_b) = a[1 - B(C/N, A/N)] \qquad (3.6b)$$

The traffic carried per channel (assuming a uniform distribution of traffic among channels) is given by:

$$L = a'/S = a[1 - B(C/N, A/N)]/ (C/N) \qquad (3.6c)$$

An important characteristic of a system is the amount of service demanded relative to that available and is called the offered load. The offered load is defined as the mean arrival rate, λ, times the mean service time, $1/\mu$. We say that arriving customers who find all servers busy are blocked. One behavior of blocked customers is that they leave or are cleared. Another is that they queue for service and are delayed. Where the calls queue, the carried load must equal offered load, although calls may be considerably delayed. We provide two examples below illustrating the basic technique for the fixed-channel assignment scheme.

Example 4

Assume there are 4, 8, 10, 20, 30, and 40 trunks per cell. Find the channel loading and the trunk utilization, assuming that each subscriber generates 0.025 Erlangs of traffic. The cell is designed for a GOS of 2%. Use Erlang B for evaluation. For a blocking probability of 2% found in Table B.1, the maximum traffic that the trunks can carry is shown in column 2 of Table 3.9. The number of subscribers per channel is obtained by dividing traffic in erlangs by the individual subscriber's traffic of 0.025 Erlangs and the number of channels. Trunk utilization is the total traffic in erlangs divided by the number of channels under consideration, as shown in column 2. The number of subscribers per channel is obtained by dividing traffic in erlangs by individual subscriber traffic of 0.02

Erlang and the number of channels. Trunk utilization is the total traffic in erlangs divided by the number of channels under consideration.

Table 3.9
Statistics for Example 3.4

Number of Channels	Traffic (Erlangs)	Channel Loading (Subscriber/Channel)	Trunk Utilization (Erlangs per Channel)
4	1.09	10.90	0.27
8	3.63	18.15	0.45
10	5.08	20.32	0.51
20	13.2	26.4	0.66
30	21.9	29.2	0.73
40	31.0	31.0	0.78

From the above computation, the channel loading and trunk utilization increases 2.9 times when the number of channels per cell is increased from 4 to 40.

Example 5

Assume channel loading of 30 subscribers. Find the total number of channels required and the channel reuse factor. The total number of projected subscribers in the first, fifth and tenth year of service is 2370, 14100, and 46890, respectively. At the channel loading of 30 subscribers, the number of channels required is: 79, 470 and 1563, respectively. Since there are a total of 395 voice channels, the channel reuse factors are 0.2, 1.19, and 3.96 in the first, fifth, and tenth year of service.

3.5.2 Monitoring the GOS for Voice Channel

The required GOS has to be equivalent to that of the public switched network, or a GOS of 0.02 or better has to be maintained throughout the life of the cellular system while traffic grows over a period of time. In order to delay an expensive hardware investment through cell splitting, facilities are provided so that a GOS better than 0.02 is achieved initially. System traffic is continuously monitored on a cell-by-cell basis. This allows the determination of when the system performance approaches the lower limit of the desired GOS. Monitoring traffic usually consists of recording the relevant data associated with subscriber's call. Statistics include: call attempts per user, number of successful calls through the system, call holding time, handoff time duration, user distribution in the system for average and peak traffic, and marginal call attempts by users trying to obtain maximum range.

The traffic monitoring unit (TMU) is a portable data acquisition unit that measures the peak plus the total system traffic. Additionally, the unit will also detect any catastrophic failure in the system. The unit generally displays the following:

1. Scrolling records, which includes the number of trunks in service, all trunks busy, number of originating and terminating blocked calls, and the peak traffic.
2. Hourly traffic records, which includes system traffic (also with day and time indication), traffic intensity, number of originating and terminating calls, originating and terminating blocked calls, and the number of trunks in use.
3. Daily traffic records.
4. Subscriber records.

From these statistics the grade of service can be computed. Once the cell with peak loading exceeds a GOS of 0.02 (guideline GOS = 0.05) for a few days in a month (typically 3–4 days out of 20 consecutive business days), action is taken to reduce blocking in the system. Before the expensive proposition of cell splitting is undertaken, the remaining available channels are assigned to the cell or channels are borrowed to prevent system blockage. These steps, if taken properly, will temporarily forestall the necessity of adding new cells or dividing a cell by using the sector transmit configuration. Therefore, the system will allow a maximum utilization of its resources at the lowest cost to the subscriber, which is one of the guidelines set forth by the FCC. Of course, in borrowing channels from other cell sites considerations are taken to prevent both cochannel and adjacent channel interference. Since borrowing channels is an interim solution, the borrowed channels are restored to their original cells when they are no longer required.

Once the GOS is not met for a few days in a business month, the busy cell is divided into two or more cells. The process of cell splitting provides the means to reuse channels, thereby multiplying the number of voice paths available to serve the subscribers. By splitting the cells, and accordingly decreasing the area of each cell, the system can adjust to a growing traffic demand density (i.e., simultaneous calls per square mile without an increase of spectrum). The expansion of the system is generally necessary for a number of reasons:

- Congestion on existing cells.
- Demand pattern may differ from original market predictions.
- Additional traffic area may come into existence, requiring the establishment of additional cells.
- It may be warranted at some stage to interconnect geographically adjacent systems by installing cells in areas where sufficient traffic does not exist otherwise.

The first two reasons can be addressed by addition of channels and/or cell splitting. These steps are taken by the cellular system operator. FCC approval is required for the third and fourth reasons for the change of a previously defined CGSA and/or new facility construction.

3.6 SIGNALING CHANNEL SELECTION AND TRAFFIC

As discussed in Chapter 2, before obtaining a voice channel for conversation it is necessary for the mobile user to find an idle channel over which it is possible to relay his or her intention to the cell site and the MTSO for call initiation. There are several known techniques that can be used for setting up a call. The three most common schemes are: all-channel signaling, scanning the voice channels, and a dedicated signaling channel.

All-channel signaling transmits low-speed data, either below or above the audio band. Sometimes a part of the speech band can be allocated for data. This scheme is generally identified as inband out-of-band signaling. Due to the limited bandwidth of the voice channel, a data rate in excess of 300 bps is hard to achieve. The problem usually shows up in the access time, which ultimately lowers the efficiency of transmission.

In the scanning scheme, the mobile user scans for the idle voice channel. Signaling is done first, followed by conversation. The sequencing process to find an idle channel is generally slow and adds delay in establishing the call, especially during periods of high traffic. Chapter 4 describes the NORDIC 450 MHz and 900-MHz systems where, for those calls originating at the mobile, the traffic channels (TCH) are used first for signaling and then for voice.

All channel signaling and scanning schemes can be discarded in favor of dedicated signaling channel for high-volume traffic (especially in view of centralized control, which is the modern trend in the design of communication systems). For larger system signaling, time is important. In other words, the signaling scheme should be such that the elapsed time between the allocation of the channel to the subscriber for signaling and the time when the channel becomes free from signaling should be minimal. If the elapsed time for signaling is greater than the rate at which the channels become idle (free of conversation), the system will build up the queue of unused channels, and consequently the channel utilization will be affected.

As the number of channels to be scanned is increased, the delay increases up to a point where an efficient operation is no longer possible and it becomes necessary to fix the signaling channel permanently. Due to the centralized control of the dedicated channel scheme, it is also possible to assign the service requests on a FIFO basis or some other priority scheme. On the other hand, with the scanning scheme there is no way to assign priorities. When a channel becomes free, the first mobile unit to lock to the channel, and consequently signal, will get the service while the other mobile units may have been waiting longer. It is strictly a function of where the scanner is with respect to the mobile desiring the service.

Thus, both all-channel signaling and scanning techniques can be discarded in favor of a dedicated signaling channel. Usually one channel per cell is sufficient for this application, even during high-traffic periods with a reasonable amount of blocking. Mobile units continuously monitor the busy/idle status of the access channel, which is the reverse channel from mobile to the cell site. If the channel is idle, mobile identification, along with the desired telephone number of the called party, is transmitted in the reverse channel. When the MTSO receives this information, it allocates an idle voice channel and informs the mobile through the cell site that the mobile can now tune to the desired channel frequency for transmission. This has been discussed in Section 2.3.2.2 under mobile-originated calls.

In this section we present the results of the access channel simulation on the AMP system reported by McDonnell and Georganas [9]. The simulation was conducted on a 40-cell system, as shown in Figure 3.9. The simulation algorithm is shown in Figure 3.10. The voice channels were assigned permanently to the cells (fixed channel assignment). Conclusions drawn from the study should hold equally for the other cellular systems. The following assumptions were made on this simulation.

Assuming a call arrival rate of λ call per second and an average holding time of 120 seconds, the traffic load (in erlangs) is given by:

$$A = \lambda 120$$

The empirical formula describing the behavior of P_{ba} (%), the access channel blocking probability for a signaling channel, as a function of traffic load, A, and the number of reattempts, R, has been stated in [16] as follows:

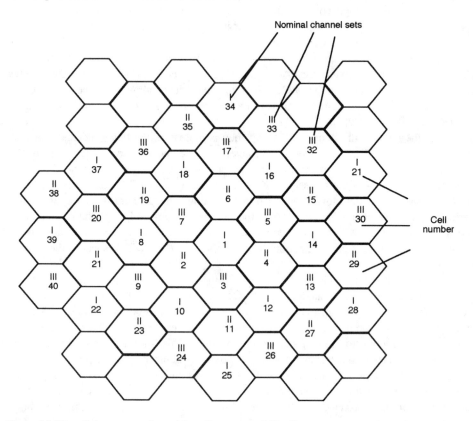

Figure 3.9 The cellular structure for mobile radio systems of 40 cells.

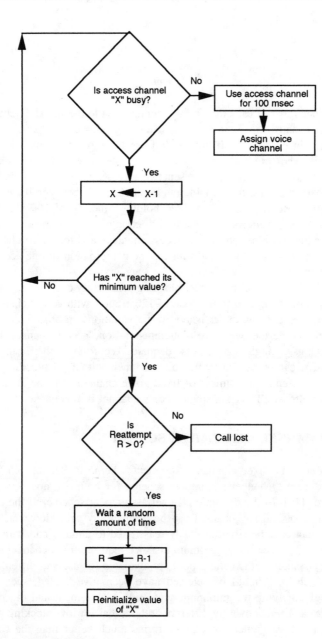

Figure 3.10 Access channel simulation.

$$P_{ba} \ (\%) = \frac{A/13}{(R+1)^2} \qquad R = 0, 1, \ldots m$$

$$= 0 \qquad R > m \tag{3.7}$$

The results of the number of calls that fail to obtain access to the signaling channel versus traffic in erlangs is shown in Figure 3.11(a). Blocking probability P_{ba} versus erlangs of traffic, are plotted in Figure 3.11(b) for two different combinations of voice and access channels per cell. The thin solid curves represent one access channel and ten voice channels, while the thick solid curves represent two access channels and ten voice channels. As stated in the assumptions, the variable random delay (D), between 0 and 200 ms has been assumed for these curves, with a fixed holding time, H, of 100 ms. The load in erlangs has been varied from a low value to 120 erlangs. The number of reattempts are zero and one for the combination of two access channels and ten voice channels, while the number of reattempts are three and four for the combination of one access channels and ten voice channels.

In Figure 3.11(a) the access channel blocking probability does not appear to exceed 2% with two reattempts, up to a traffic load of 120 erlangs with access channels and ten voice channels per cell. Thus, under heavy traffic situations an objective of 2% access channel blockage can be met with two reattempts and one access channel. For a traffic load up to 80 erlangs, a single reattempt is adequate. According to the Erlang B formula, 80 erlangs of traffic can be carried by 91 voice channels with a 2% blocking probability. For most cellular systems, the number of fixed voice channels will be far less than this number. Thus, in almost all cases a single access channel is adequate.

3.7 HYBRID CHANNEL ASSIGNMENT SCHEME

We have considered the fixed-channel assignment scheme in Section 3.5.1. Jakes has considered the dynamic channel assignment scheme for both one and two-dimensional cellular structures [19]. In the dynamic channel assignment scheme all the channels are kept in a central pool and any channel can be used in any cell. However, the channel used in one cell can only be simultaneously reassigned to another cell if the separation between the two cells exceeds the minimum distance required for cochannel interference. Usually several channels in the central pool are available for use and different strategies for choosing which one should be selected have been investigated, such as the first available channel satisfying the minimum separation requirement and the ring strategy (round robin). It has been shown by different authors that for low blocking probabilities the dynamic channel assignment scheme performs much better than the fixed-channel assignment system [4]. However, for high blocking probabilities, associated with high traffic, the fixed-channel assignment scheme performs better.

The hybrid channel assignment is a mixture of both the fixed and dynamic channel assignment scheme, and thus, should have the properties of both the fixed and dynamic

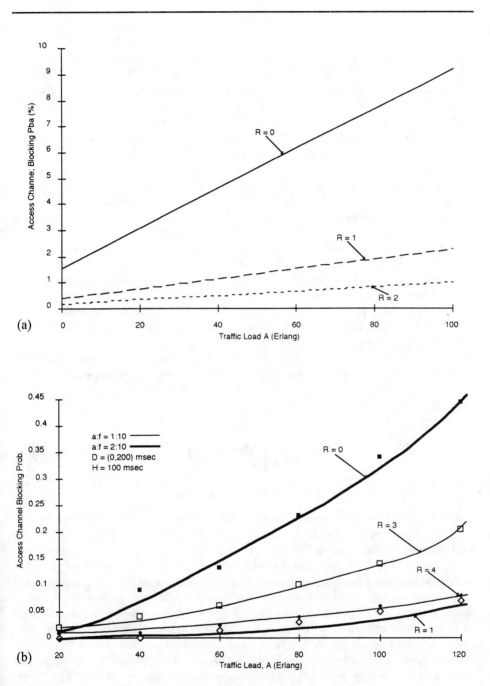

Figure 3.11 (a) Number of calls that fail to access the signaling channel versus traffic; (b) response of access channel blocking probability to increasing traffic load. Source: [24].

system. In the hybrid scheme, a total of T channels are divided into two sets A and B, which do not necessarily contain an equal number of channels. Set A contains channels that are used in the cells using fixed-channel assignment schemes. Set B contains channels that are used in any cell within the system using the dynamic channel assignment [4]. Unlike the dynamic channel assignment scheme, in the hybrid scheme the intersection point of improved performance over the fixed channel system can be moved towards lower or higher loads by changing the number of dynamic channels.

The hybrid channel assignment scheme impacts both software and hardware. By using the hybrid channel scheme, it is expected that the number of software operations on a per-call basis should lie between the fixed and dynamic channel assignment schemes. Assuming that the software real-time load to the processor is dependent on the average number of functions needed to complete a call, it can be intuitively seen that this number should be less for the fixed-channel assignment scheme than for the dynamic system. Since the hybrid channel scheme uses both fixed and dynamic channel assignment schemes, the average number of functions performed per call should be between that of a fixed and a dynamic system. Unlike the minimization of software functions, the hardware complexity remains at the level of the dynamic assignment scheme. The reason for this is that once it is necessary to borrow a channel, the steps and the decisions the processor goes through remain the same as that of a dynamic system.

For those cases where the traffic is different in different cells, the dynamic channel assignment scheme exhibits considerably more service deviation between cells than the fixed-channel assignment scheme. The measure of traffic variation between cells can be expressed in terms of service deviation (SD).

$$
SD = \left[\frac{\sum_{i=1}^{N} (B_i - \overline{B})^2}{N - 1} \right]
\tag{3.8}
$$

and, $\overline{B} = 1/N \sum B_i$ where $\overline{B}$ is the average blocking probability in the entire system, B_i is the ratio of calls blocked in cell i to offered calls in a given time interval, and N is the total number of cells in the system.

A well-designed mobile system has a service deviation value close to zero. In other words, the blocking probability of the overall system should equal the blocking probability in an individual cell. The hybrid channel assignment scheme will be able to respond to spatial shifts in offered traffic and therefore prevent a large number of calls from blockage in an individual cell while the channels are available in some other cells. On the other hand, the allocation algorithm that has a fixed channel allocation is simpler to analyze than the dynamic algorithm, which distributes channels freely within the system. Hybrid channel allocation is dynamic and is difficult to analyze analytically, and thus simulations are generally performed for performance prediction. In this section, we first discuss an algorithm and the results for hybrid channel assignment done by Kahwa and Georganas.

For measuring the performance of the hybrid scheme, a GPSS simulation has been carried out by Kahwa and Georganas on a large 40-cell layout, as shown in Figure 3.9. Here the system layout is similar to the one described for access channel evaluation. The reason for the large system consideration is the avoidance of edge effects. For small system cells where the boundaries will not have neighboring cells to borrow will lead to higher blocking for the center cell than the cells located at the boundary. In order to avoid the unequal traffic, large systems have been considered in simulation and only the samples taken on the central cells were accounted for in traffic.

Apart from the Poisson arrival rate and the exponential distribution of interarrival times, a holding time of 120 seconds is assumed. The other assumptions made in the simulation are:

- The first available channel that satisfies the cochannel interference spacing criteria $\sigma = D/R$ is borrowed.
- Mobile location is known at all times at the cell site and the MTSO.
- Mobile can tune to any of the 333 (total number of channels in AMPS based on original 10-MHz channel allocation) channels.

With these assumptions, a simulation was first carried out by assuming a fixed-channel assignment scheme as the baseline, using the Erlang B traffic formula for the desired grade of service. Channel simulations with 10, 18, 28, and 35 channels per cell were carried out. These channels were then divided among the fixed and the dynamic channels for each case. The case of 10 and 18 total channels for simulation is reproduced here in Table 3.10. Let the ratio of fixed-to-dynamic channels be represented by $K_1 : L_1$; where K_1 is the average number of fixed channels per cell, and L_1 be the average number of dynamic channels per cell. Then the total number of channels per cell is $K_1 + L_1$.

The results of the simulations are shown in Figures 3.12 and 3.13 and have been plotted as percentage of increase in load versus average blocking for the middle 20 cells.

Table 3.10
Channel Simulations in Uniformly Loaded System With Fixed Assignment

Average Channels Per Cell	Channel Partitioning		Traffic Loading (Erlangs)
	K_1	L_1	
10	8	2	5, 6, 7, 8, 9, 10*
	5	5	
	10	0	
18	18	0	11, 12, 13, 14, 15, 16†
	16	2	
	14	4	
	12	6	

*Results in Figure 3.12.
†Results in Figure 3.13.

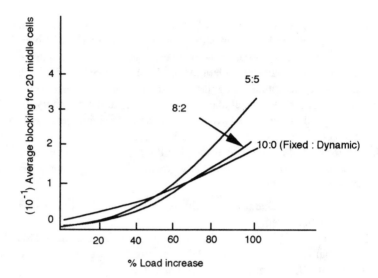

Figure 3.12 Simulation result with 10 channels.

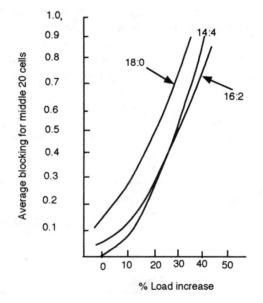

Figure 3.13 Simulation result with 18 channels.

In Figure 3.13, the base load with a fixed number of channels is assumed to be 5 erlangs with 10 fixed channels. The resulting blocking is 0.018 with a constant holding time of 120 seconds. The traffic rate was increased from 5 erlangs to 10 erlangs in increments of 1 erlang (20% increase in load with each additional Erlang of traffic). The simulation result for fixed channel was identical to the Erlang B case. The channels were then divided between the fixed and the dynamic in ratios of 8:2 and 5:5. From Figure 3.12, it is seen that the 5:5 division is better than the 8:2 division up to a 15% increase in load. Beyond a 15% increase in load and up to about an 80% increase of load, an 8:2 division is found to outperform the 5:5 division. This is justified since an 8:2 division is much closer to a fixed channel assignment than 5:5 division. Beyond an 80% increase in traffic load, fixed channel assignment is a better choice. Figure 3.13 has a uniform load of 11.4 erlangs on an average per-cell basis. For a load increase of up to 15%, channel partitioning of 12:6 and 14:4 provides the lowest probability of blocking. Beyond a 17% increase in the load, the 14:4 and the 16:2 give better results. The 14:4 channel partition gives the best results with up to about a 30% load increase. Beyond 30% the 16:2 gives better results. This is once again justified since a 16:2 division is much closer to a fixed channel assignment than 14:4. Simulation was only carried out in this case with up to a 42% increase of load. It is expected that beyond a 50% increase in load, the fixed channel assignment will outperform the hybrid channel schemes, as in the last simulation. From both these figures one can conclude that the use of a hybrid channel assignment scheme gives a better grade of service than the system that uses the fixed-channel assignment scheme with up to about a 50% increase of the base load. One explanation for this behavior is that at high loads hybrid allocation cannot maintain the minimum reuse distance and that this deficiency becomes apparent when compared to fixed allocation, which always uses the minimum reuse distance. This deficiency is less important at low loads because conflicts seldom arise, but as the load increases conflicts becomes more frequent and the scheme cannot allocate channels to cells in an optimal way.

In the next section we discuss the same hybrid assignment scheme with a reassignment of channels such that the performance of the system can be improved further.

3.8 HYBRID CHANNEL ASSIGNMENT WITH REASSIGNMENT

In the last section, we saw that the hybrid channel assignment scheme improves the performance of the system only up to about a 50% increase in load, beyond which the fixed channel scheme performs better. In this section we shall discuss simulation results using a flexible fixed channel assignment with borrowing and channel reassignment in a way that minimizes the blocking probability [16–18]. A group of channels is assigned to each cell according to the fixed-channel assignment scheme so that every cell has a list of nominal channels. A call request will try to assign a nominal channel if possible; otherwise a channel is borrowed from the neighboring cells instead of blocking the call. The borrowing starts by counting the number of channels available in the six surroundings cells. A channel is available if it is not being used at that instant in any other cell where it can cause cochannel interference. The channel is borrowed from the cell having the maximum number of available channels as compared to the other five adjacent cells. Finally, the channel borrowed is the last channel in the list of available channels so that it has a good chance of not being used if the channels are left alone. When a call using a borrowed channel is terminated, the borrowed channel is returned to its original cell. The released channel is also set free in the three other cochannel cells. The basic scheme of channel borrowing is shown in Figure 3.14. Here, channel x is borrowed from cell 2 (shown by an arrow) to be used by a call in cell 1. For the whole duration that the channel is used in cell 1, channel x is locked into three other cochannel cells. Note that due to this reassignment of the channel, the spatial distance from the three other cells has increased. Thus, this channel can be used by those cells, which means that only three cells have to be watched for assigning the same frequency instead of all six cells. Three examples of the borrowing process are shown in Figure 3.15. Assuming 10 nominal channels are assigned per cell, channels 1, 2, 3, ..., 10 are assigned to cell 1, channels 11 through 20 are assigned to cell 2, channels 21 through 30 are assigned to cell 3, and channels 61 through 70 are assigned to cell 7.

Example 6

Nominal channels 1 through 8 are occupied and no channels have been borrowed. If the call served by channel 5 is terminated, the call served by channel 8 is switched to channel 5 and channel 8 is set free. Thus, higher order channels are only used when all the lower order channels are already in use. This scheme is shown in Figure 3.15(a).

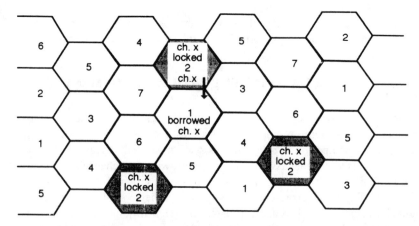

Figure 3.14 Cell 1 borrows channel x (ch. x) from adjacent cell 2. Channel x is locked in three hatched interfering cells. Source: [17].

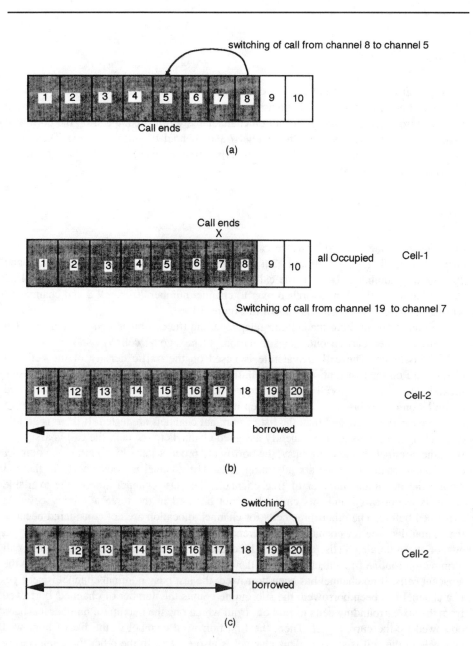

Figure 3.15 Five switching strategies: (a) Example 6; (b) Example 7; (c) Example 8. Source: [17].

Example 7

Assume that there are twelve calls in cell number 1. Thus, all 10 nominal channels are occupied. Let us also assume channels 19 and 20 (the highest two channels) have been borrowed from cell 2. Now, if the call on channel 7 is terminated, a call on the borrowed channel 19 will be switched over to channel 7 and channel 19 will be set free. Thus, once again a lower number channel(channel 19) is set free for use in cell 2. This scheme is shown in Figure 3.15(b).

Example 8

Consider the same case with twelve calls, with calls 1 through 10 being served by the nominal channel and the 11th and 12th calls being served by borrowed channels 19 and 20 from cell number 2. Now, if the call served by channel 20 is terminated, the call on channel number 19 will be switched over to channel number 20 of cell 2 and channel 19 will be set free. This scheme is shown in Figure 3.15(c).

Simulations on both the cellular telephone and the combined cellular and dispatch system have been carried out. For simulation, 49 hexagonal cells (as shown in Figure 3.16) is assumed. The call arrival rate is based on the traffic density of the cell and assumes a Poisson distribution. For call initiation and termination, a time interval of one second is considered. If a call initiation request occurs in a certain cell, the call is processed by initiation of a subroutine. The first step in the subroutine is to increase the number of calls requested by one, and then the list of nominal channels are tested. If there is an idle channel among the list of permanently assigned channels to the cell, the call is assigned. If all the nominal channels are busy, the borrowing process starts by counting the number of available channels in the six adjoining cells. The channel is borrowed from the cell having the maximum number of free channels. The last channel among the available channels is borrowed and this channel cannot be used in the three adjoining cells, as explained before. The other three cells for channel allocation are not considered because the spatial distance for cochannel interference is increased. If no channel is available on any of the adjoining cells, the call is blocked. When the call is terminated, the call termination routine first checks to see if the current cell has borrowed a channel from the adjacent cells. If no channel has been borrowed, the last busy nominal channel is set free. If a channel has been borrowed, the subroutine counts the number of channels borrowed from the six surrounding cells to find out from which one the maximum number has been borrowed by the current cell. Then, the first borrowed channel in the list of borrowed channels in the cell is set free. This channel is also set free in the other three interfering cells. The simulation for fixed channel assignment uses the Erlang B discipline with an average blocking probability of about 0.02. This is plotted in Figure 3.17 for comparison against the new algorithm proposed in this section. In this same figure, the behavior of the hybrid channel assignment, algorithm I discussed in Section 3.7, is also drawn. The

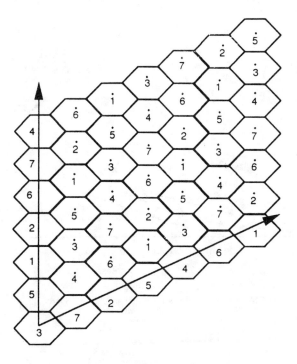

Figure 3.16 Cellular system used for simulation of hybrid channel assignment with reassignment strategy. Source: [17].

result of the new algorithm discussed in this section shows uniform improvement throughout a range of up to a 100% increase in load. It should be noted that both the hybrid channel assignment scheme and the hybrid assignment scheme with reassignment make the system complexity high when compared to fixed channel assignment.

The average delay for dispatch calls is shown in Figure 3.18. In simulation, the basic traffic to each cell has been kept to 100 telephone calls and 100 dispatch calls per hour with independence of calls assumed. Also, Poisson call arrival has been assumed. The holding time for dispatch calls in the simulation was assumed to be 20 seconds. Based on the recommendations of a Bell Laboratories technical report, 5 erlangs of telephone traffic and 5/9 erlangs of dispatch traffic has been assumed. From Figure 3.18, it can be inferred that the average delay for dispatch calls is decreased up to a 100% increase in traffic load.

3.9 SUMMARY AND CONCLUSIONS

We started this chapter by asking a fundamental question: How one can arrive at the number of subscribers per cell before even considering the traffic design for the cellular

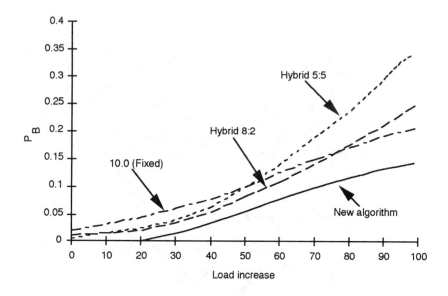

Figure 3.17 Comparison of blocking probabilities for different schemes. Source: [17].

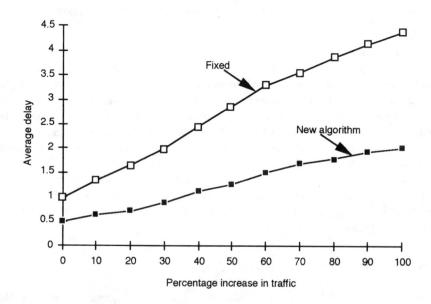

Figure 3.18 Average delay for dispatch calls. Source: [17].

system? We provided the answer by going through an example of market survey and demographic study. Following this we discussed the three basic formulas: Erlang B, Poisson, and Erlang C and their underlying assumptions. Assumptions are rather important for a judicious choice of the formula. This is important from two different viewpoints. First, a through understanding of the subject is required. Most importantly, since cellular is still growing and all the facts are not completely known, a greater understanding at the beginning will better prepare you for future changes. Three well known studies (AT&T study, Rydax automatic mobile telephone system study at Honolulu, and AUTOTEL pilot system study at Vancouver, BC, Canada) were discussed for arriving at the erlangs per subscriber and the average holding time. The importance of monitoring the GOS, which allows for system expansion in a timely fashion, was discussed in Section 3.5.2. The basis for choosing one signaling channel per cell site was established in Section 3.7. The second half of the chapter dealt extensively with the hybrid channel assignment scheme and it was concluded that the hybrid scheme performs better than the fixed channel assignment up to about a 50% increase in load, beyond which a fixed channel assignment performs better. The problem beyond a 50% increase in traffic was circumvented by modifying the hybrid assignment approach by subsequent reassignment of the channel. The result here shows a uniform improvement of up to 100% in traffic load. However, the scheme is far more complex than the hybrid assignment technique and one has to judge this for its practicality.

PROBLEMS

3.1 If C is the market penetration probability constant and W is the number of working lines/mile^2, then the density of the offered traffic in Erlangs per square mile can be given by:

$$E = C \; W^{0.68}$$

A mix of MTS and ADS subscribers are assumed in the above equation [1]. For a cell radius of 8 miles, compute the value of E if the market penetration constant is assumed to be 10%. Assume the number of channels in the cell to be 20.

3.2 For a group of 10 trunks find the divergence between the Erlang B, Poisson, and Erlang C formulas in terms of the load carried per trunk when the blocking probabilities are 1%, 5%, 10%, and 20%. What conclusions you can draw from these?

3.3 For Example 3.4, compute the channel loading and trunk utilization assuming GOS to be 0.01, 0.05 and 0.1. Draw the curves for number of channels versus channel loading for these cases. What conclusions can one draw from this example?

3.4 The result of a market survey for the first ten years of cellular service predict the number of subscribers to be 2,370; 4,710; 9,390; 11,730; 14,070; 18,750; 23,430; 28,830; 37,470; and 46,830. Assume the channel load to be 30 subscribers per channel and find the voice frequency reuse factor. Assuming that GOS is never allowed to degrade below 0.05, which year will you propose cell splitting?

3.5 Assuming that a group of 10 trunks each serve mobile and dispatch users separately, find the traffic carried by each group using Erlang B discipline and assume the GOS to be 0.001 and 0.2. If we now combine both the groups, find the resulting blocking probability. Find which group benefits by this process. Does a group suffer by this combining also?

3.6 Suppose in a certain cell it is found that the dispatch mobile unit generates on the average 1/120 erlangs of traffic during the busy hour, and that the average holding time is 15 seconds. Regular mobile users generate 1/30 erlang during the busy hour, with a holding time of 3 minutes. If the number of radio channels available in a cell is 40 and the number of mobile and the dispatch mobile subscribers is the same, show that 32 channels can be used in this case for 810 mobile telephone users with a GOS of 0.05. The remaining 8 channels can be used for an equal number of dispatch users with the GOS less than one holding time average delay.

3.7 The approximate value of induced delay, τ, (that is, the holding time of the register at the head of the queue) needed to give the Erlang C source an overall mean delay of t in the queue may be found by using the following relation:

$$E = \tau\lambda$$

Where λ equals the arrival rate of the dispatch users. The average delay in holding times function $f(E)$ for a single server is:

$$f(E) = E/(2 - 2E)$$

The relationship between average delay is given by $t = \tau f(E)$. Show that the traffic intensity offered to the server group is:

$$E = -tl + (t^2\lambda^2 + 2t\lambda)^{1/2}$$

Using the data of problem 3.6, show that the average arrival rate is 0.45 arrivals/sec, and the value of the induced delay is $\tau = 2.1$ sec.

3.8 The number of calls completed per day follows the Poisson distribution with the random variable a. Let the probability of an incomplete call be p and the number of incomplete calls per day be n, then show that $E(n) = pa$ and

$$P(n = k) = \sum_{n=k}^{\infty} \frac{a^m}{m!} p^k q^{m-k}$$

3.9 Traffic in erlangs for a 21-cell cellular system, with the total available channels of 395, is shown below in Figure 3.19. Assuming that the service has to be provided with 2% blocking probability, compute the number of channels required at each cell for a fixed channel assignment. Use the block call abandoned discipline. Make adequate assumptions for the channel reuse factor if necessary.

3.10 Find the increase in traffic carrying capacity if 24 channels are placed in one group compared to 15 channels in one group. Use the Erlang B table.

3.11 The basic reuse pattern for a seven-cell system using omnidirectional antenna is A1/ B1/ C1/ D1/, and the number of subscribers associated with these cells is shown in Figure 3.20. Each user generates traffic of 1.08 CCS. Find the number of channels required in each cell when the desired grade of service is 0.05.

3.12 In the Illinois Bell market trial system, the average size of a cell is 210 square miles. There are a total of 10 sites, each of which illuminates three cells. Dividing the total 312 voice channels into 30 frequency groups, with a frequency reuse factor of 1.44, which provides 15 channels per group. Find the total number of subscribers that the system can serve and the number of subscribers per square mile.

3.13 Derive the blocking probability, P_b, the carried traffic per cell, a', and the traffic carried per channel, L, for Poisson and Erlang C cases. (Hint: rederive (3.6a) through (3.6c).)

3.14 Traffic in erlangs for a 12-cell cellular system (total available number of channels 395) is shown in Table 3.11. Assuming that the design GOS is to be 0.01, find the number of channels required to satisfy this traffic. Further, if we assume that the average traffic is the same throughout all the cells, find the new value of the number of channels per cell and the total number of required channels. Assume Erlang B formula.

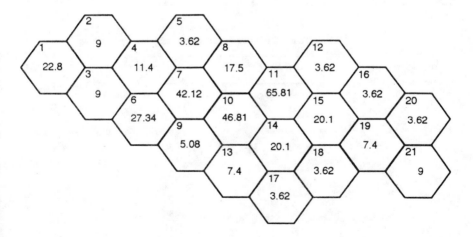

Figure 3.19 Traffic distribution for 21-cell system.

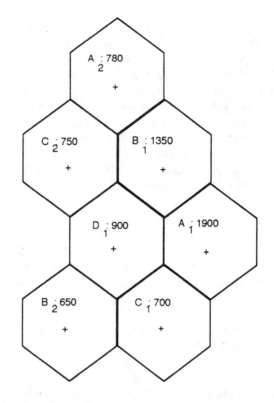

Figure 3.20 Seven-cell system using omniantenna.

Table 3.11
Traffic for a Twelve Cell Cellular System

Cell Number	Traffic (Erlangs)
1	12
2	11.2
3	3.78
4	13.7
5	4.46
6	9.65
7	11.2
8	3.78
9	3.78
10	5.88
11	5.11
12	10.4

REFERENCES

[1] High Capacity Mobile Telephone System Technical Report, Bell Laboratories, 1971.

[2] Anderson, L. G. "A Simulation Study of some Dynamic Channel Assignment Algorithms in a High Capacity Mobile Telecommunications Systems," *IEEE Trans. on Vehicular Technology*, Vol. Com-21, No. 11, November, 1973.

[3] Choudhari, G. L., and S. S. Rappaport. "Cellular Communication Schemes Using Generalized Fixed Channel Assignment and Collision Type Request Channels," *IEEE Trans. on Vehicular Technology*, Vol. 2, No. 2, May, 1982, pp. 53–65.

[4] Sengoku, M. "Telephone Traffic in a Mobile Radio Communication System Using Dynamic Frequency Assignments," *IEEE Trans. on Vehicular Technology*, Vol. VT-29, May 1980.

[5] American Telephone and Telegraph Company. *Switching Systems*, 1961.

[6] "Traffic Handling Capacity of Trunked Land Mobile Radio Systems," *International Conference on Communication*, 1979, 57.2.1–57.2.5.

[7] Cox, D. C., and D. Reudnik. "Increasing Channel Scale Mobile Radio System: Dynamic Channel Reassignment"; *IEEE Transactions on Communications*, Vol. Com-31, No. 11, November, 1973.

[8] Reeves C. M. "An Overview of Trunking Techniques in Mobile Radio Systems," *IEEE Vehicular Technology Conference*, 1980, pp. 1–4.

[9] Nehme, and N. D. Georganas. "Spectral Efficiency of the Hybrid Channel Assignment Scheme as Compared to the Fixed One, in Cellular Land-Mobile Systems," IEEE. 1981.

[10] Sin, J.K.S., and N. D. Georganas. "A Hybrid Channel Assignment Scheme for Cellular, Land-Mobile Radio System with Erlang C. Service," *IEEE National Telecommunication Conference*, 1979.

[11] Box, Frank. "A Heuristic Technique for Assigning Frequencies to Mobile Radio Nets," *IEEE Transactions on Vehicular Technology* Vol. VT-27, No. 2, May 1978.

[12] FCC Filing for Kansas City SMAS by McCaw Communications.

[13] "Memphis SMAS Population and Land Characteristics by County," U.S. Department of Commerce, Bureau of Census, 1980.

[14] Haccoun, P.C.D., and H. H. Hoc. "Traffic Analysis for Different Classes users of Land Mobile Communication Systems," *IEEE Vehicular Technology Conference*, pp. 283–285.

[15] Callendor, M. H., and C. I. Donald. "Network Design for a Fully Automatic Wide Area Radiotelephone Service," *29th IEEE Vehicular Technology Conference*, Arlington Heights, Ill., March 27–30, 1979.

[16] Singh, R. "Channel Assignment Scheme and Traffic Capacity of Mobile Radio Systems," *IEEE Vehicular Technology Conference*, 1981, pp. 281–284.

[17] Elnonbi, M., R. Singh, and S. C. Gupta. "A New Frequency Channel Assignment Scheme Algorithm in High Capacity Mobile Communication systems," *IEEE Transactions on Vehicular Technology*, Vol. VT-31, No. 3, August 1982.

[18] Elnonbi, M., R. Singh, and S. C. Gupta. "A New Channel Assignment Scheme in Land Mobile Radio Communications," *IEEE National Telecommunication Conference*, 1979.

[19] Siemens Aktiengesellschaft. "Telephone Traffic Theory Tables & Charts", 2nd Edition, 1970.

[20] Sallberg, K., B. Stavenow, and B. Eklundh. "Hybrid Channel Assignment and Reuse Partitioning in a Cellular Telephone System," *IEEE Vehicular Technology Conference*, 1987.

[21] Haslett, N. J., and A. J. Bonney. "Loading Considerations for Public Safety on Trunked Radio System," *IEEE Vehicular Technology Conference*.

[22] Bakry, S. H., and M. K. Samarkandy. "Computer Investigation of Telephone-Traffic Capacity for Cellular Mobile Radio Systems," *IEEE Vehicular Technology Conference*, 1989.

[23] Briley, B. E. *Introduction to Telephone Switching*, Addison-Wesley Publishing Company, Inc.

[24] McDonnell, M., and N. D., Georganas, "Simulation Study of a Signaling Protocol for Mobile Radio Systems," *National Telecommunication Conference IEEE*, 1981 p. 23.6.4.

Chapter 4
Different National and International
Cellular Systems

4.1 INTRODUCTION

One of the many innovations in the field of communications over the last decade is cellular mobile radio technology. Cellular telephones are being installed in almost every major city throughout the western world. Cellular systems are being considered for third world countries that are essentially lacking in telephone infrastructure and where cellular technology offers a relatively economical means of establishing their telephone systems. The basis for the success of cellular telephones lies in efficient spectrum utilization by repeating channels with sufficient geographical separation so that cochannel interference can be effectively controlled. Presently, analog voice can be supported on a continuous basis with the help of the handoff technique. Digital cellular radio system, which will eventually replace the present analog system, will not be discussed in this chapter because it is not fully operational today and an exhaustive discussion is provided in Chapter 7. This chapter is devoted to the analog cellular services offered in the United States, Canada, Japan, the Nordic countries, and Germany. These are the prëeminent systems of the world, systems that have been adopted by many other countries. Section 4.2 briefly describes some important technical parameters the reader will encounter in Sections 4.3 through 4.9. Each of these sections is devoted to a particular system, where we first describe the associated national history of mobile communication, then deal with system characteristics and parameters, and close with an overview of that country's system. The chapter is concluded by a comparative study of these systems.

4.2 TECHNICAL PARAMETERS

In this section, we briefly describe those parameters that are important and common to systems described in this chapter. Parameters specific to a particular system are discussed in the section that deals with that system. The interested reader should understand these parameters clearly before going through different systems. This will enable the reader to understand the systems and appreciate their differences. To achieve this objective, we have selected the following subset of cellular system parameters: BCH codes, cell repeat pattern and channels per system, channel spacing, compandor (compressor/expandor), dotting sequence and word synchronization code, frequency deviation for FM voice, frequency separation between the transmit and receive parts of a channel, frequency shift keying (FSK) and fast frequency shift keying, grade of service, handoff process, IF bandwidth, Manchester-encoded data, phase modulation, receiver noise figure, and speech quality measurement, and spurious emission.

4.2.1 BCH Codes

In the case of block codes, the data is segmented into blocks of k messages or information bits: each block can represent any one of 2^k distinct messages. The encoder adds $(n - k)$ bits and forms a block n bits long, which are called code bits. These $(n - k)$ added bits are known as redundant bits, parity bits, or check bits, and they carry no information. The code is referred to as (n, k) code. The ratio $(n - k)/k$ within a block is called the redundancy of the code, and the ratio of data bits to total number of bits, k/n, is called the code rate. The code rate is the portion of a code bit that constitutes information. Thus, for a 3/4 code rate, there are three bits of information for every four bits of code word, and the redundancy in this case is 33% and the bandwidth expansion is 4/3. Besides the code rate k/n, an important parameter of a code word is its weight, which is simply the number of nonzero elements that it contains. In general each code word has its own weight. When all the M code words have equal weight, the code is called a fixed-weight code or a constant-weight code. Let us consider two code words, C_j and C_i, in an (n, k) block code. The distance between two code words is defined as the measure of the difference between them or the number of positions where they differ. This measure is called the Hamming distance $d(j, i)$. Clearly, $d(j, i)$ for $j \neq i$ satisfies the condition $0 < d(j, i) < n$. The smallest distance value of $d(j, i)$ among all code words is called the minimum distance of the code and is denoted as $d_{min}(j, i)$. Minimum distance is important as it represents the weakest link in a chain and provides the minimum capability, which provides the strength of the code. In order to maintain reasonable throughput under multipath fading, block coding is used in most cellular systems of the world. In the United States advanced mobile phone service (AMPS), the word length for the forward signaling channel is 40 bits long. Each 40-bit encoded word contains 28 bits of data and 12 bits of check bits, and forms a (40, 28; 5) BCH code. Here the distance between code words is five. In the reverse control channel the word is formed by encoding 36 data bits into a (48, 36) BCH code word that has a distance of five also, (48, 36; 5). Both in the forward and reverse channels the left most bit is designated as the most significant bit. As described in Chapter 2, the data is transmitted in the voice channel also by a technique known as blank and burst. The word is formed in the forward voice channel by encoding the 28 bits into a (40, 28; 5) BCH code. Similarly, the word is formed in the reverse voice channel by encoding 36 bits of data into a (48, 36; 5) BCH code. Coding in other major systems of the world are very much like AMP and are discussed in their respective sections.

4.2.2 Cell Repeat Pattern and Channels Per System

The total number of channels in a system is fixed, based on the total availability of spectrum and the channel bandwidth. For example, if the maximum allotted spectrum is 10 MHz in each direction of transmission with a channel bandwidth of 30 kHz, there are a total of 333 channels. Two simplex channels are required for the full-duplex operation between cell sites and the mobile. The cell site transmission takes place over a forward channel, while the mobile transmits over a reverse channel. Similarly, the data channel transmitting from the cell site is known as the forward or paging channel, while the data channel from the mobile to cell site is known as the access or reverse channel. These channels are allocated to the basic cluster set, N; outside the cluster, channels are repeated. The common repeat pattern in analog cellular system is four cells in the United Kingdom and seven and twelve-cell patterns in the United States.

4.2.3 Channel Spacing

Channnel spacing is the frequency separation between two adjacent cellular channels. In the AMP system the channel separation is 30 kHz, while in most other systems of the world the separation is 25 kHz. The channel separation in C-450 (450 MHz) is 20 kHz. Lower channel separation implies less overall spectrum requirement for a specified number of channels. A frequency modulated signal requires a higher carrier-to-noise ratio for the same voice quality as the channel bandwidth is reduced.

4.2.4 Compandor (Compressor/Expandor)

A compandor is used in a signal processor for improving the quality of voice in multipath fading. The compandor controls the effect of speech-level variability on clipping distortion and frequency deviation generated by the modulator. The total dynamic range of speech is high. Low-level speech produces a low signal-to-noise ratio, at the FM demodulator output as the transmitter frequency deviations, which are proportional to speech level at the modulator input, are reduced. On the other hand, a high level of speech will produce overdeviation, resulting in distorted output. The speech variability superimposed by rapid Rayleigh fading requires some control in the transmitter and receiver such that the required signal-to-noise ratio at the receiver output can be maintained. A safeguard against this speech variability is obtained by syllabic companding. The normal reference level of the compressor corresponds to a 1,000 Hz acoustic tone at the expected normal speech voltage. The normal level is supposed to produce a peak frequency deviation of ± 2.9 kHz of the carrier. The transmitter includes the compressor portion of a 2:1 syllabic compandor. Here, for every 2-dB change in input level to a 2:1 compressor within its operating range, the change in output level is 1 dB. Expansion takes place in the receive processor, where a 1-dB level change at the input produces a 2-dB level change at the output of the expandor. The characteristics of the compandor are shown in Figure 4.1(a). In the AMPS system, the compressor has a nominal attack time of 3 ms and a nominal recovery time of 13.5 ms as defined by the CCITT. (CCITT Recommendation G162, Geneva, May-June 1964, *Blue Book*, Vol. 111, p. 52). The compressor, is followed by the preemphasis network in the transmitter processor and the expandor follows the deemphasis network.

One way to reduce the effect of noise at the receiver output is to include a filter after a discriminator, which will deemphasize the noise at a $1/\omega^2$ rate. This is desirable, but if nothing else is done this deemphasis filter will degrade the high-frequency performance of speech further. The way to circumvent this problem is to include a preemphasis network in the transmitter, which will emphasize the speech and correspondingly deemphasize the speech in the receiver in a way that the composite effects of both networks are nullified. Thus the transmitter filter must have a transfer function, which is the inverse of the receiver transfer function, that is,

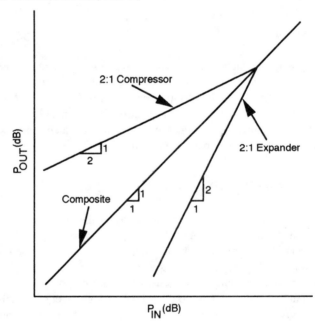

Figure 4.1(a) Characteristics of 2:1 compandor.

$$H_d(\omega) = \frac{1}{H_p(\omega)} \tag{4.1}$$

The characteristic of $H_p(f)$ has a rising response of +6 dB per octave between 300 Hz and 3 kHz, as shown in Figure 4.1(b). The corresponding deemphasis network response is dropping with the same slope of −6 dB per octave. Thus, these networks simply boost the response in the transmitter and correspondingly reduce the response in the receiver. This arrangement of filters can easily provide signal-to-noise improvement nearly equal to 5 dB.

4.2.5 Dotting Sequence and Word Synchronization Code

In cellular systems, both the forward and the reverse control channels contain the alternating sequence (10101010), known as the dotting sequence for bit synchronization, and the code word (Barker Code) for word synchronization. The dotting sequence is recognized

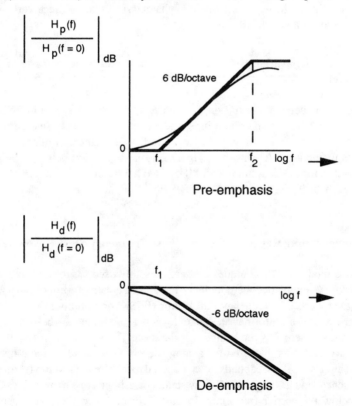

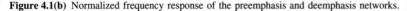

Figure 4.1(b) Normalized frequency response of the preemphasis and deemphasis networks.

as a fixed-frequency tone, which initializes the phase of the clock at the receiver. In the United States, the forward control channel contains a 10-bit dotting sequence and the 11-bit Barker code (11100010010) for word synchronization. In the reverse control channel, (from mobile to cell site) the dotting sequence is 30 bits long, while the code (Barker Code) length is kept the same. Since the reverse control channel is asynchronous, the dotting sequence is of greater length.

4.2.6 Frequency Deviation for FM Voice

For FM systems, the instantaneous frequency deviation from the carrier frequency, fc, is

$$\Delta f = f - f_c = (k/2)m(t) \tag{4.2}$$

where $m(t)$ is the modulating signal and k is a constant. Thus the instantaneous frequency is directly proportional to the modulating signal $m(t)$. In order to reproduce the signal $m(t)$ correctly, the frequency discriminator characteristics at the receiver must remain linear within the limits of this deviation.

4.2.7 Frequency Separation Between the Transmit and Receive Parts of a Channel

FM cellular radio operates in a full-duplex mode, where the transmission from mobile to cell site and from cell site to mobile is independent and simultaneous. In order to minimize interference between the transmit and receive parts of a channel, a minimum frequency separation is maintained. As shown in Figure 2.1, the separation between the two halves of the channel in the AMP system is 45 MHz. Two halves of a channel are also automatically selected, that is, for a selected channel from cell site to mobile (forward channel) there is a corresponding fixed channel from mobile to cell site (reverse channel).

4.2.8 Frequency Shift Keying (FSK) and Fast Frequency Shift Keying (FFSK)

There are two modulation techniques that are presently used for the data transmission in analog cellular systems: noncoherent binary FSK and a coherent form of binary frequency shift keying known as fast frequency shift keying (FFSK) or minimum shift keying (MSK). The binary FSK uses an information stream to switch the frequency between two carriers every T seconds, where T is the symbol for duration. The receiver can also use the same discriminator detector as that used for analog speech. Generally, a "logic one" into the modulator (high state) corresponds to a nominal peak frequency deviation above the carrier frequency, and a "logic zero" (low state) corresponds to a nominal peak frequency deviation below the carrier frequency. In other words, 1 and 0 are represented by the following equations at the modulator output.

$$S_1(t) = A \cos(\omega_1 t) = A \cos(\omega_o + \Delta\omega)t$$
$$S_2(t) = A \cos(\omega_2 t) = A \cos(\omega_o - \Delta\omega)t \tag{4.3}$$

The value of $\Delta\omega$ in the AMP system is $2\pi 8 \times 10^3$ Hz. The frequency separation between two tones is $1/T$.

FFSK (or MSK) is an adaptation of offset phase shift keying (OQPSK) in which the baseband pulses are sinusoidal instead of rectangular. One representation of MSK waveform is

$$S(t) = \alpha_I(t)\cos(\pi t/2T)\cos(2\pi f_c t) + \alpha_Q(t)\sin(\pi t/2T)\sin(2\pi f_c t) \tag{4.4}$$

Thus, the phase of this quadrature-represented signal, $\theta(t)$, is

$$-\tan^{-1}\frac{\alpha_Q(t)\sin(\pi t/2T)}{\alpha_I(t)\cos(\pi t/2T)} \tag{4.5}$$

The above equation can alternately be represented by

$$S(t) = \cos[2\pi f_c t + \Phi_k + a_k(\pi t/2T)] \tag{4.6}$$

Here a_k assumes the value of $+1$ when both $\alpha_I(t)$ and $\alpha_Q(t)$ differ and -1 when the bits are same. The instantaneous phase of the waveform is given by $a_k(\pi t/2T)$. Thus, the phase varies linearly during the bit interval and is regarded as continuous. The modulation is considered to be coherent FSK with frequency spacing of $1/2T$ Hz. This is also the minimum spacing required for orthogonal waveforms. Time domain waveforms for FSK and FFSK (MSK) is shown in Figure 4.2(a,b). For FSK, there are two carrier cycles at the higher frequency because the separation between tones is $1/T$. For MSK, the corresponding number of cycles at state one is 1 1/2

4.2.9 Grade of Service

The grade of service, as described in Section 3.3.2, is the measure of insufficiency of the number of channels available to the mobile users. The cellular system design is usually based on a GOS of 0.02 or better. A GOS of 0.02 means that, on average, 2 out of 100 calls by subscribers will be blocked during the peak traffic hours of a normal day. Naturally, the GOS will improve during the nonbusy hours of the day. The cellular system is also designed to provide a GOS of 0.05 or better during the life of the system, as discussed in Chapter 3.

4.2.10 Handoff Process

Handoff is the process whereby the cellular system redirects the mobile/base station radio path from one base station to another, thus effecting the transfer of the cellular subscriber

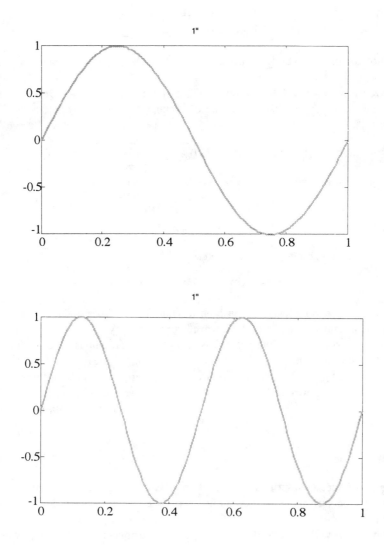

Figure 4.2(a) Time domain waveforms of FSK signals.

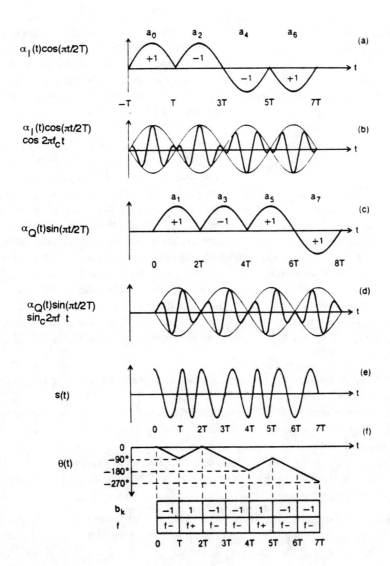

Figure 4.2(b) Time domain waveforms of FFSK signals.

from one cell site to another. Note that the mobile has to retune its channel frequency as it changes cells. The objectives are to achieve this process without the knowledge of the user, and once a handoff is made the mobile should not be immediately handed off again. The threshold value at which this switchover takes place is initially set at the time of system installation. This initial parameter setting is based on the desired performance of the system and cannot be changed arbitrarily. There are three basic reasons why handoff is necessary: to maintain a high signal quality, to balance traffic among cells, and for recovery in the event of failure of a control channel.

Signal quality refers to the minimum required carrier-to-interference ratio of 18 dB or more (AMP, with 30-kHz bandwidth), which is necessary at the mobile receiver for good quality reception. The carrier-to-interference ratio of 18 dB is arrived at based on subjective quality tests of customers where 75% of users find the voice quality either "good" or "excellent." The handoff process essentially starts when the mobile signal level, received at the cell site, falls below the preassigned threshold. Once this condition is reached, the serving cell site notifies the MTSO to ask all other adjoining cells to make a determination of the specific cell where the mobile signal can best be received. Once the best new cell is identified, the serving cell site is notified of the new channel number where the mobile should tune to for handoff. As the mobile retunes to the new channel, handoff has essentially taken place. The handoff process has been discussed in Chapter two. Here we only want to bring up some additional important considerations for an effective handoff. For a high-quality handoff system, one must take into account the following:

- Handoff must take place before a noticeable degradation of the signal quality occurs at the mobile.
- Before the handoff decision, the signal received from the mobile must be integrated for a sufficient time to assure that the quality degradation is real and not due to multipath fading.
- The new channel to which the mobile is being switched should be of a sufficiently higher level(high signal-to-noise ratio) so that the new channel is not immediately switched again.

Unlike the signal quality–directed handoff, which is done to accommodate the user, the traffic balancing handoff is done to accommodate the system. Its main purpose is to balance the load among the various cells so that an overload condition is not experienced in any one cell. Load balancing is most effective when there is a significant amount of overlap between adjacent cells and is achieved via a technique known as "directed handoff." A directed handoff is initiated when the cell load reaches a preset percentage of the full load. Cells encountering this condition will endeavor to find alternate cells and part with some of their load. The obvious underlying assumption for load balancing is that simultaneous overload conditions do not occur in adjacent cells. With proper system design the load balancing algorithm will provide help in handling peak load conditions, but it will not compensate for an inadequate quantity of voice channels. The algorithm for traffic balancing must consider the following:

- The new cell must be able to handle the call without the need for immediate handoff due to a low signal-to-noise ratio and,
- The new cell itself must not be in an overload state.

A candidate for load balancing handoff is the Japanese cellular system around Tokyo, discussed in Section 4.5, where the peak load hours are geographically separated.

The third objective for handoff is due to a failure of the control channel when a voice channel is used as a backup control channel. Systems designed with this feature will require the mobile to be switched over to another channel if it is using the designated backup control channel for voice communication. The main objective for failure driven handoff is to vacate the channel so that the channel is free of voice traffic and ready to be assigned as a control channel.

We shall now describe the basic process through which the handoff takes place in a typical cellular system. Before we actually describe the process, a general outline of the problem is in order. As described in Section 7.2, the 39-dBu contour (average level) corresponds to a signal level of −125 dBW. This is the minimum median signal [F(50, 50)] level for acceptable voice quality. F(50, 50) represents loss for 50% of the time and 50% of the locations. Signals below this level may be usable but are noticeably noisier than the average landline telephone and thus may provoke customer complaints. Cells are designed with overlap so that the median signal levels associated with the serving cell exceed -95 dBm at the cell boundary. For a typical cell, with the base antenna height of 150 feet above the average terrain and with the maximum radiated power of 100W, this occurs roughly at a distance of about 8 miles provided the cell terrain is of the regular type. Regular, as used here, means that the area does not have significant hills and valleys. Figure 4.3 shows the two cases of cell overlap with a cell boundary −95 dBm. The locus of equal signal-strength points lies on a straight line where the signal strength is greater than −95 dBm. A mobile served from cell A and moving in the direction of cell B will switch over to cell B at a point to the right of the equal strength line. In this case the mobile signal quality will improve after switchover. It is generally, preferable to hand off in a region of moderate signal strength to a cell with somewhat stronger signal strength. This will provide hysteresis and minimize any perceivable effects of handoff. Even in areas of little or no overlap, if the signal from the serving cell is noisy and the new cell presents a similarly noisy signal, the user will perceive no effect from the handoff procedure. If the signal at the point of equal signal level between two cells is below −95 dBm, then the RF coverage is simply inadequate. It should be noted once again that no amount of handoff adjustment will overcome the problem of poor coverage. The smaller radius of Figure 4.3(b) is simply achieved by reducing the cell-site transmit power. If the mobile user is crossing the boundary from cell A, he or she may be asked to reduce transmit power so as to balance the two sides of the link, from base to mobile and from mobile

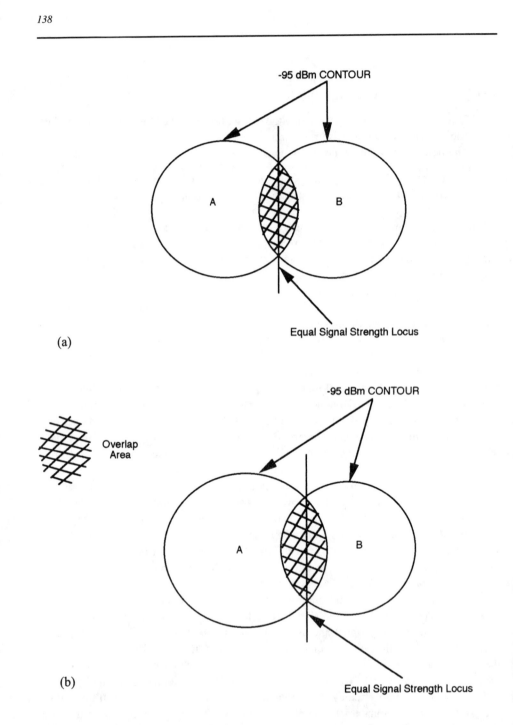

Figure 4.3 Overlapped coverage of two cells: (a) equal radii; (b) unequal radii.

to base. Conversely, the mobile user may be required to raise transmit power while crossing from cell B toward cell A.

The received signal levels at the mobile, plotted as a function of the distance from the cell sites of two cells transmitting equal power, are shown in Figure 4.4. The intersection point is the equal level point on two curves. From the figure it can be seen that as the intersection level moves higher, the overlap increases. The two cases of overlap, extensive and minimal, are shown in Figure 4.5(a,b). For Figure 4.5(a), if the mobile is traveling from cell A towards cell B, the switchover threshold must be to the right of the intersection point. This is the only condition where the mobile is switched from a marginal signal level to an improved signal quality, which satisfies the handover criteria. Here, 1 and 1' are appropriate switchover points for mobile units within cells A and B. The offset 11' can be regarded as a natural hysteresis. The second case, where the threshold is at −90 dBm to the left of the crossover point, is rather interesting. In this case, switchover from cell A to cell B will be detrimental as the mobile will have a lower signal-to-noise ratio at cell B. Rather, switchover should be put on hold until the mobile reaches point 2 where the signal-to-noise ratio improvement is greater than 20 dB. This holding of the switching point can be built into the software.

4.2.11 IF Bandwidth

In cellular systems in the United States, the IF bandwidth of 30 kHz is based on the channel frequency peak deviation of 12 kHz and the maximum frequency content of 3 kHz for voice. Carson's rule bandwidth, which conserves about 98% of the power within the band is

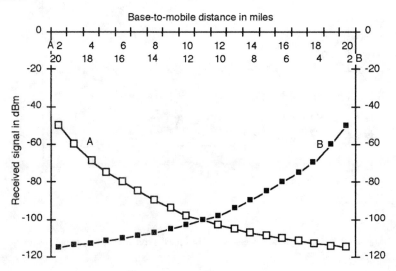

Figure 4.4 Signal level at the mobile from cells A and B.

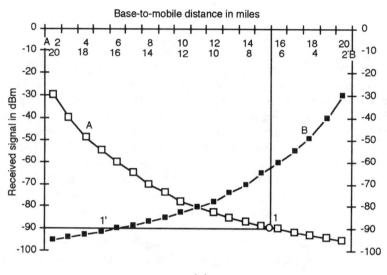

(a)

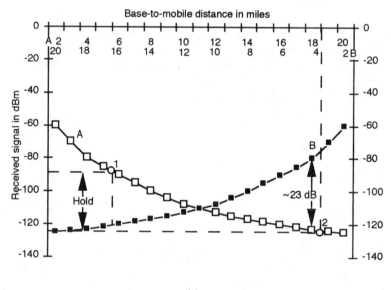

(b)

Figure 4.5 Signal-level variation between cells A and B: (a) extensive overlap; (b) minimal overlap.

$$B = 2(\Delta f + f_m) \tag{4.7}$$

Substituting $\Delta f = 12$ kHz and $f_m = 3$ kHz, the computed value of channel bandwidth is 30 kHz. Systems with lower values of frequency deviation, such as TACS, have a lower IF bandwidth.

4.2.12 Manchester-Encoded Data

Manchester encoding of data is applied to both control and voice channels. Wideband data streams are encoded such that each nonreturn-to-zero binary one is transformed to a zero-to-one transition, and each nonreturn-to-zero binary zero is transformed to a one-to-zero transition before modulation. At the receiver, an inverse recovery takes place where Manchester-encoded data is converted to NRZ data. The normalized spectrum density for $B_{i\phi}$ data is

$$\frac{S(f)}{E_S} = \frac{\sin^4 \pi f T_s/2}{(\pi f T_s/2)^2} \tag{4.8}$$

A normalized plot of $s(f)/E_s$ versus fT_s for $Bi\phi$ signals is shown in Figure 4.6. From the figure it can be seen that choosing 10-Kbps signaling over a voice band of 3 kHz, that is, $fT_s = 0.3$, will not interfere with the speech since most of the energy is beyond the normalized $fT_s = 0.3$ and the peak of $s(f)/E_s$ occurs at $fT_s = 0.75$. It should be also be noted that the Manchester encoding provides a fruitful solution to the baseband dc wander problem typically encountered in digital communication systems since there is no dc component to the waveform.

4.2.13 Phase Modulation

In exponential modulation, the modulated wave in phasor form is represented as:

$$V(t) = Re[A_c e^{j\theta_c(t)}] = A_c \cos \theta_c(t) \tag{4.9}$$

where $\theta_c(t)$ is a linear function of message, and A_c are constants. There are two forms of angle modulation, namely phase modulation (PM) and frequency modulation (FM). These modulations are best described by a time varying frequency or phase. To explain the differences between these two modulations, first consider:

$$\theta_c(t) = 2\pi f_c t + \phi(t) \tag{4.10}$$

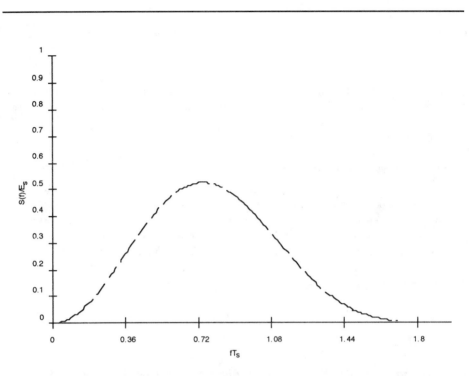

Figure 4.6 Spectral density of Manchester-encoded waveform.

where f_c is the carrier frequency and the second term $\phi(t)$ is the relative phase angle in the sense that the phasor $e^{j\theta_c(t)}$ differs in angular position from $e^{j\theta_c(t)}$ by $\phi(t)$. Also, the angular frequency can be defined as the time derivative of the angular phase. Thus we define the instantaneous angular frequency $\mathscr{F}(f)$ by

$$\mathscr{F}(f) = \frac{1}{2\pi}\frac{\theta_c(t)}{dt} = f_c + \frac{1}{2\pi}\frac{d\phi}{dt} \tag{4.11}$$

where $(1/2\pi)d\phi/dt$ can be interpreted as the instantaneous frequency deviation relative to f_c. From the above equation we can relate the phase $\theta_c(t)$ to instantaneous frequency, $\mathscr{F}(t)$, as

$$\theta_c(t) = 2\pi\int_{-\infty}^{t}\mathscr{F}(\lambda)d\lambda \tag{4.12}$$

The lower limit of integration represents a constant phase term, which can be dropped without loss of generality.

With the help of (4.12), we can now define the phase modulation as a process whereby the relative phase $\phi(t)$ is proportional to the message. In frequency modulation, the instantaneous frequency deviation, $(1/2\pi)d\phi/dt$, is proportional to the message. Thus,

$$x_c(t) = A_c\cos[w_c t + K_p x(t)] \tag{4.13}$$

where K_p is the phase deviation constant, that is, the maximum phase shift produced by $x(t)$. We note that $|x(t)| \le 1$. Similarly, the instantaneous frequency of the FM wave is

$$\mathcal{F}(t) = f_c + \Delta f x(t) \tag{4.14}$$

where Δf is the frequency deviation constant, or

$$x_c(t) = A_c\cos\left[2\pi f_c t + 2\pi\Delta f \int_{-\infty}^{t} x(\lambda)d\lambda\right] \tag{4.15}$$

We note that the message has no dc component, that is, $\langle x(t)\rangle = 0$, Otherwise, the integral will diverge as $t \to \infty$.

4.2.14 Receiver Noise Figure

This parameter provides an estimate of the signal degradation due to the noise generated within the receiver. The noise is largely controlled by the low noise amplifier of the receiver front end. Noise generated after the low noise amplifier of the receiver does not contribute much toward degradation of the signal due to the high gain of the amplifier. The value of the receiver noise in UHF band is on the order of 6 dB and is mostly due to the noise temperature of the low noise amplifier.

4.2.15 Speech Quality Measurement

Speech quality for FM in the presence of Rayleigh fading is measured in terms of the sound articulation index (AI). Sound articulation of 80% or more is generally maintained throughout the conversation. The minimum intelligibility tolerance is ascertained through the use of intelligibility testing and is defined by the American National Standard Institute (ANSI) and is shown in Figure 4.7. As shown in this figure, the AI of 0.3 corresponds to an intelligibility level of approximately 94% for sentences known to listeners. An important class of intelligibility tests in speech work are known as rhyme tests, where the listener is supposed to recognize one of several possible consonants in the closed rhyming set, such as feat, heat, meat, beat, and so forth. Tests can also be done by phonetically balanced (PB) stimuli rather than the rhyming word. However, in PB testing

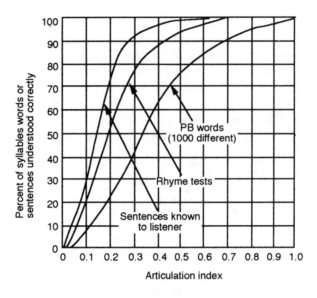

Figure 4.7 Measure of speech intelligibility. Note: these relations are approximate they depend on the type of material and the skill of talkers and listeners. Source: [27].

the intelligibility is highly dependent on the number of words in the test set. It is generally agreed that for good radio system design, AI should be greater than 0.5, which leads to a conversational intelligibility level greater than 97%. The experimental results of sound articulation between landline and automobile telephone for signals received at the base station and at the mobile station are shown in Figure 4.8. For an 80% or higher AI, the received level at the base station should exceed 7.5 dBμV. The corresponding voltage level at the mobile should exceed 14 dBμV [17].

4.2.16 Spurious Emissions

Spurious emission is the unwanted signal energy falling in other channels. There are two types of specifications that control the mobile radio emission. First, the emission from the mobile transmitter has to be controlled. Additionally, the emission generated by the mobile receiver is also controlled. The emission level generated from the mobile transmitter is controlled according to the levels specified in the Table 4.1. Additionally, the emission levels generated by the transmitter in each 30-kHz mobile receive band (U.S.) must be lower than −80 dBm. This level is specified at the transmit antenna connector. The emissions from the mobile receiver falling in the mobile receiver band must also be below −80 dBm, measured at the antenna connector. In the Japanese cellular system, this number is specified with respect to the mobile receiver sensitivity and will be discussed in Section 4.5.

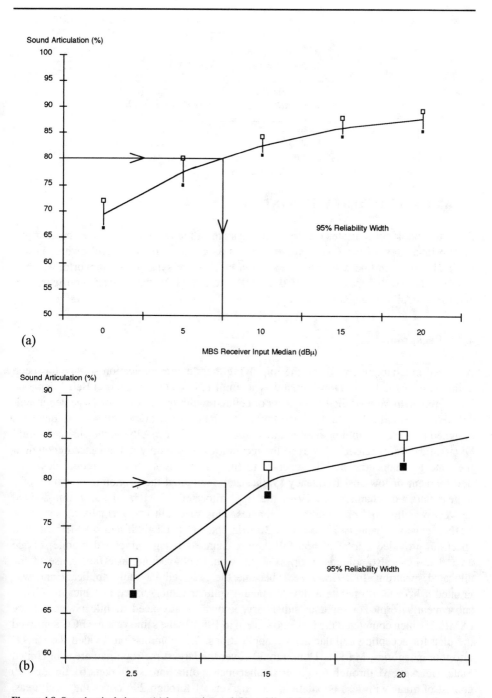

Figure 4.8 Sound articulation at (a) base station and (b) mobile station.

Table 4.1
Mobile Transmitter Emission Level

Frequency Away From Center of Channel (kHz)	Spurious Level Below Carrier (dB)
±20	< −26
±45	< −45
±90	< −60

4.3 AMPS—UNITED STATES SYSTEM

Since this book concentrates on the AMPS system of the United States, we consider this system first. Most of the AMPS system and its parameters are covered elsewhere in this book. However, in line with our examination of different systems of the world, we shall briefly examine the background of the AMPS system [1,2], the network configuration, and the system's parameters.

4.3.1 Background

As stated in Chapter 1, work on the improving mobile communication service has been going on since the 1930s. However, it was not until 1947 that researchers at Bell Telephone Laboratories discovered the key concept of cellular radio: by dividing the large geographical area into small cells, and by using low-power transmitters at the cell sites, frequencies assigned in one cell could be reused and thereby the traffic capacity on the channels could be significantly increased. However, the technology was simply not advanced enough at the time to enable the mobile receiver to tune to different channel frequencies. The development of low-cost frequency synthesizers facilitated the economic operation of a large number of channels. In 1974, the FCC allocated 40 MHz of spectrum, divided between wireline and nonwireline companies. This spectrum was formally allocated for VHF television channels 70 to 83. A 30-kHz channel bandwidth and 40 MHz of total spectrum provided a total of 666 full-duplex channels per geographical mobile service area. The FCC also solicited proposals that would demonstrate an efficient use of the allocated spectrum. In 1975, AT&T became the first cellular radio applicant and was granted a license to operate a developmental cellular radio service in Chicago. AT&T subsequently formed a separate subsidiary, known as advanced mobile phone service (AMPS), which completed the Chicago system in 1982. In the same year the FCC approved the plan for accepting cellular license applications. The commission divided the market into four categories. Markets 1 through 30 cover the largest metropolitan areas in United States, markets 31 through 60 cover smaller metropolitan areas, markets 61 through 90 consist of areas with less population, and markets 91 through 296 cover the rural areas. Appendix A provides a list of the first 120 CGSAs with operational cellular systems.

4.3.2 The Chicago Developmental Cellular System

In March 1977, the FCC authorized Illinois Bell Telephone (IBT) to construct and operate a developmental cellular mobile telecommunications system in the Chicago area [3, 4]. The system was configured as a startup AMPS system employing large cells and omnidirectional antennas to minimize the initial equipment required to cover the serving area. Approximately 2,100 square miles of the urban and suburban Chicago area were covered by a 10-cell site system, employing a total of 136 voice channels controlled by an MTSO located at Oak Park, Illinois. To minimize real estate procurement and zoning problems, sites were selected on the Bell system property as much as possible. The coverage area included the skyscrapers of downtown Chicago as well as the wooded areas of the suburbs. Technical and marketing evaluations of the system were accomplished during a two-phase program. An equipment-test phase using approximately 100 mobile units (about 90 of which were Bell System units) assigned to AMPS development personnel was followed by a service-test phase where IBT was authorized by the FCC to provide tarriffed mobile service to a maximum of 2,500 mobile users. The equipment-test phase commenced in mid 1978 and the service test phase started in late 1978. For the service-test phase, 2,100 mobile sets were procured from three suppliers and the system served over 2,000 trial customers. The quality of the service was so good that the people at the land end of the mobile conversation usually found it difficult to believe that they were talking to someone in an automobile. The Chicago developmental system proved that cellular technology could provide a new quality of mobile telephone communications.

Among the principal purposes and objectives for the Chicago AMPS development trials were:

- To complete all system shakedown and debugging activity necessary to ensure a high-quality reliable system;
- To verify the quality and the reliability of service;
- To test prototype designs;
- To check the engineering procedures used to lay out the systems and improve these tools based on the experience gained;
- To gain experience in installing and operating a cellular system;
- To confirm the viability and worth of AMPS by demonstrating that the public's mobile communication needs could be met at a reasonable cost;
- To develop and validate methods for estimating the average traffic generated per mobile and the geographical distribution of mobile traffic;
- To determine customer reactions and sensitivities to the basic service, the mobile installation, the maintenance procedures, and the vertical service provided;
- To verify acceptability of system recovery procedures, call processing sequences, and the overall signaling plan; and
- To successfully operate and maintain the AMP system.

The AMPS developmental system in Chicago was engineered as a representative startup cellular system. Figure 4.9 shows a 2,100-square-mile area covered by a system of 10 cell sites, the locations of which are indicated by crosses and three-letter abbreviations. The size of these circles depends on the height of the antennas at different cell sites: the higher the antenna height, the larger the coverage area. However, the cable loss also increases as the antenna mounting height is increased. A circle represents the estimated ideal 39-dBu contour. The 39-dBu contour represents the minimum power level of -125 dBW at the cell boundary.

As stated above, existing structures were chosen for the cell locations while minimizing site-location deviations from the ideal hexagonal grid. Developmental system cell-site locations and the height of the antenna are shown in Table 4.2. The layout required only three new antenna masts out of a total of ten cell sites. Antennas for the seven cell sites were placed on existing structures. With 136 channels, the system was engineered to serve approximately 2,500 mobile customers with a blocking probability of 2%. A total of 136 voice channels were used.

Many different system-level tests were conducted. In one protracted test, signal strengths were measured at selected locations throughout the entire service area to define the actual system boundaries and the general adequacy of the signal strength within the service area. Testing and evaluation of the forward and reverse blank-and-burst function of data transmission over the voice channel were accomplished, as well as a test and evaluation of the data transmission over the setup channels. In another test, an evaluation of the location and handoff algorithm was undertaken using the mobile-equipped vehicles. The response of the system to the loads encountered as commercial customers started to use the system was tested and evaluated. System reconfiguration tests were done to evaluate the ability of the generic MTSO software to isolate and reconfigure any voice/ data trunk group, or other redundant equipment groups in the data terminal equipment or the cell sites, by simulating a catastrophic failure of the equipment and noting the performance and the integrity of programs at the MTSO. The effects of outages and of automatic reconfiguration on the overall service were evaluated as part of these system tests.

System tests were conducted in the Chicago developmental system to verify that no frequency reuse problems would be encountered in this startup configuration. The technical evaluation of the AMPS developmental system installed in Chicago was completed in 1979.

In addition to the technical evaluation of the cellular system, preparation was also done for a market trial (service test). The marketing plan was designed to demonstrate the demand for public mobile phone service, verify sales predictions and market research studies, and assess customer reaction to the AMPS service. An assessment of the demand for AMPS service was based on the sales figures in the service test. Interviews conducted with subscribers were analyzed to evaluate user reaction to the service. The results showed that sales were significantly greater than predicated and that the majority of customers were very enthusiastic about the service. After being asked to identify their reasons for subscribing, almost all of the test participants mentioned saving time and convenience as

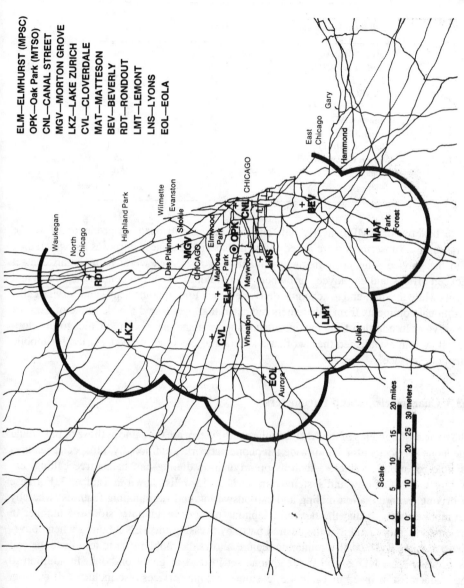

Figure 4.9 Chicago cell-site locations. Source: [28].

Table 4.2
Cell-Site Locations and Duplex Voice Channels in Illinois

Cell Site (Abbreviation)	Number of Voice Channels	Antenna Height (ft)
Beverly (BEV)	16	150
Canal Street (CNL)	26	550
Cloverdale (CVL)	15	325
Eola (EOL)	8	310
Lake Zurich (LKZ)	9	285
Lemont (LMT)	8	250
Lyons (LNS)	16	150
Matteson (MAT)	12	260
Morton Grove (MGV)	18	185
Rondout (RDT)	8	150
Total	136	

major benefits of AMPS; nearly half cited an increase in business productivity. In the same survey, subscribers were asked to list the specific business purposes for which they used the service. The most frequently mentioned responses were: to maintain contact between offices and employees or owners who are on the road; to contact customers and clients while driving; and to schedule deliveries, pick ups, appointments, and service calls. Additional comments from subscribers indicated that they found a wide range of uses for the service. In addition to being an outstanding technical success, the Chicago developmental system established the fact that a substantial customer market existed for mobile telephone service.

4.3.3 Characteristics and Parameters

The basic requirements of the AMP cellular radio system are the same as those of individual telephone services in the nationwide telephone network. Additionally, all special features suitable for mobile vehicles, such as preorigination dialing and hands-free calling, are provided. A list of user and system characteristics [5–7] is shown in Table 4.3. For ease of driving, preorigination dialing and call-answering and -terminating features (whereby the mobile user can dial the desired telephone number in advance, store the number in the unit's memory, and dial the desired party by pressing the SEND button when convenient) are added. The call-termination feature allows the mobile user to terminate the call by pressing the END key in the telephone set instead of going on-hook. In addition to the characteristics listed in Table 4.3, customer calling services also include call waiting, speed calling, and three-way calling. Call waiting, as the name suggests, alerts the mobile user of an incoming call while he or she is having a conversation with someone else. Speed calling permits the subscriber to originate calls to frequently called numbers by

Table 4.3
User and System Characteristics of AMP

Characteristic	Description
Efficient use of spectrum	Channels are repeated in geographically separated cells
Large customer capacity	System is capable of serving up to 100,000 customers
Nationwide accessibility	Customer belongs to one service area and is able to obtain service in other service areas without operator intervention
Telephone grade of service	A 2% blocking probability at peak traffic; system designed to have telephone-quality voice
Handoff	Standard feature; entire process takes less than 200 ms
Preorigination dialing	User can store the called number in a register before pressing the SEND key
Call answering and termination	Mobile user answers call by removing handset from cradle and terminates call by returning handset to cradle; calls are also terminated by pressing the END key

simply pushing one or two buttons (activation of previously stored telephone number in its memory). Three-way calling permits a user who is already connected to another phone to originate a call to a third party, to switch back and forth between connections, to hold a three-way conversation, or to connect the other two parties for continued conversation while disconnecting his or her unit.

The technical parameters of AMPS [5–7] are grouped into three categories: system related, communication system parameters, and mobile unit parameters. These parameters are shown in Table 4.4. The system-related parameters include all major system requirements, based on which the system design is implemented. The communication system parameters include those related to speech quality. The mobile unit parameters include the relevant requirements based on which the mobile transceiver and control is designed.

As stated above, the FCC originally allocated 40 MHz of RF spectrum for cellular mobile radio systems, divided into two groups of 20 MHz each. The upper half of the 40-MHz band was allocated to the local telephone company and the lower half to the radio common carrier (RCC). Frequencies of 870–890 MHz are used by the cell site for their transmissions, while frequencies of 825–845 MHz are used for mobile transmission. A pair of duplex channels is separated by 45 MHz, while the channel separation is 30 kHz. In 1989, the FCC allocated an additional 10 MHz of spectrum, divided between the FCC and the RCC. The additional spectrum provides a total of 166 full-duplex channels, equally divided between the two operators. In order to meet this requirement, the IF bandwidth of the receiver is kept at 30 kHz. The mobile receiver is about 60-dB down, 15-kHz away from the center of the channel. For speech modulation, FM with a peak frequency deviation of ±12 kHz is used. Frequency preemphasis and amplitude companding, as described in Section 4.1, is used for voice transmission. Continuous data on the paging channel is transmitted from cell site to mobile at a rate of 10 Kbps. The modulation is FSK with a peak frequency deviation of ±8 kHz. With voice channels, discontinuous

Table 4.4

Parameters of the AMPS System

Category	Parameters
System-related:	
Number of channels	832, 2 groups of 416 channels; each group includes 21 signaling channels
Cell radius	2–20 Km
Mobile-receiver frequency range	869–894 MHz
Mobile-transmit frequency range	824–849 MHz
Channel spacing	30 kHz
Mode of transmission	Full duplex
Voice transmission	Frequency modulation with peak frequency deviation of ±12 kHz
Data transmission	Signaling-channel data is frequency modulated; Manchester-encoded binary at 10 Kbps; peak frequency deviation is ±8 kHz; on voice channel, burst-type digital messages are transmitted at 10 Kbps
Number of cells	50 (typically, on a fully developed system)
Maximum base station ERP	100W per channel
Exchange	Fully automatic with DTMF receiver
Frequency separation	45 MHz between transmit and receive channels
Error protection coding	Shortened, 63:51 BCH, repeated 5–11 times
Base-to-mobile	40:28 BCH
Mobile-to-base	48:36 BCH
Communication system:	
Speech quality	Toll quality
Speech processing	2:1 syllabic compandor
Grade of service	2% blocking probability
Mobile unit:	
Tx RF output	3W (nominal)
Tx RF power control	10 steps of 4-dB attenuation each; minimum power: −34 dBW
Rx spurious response	−60 dB from center of the passband
Number of synthesizer channels	832; tunable to any channel
Rx noise figure	6 dB measured at the antenna port
Rx sensitivity	−116 dBm from a 50W source applied at antenna terminals should produce 12-dB SINAD (C-message weighting)

transmission at 10 Kbps takes place. A technique known as blank-and-burst signaling is used, whereby the speech is blanked out for a period of 50 ms and data is transmitted. The customer does not notice any disruption in speech due to the 50-ms blanking. The data is Manchester encoded such that its spectrum lies above the voice band.

Both speech and data are passed through limiters and FM discriminators in the receiver. Thus, both data and speech use a common receiver circuit up to the discriminator output. Discriminator output is split into two parts; one goes for speech processing and the other for data demodulation and decoding. For protecting the data under multipath fading, data is BCH encoded and is repeated 5–11 times. Majority decisions are made in the receiver. Data consists of two word groups that are directed toward two different groups of mobiles. Half of the mobiles in a service area are supposed to read the words for group A, while the other half of the mobiles are supposed to read the words belonging to group B. These interspread words and their repetition reduce the multipath effects due to partial decorrelation of the transmitted data. A continuous data frame in the forward channel consists of a 10-bit dotting sequence for bit synchronization, an 11-bit Barker code for frame synchronization, and 40 bits of data from users A and B, repeated five times. This is shown in Figure 4.10(a). Additionally, a single-bit busy/idle status indicator is inserted every 10 bits to reduce the probability of collision due to random access of the users in the reverse access channel. The reverse channel, shown in Figure 4.10(b) and used by the mobile, consists of 48 bits of seizure precursor plus five repeats of 48 bits each from subscribers A and B. Due to discontinuous message transmission, a bit synchronization word is longer than the forward paging channel. The data format used over voice is shown in Figure 4.10(c). As stated above, the technique described as "blank-and-burst" is used over this channel. Here the messages are repeated eleven times in the forward direction but only five times in the reverse direction. The reason for this is that the forward direction message is more critical due to handoff messages. The blank-and-burst message begins with an initial 100-bit dotting sequence, which generates a 5-kHz tone that is used to detect the beginning of the message. Each repeated word also contains 37-bit dotting sequence followed by an 11-bit Barker sequence.

The mobile adjusts its power based on instructions from the cell site. The power adjustment is based on the periodic signal-level measurements by the cell site upon request from the MTSO. If the voice signal-level measured is below a certain level, the cell site will send out a power-level adjustment order over the forward voice channel using blank-and-burst signaling. This will allow the mobile transmitter to turn down its power level in steps of 4 dB. There are total of seven steps with a minimum attainable power level of −39 dBW(nominal, class IV for dual-mode subscriber) at the mobile transmitter. On the other hand, if the mobile power level received at the cell site is above a certain level, the MTSO may direct the cell site to send out an order instructing the mobile unit to decrease its power level. Internally generated receiver noise is limited by its 6-dB noise figure. The minimum sensitivity of the receiver is specified to be −116 dBm at the antenna input terminal and provides an output signal-to-noise ratio of 12 dB.

4.3.4 System Overview of AMPS

The network configuration of the AMPS, along with the functions of three control components, is shown in Figure 4.11. The MTSO is implemented by the 1/1A family of electronic switching systems (ESS). It is a control center and interfaces with the land telephone network. An MTSO services a large geographic coverage area, and all the mobile calls are switched through the MTSO. The 1/1A ESS consists of processors, memory, and

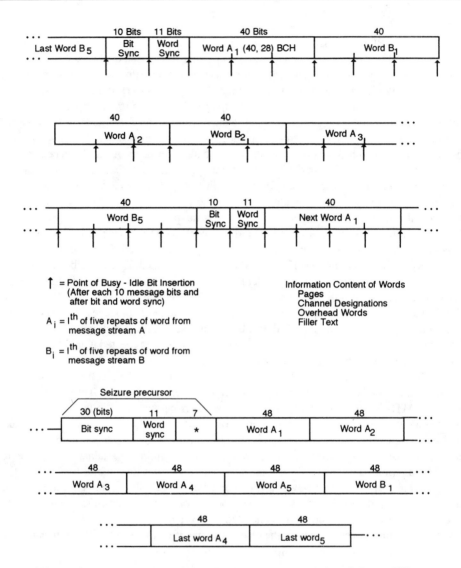

Figure 4.10 Data format on the (a) forward-control and (b) reverse-control channel. Source: [29].

service circuits. Programs stored in the switching system memory provide the logic to control telephone calls. The processors and memory are duplicated for greater reliability. Reed switches are used as matrix elements. The various control functions performed by the MTSO, cell sites and Mobile unit are narrated in Figure 4.11.

Lines from local subscribers terminate on the switching network at the MTSO. Likewise, trunks interconnecting with other switching offices terminate on the network

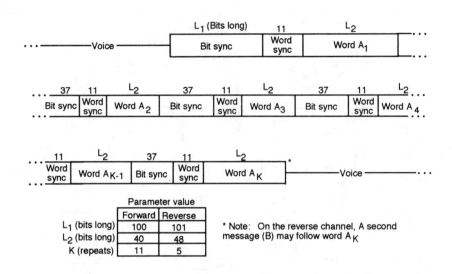

Figure 4.10(c) Data format used on forward voice channel. Source: [29].

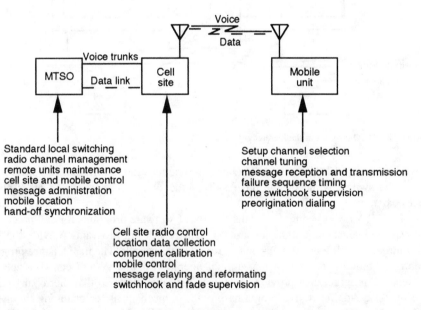

Figure 4.11 Partitioning of control functions among AMPS control elements. Source: [29].

through trunk interface circuits at the MTSO. The reeds are switched under the control of the central processor to produce a metallic voice connection or path between them. The switching network is configured to connect any two lines and/or trunks together with an engineered probability of blocking. The positioning of the MTSO in the hierarchy is

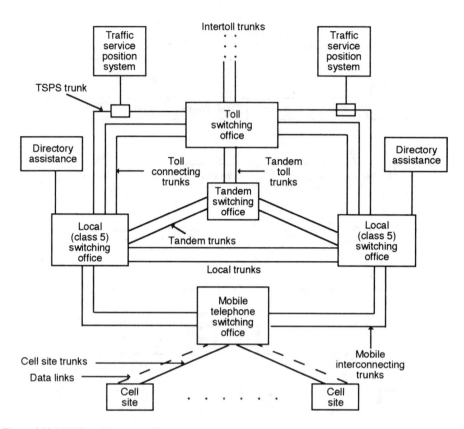

Figure 4.12 MTSO position in the hierarchy. Source: [30, p. 74].

shown in Figure 4.12. Here, the MTSO provides connection to different cell sites on one side of the switch, while on the other side of the switch connections to other class 5 offices exist.

The 1/1A ESS operates under the control of software programs for call processing, hardware maintenance, and administration. Call-processing programs provide the logic that controls the call setup and disconnect for all types of calls. The maintenance programs reconfigure failed hardware so that normal operation of the system can continue. The maintenance program also provides diagnosis for failed units so that the repairs can be done quickly. Administrative programs provide a mechanism for changing the system database. The database includes customer records, billing data, trunk records, and traffic counts.

In the Chicago equipment test, the MTSO occupies a position in the switching hierarchy below a class 5, or local, office. The MTSO can be interconnected with one or more local offices over standard trunk facilities. The MTSO interconnection arrangement is similar to the private branch exchange (PBX) and makes use of existing land lines to other class-5 offices. Calls originated from the mobile subscriber are out-pulsed from the MTSO using Touch Tone, a registered trademark of AT&T, or dial pulse signaling. The MTSO selects and sizes the trunk to the local class-5 office. It begins sending the dialed information after it receives a wink signal from the local class-5 office. A wink signal is a momentary battery reversal on the trunk that confirms the readiness of the local office to receive the digits. There are three possible calls within the system: mobile-to-mobile, mobile-to-landline, and operator-assisted and service calls from the mobile. For a mobile-to-mobile call, the MTSO receives digits from the cell site, determines that the call is meant for another mobile within the same cellular system, and completes the connection to another mobile. Other class-5 offices are not involved in this call. This interconnection is shown in Figure 4.13(a). For mobile-to-land calls, upon receiving the digits from the cell site the MTSO transfers the called number to the proper class-5 office. The mobile interconnecting trunk provides access to the called subscriber through other class-5 switching offices. The connection may be completed with or without going through the tandem office. This case is shown in Figure 4.13(b). Calls for service from the mobile are identified at the local class-5 office, where it connects to the operator and the service position trunk, as shown in Figure 4.13(c). For further details of these calls, the interested reader is referred to Section 2.3.2.1 where both national and international calling sequences are discussed.

4.4 THE CANADIAN SYSTEM

As in the United States, the demand for cellular radio in Canada lies primarily in professions such as real estate, construction, sales, and medicine. The cellular system developed in Canada is completely compatible with the system in United States. The reason for this compatibility due to the large number of roamers from the United States.

4.4.1 Background

Cellular mobile telephone service in the of 800-MHz band was first examined by the Department of Communication (DOC) in September 1981, when they issued a paper entitled "Radio Licensing Policy for Cellular Mobile Radio Systems and Preliminary Mobile-Satellite Planning in the Band 806–890 MHz." This paper invited comments and suggestions concerning the guidelines for licensing cellular mobile radio and the planning of mobile-satellite networks. Following a review of comments, the DOC issued a notice on cellular mobile radio policy and called for license applications in October 1982. In this notice, the DOC invited applications for the first 23 big metropolitan sites throughout Canada. The major sites covered in this notice are shown in Table 4.5 [9]. The DOC further declared that each area where the cellular radio telephone system was to go into service would essentially be served by two operating companies: one a local telephone company serving the area, and the other a nontelephone operating company. Half of the

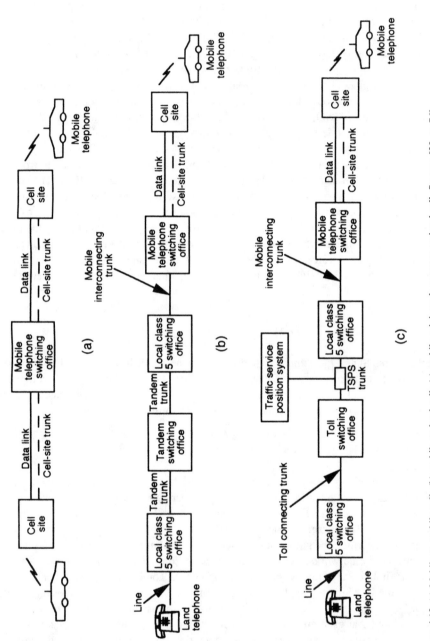

Figure 4.13 (a) Mobile-to-mobile call; (b) mobile-to-land call; (c) mobile-originated, operator-assisted call. Source: [30, p. 76].

Table 4.5
Cellular Radio's First Canadian Sites

Cities	*Province*
Calgary, Edmonton	Alberta
Vancouver, Victoria	British Columbia
Winnipeg	Manitoba
St. John	New Brunswick
St. John's	New Foundland
Halifax	Nova Scotia
Hamilton and Kitchner, London and Oshawa, St. Catharines Niagra, Sudbury, Thunder Bay, Toronto, Windsor, and Ottawa	Ontario
Montreal, Quebec City, and Chicoutimijonquiere	Quebec
Regina, Saskatoon	Saskatchewan

total spectrum would be given to the operating company, and the other half of the spectrum to the nontelephone company. The system had to be interconnected to the telephone network of the country.

A majority of the companies involved in the installation of the cellular mobile telephone system favored the interconnection of cellular systems to the public switched telephone network. It was obvious that this interconnection would not only enhance the value of cellular mobile radio system to users, but would also lead to technological innovations through competition. Among the major companies who now provide service in the cellular market in Canada are Bell Cellular, British Columbia Cellular, and Contel. Bell Cellular, a subsidiary of Bell Canada Enterprises is based in Quebec and provides service to the major cities throughout Ontario and Quebec (see Table 4.5). These services were initiated in July of 1985. British Columbia Cellular, which is a subsidiary of B.C. Telephone Company, initiated cellular service in the Vancouver area in January 1986. Contel (a joint venture of Roger Telecommunications Limited of Toronto, Telemedia Enterprises of Montreal, and First City Telecommunication of Vancouver) started offering cellular mobile service in July 1985. Initially the offering of cellular service was limited to Ontario and Quebec. In 1986 they expanded their service to the cities of Victoria and Vancouver in British Columbia and Calgary and Edmonton in Alberta.

4.4.2 Characteristics and Parameter

The characteristics and parameters of the cellular systems in Canada are the same as for those in the United States and thus need not be discussed. However, we will describe a particular system developed by the Westech Corporation that was implemented in Alberta, Canada. We turn briefly the technical features of the system.

4.4.3 Overview of the Alberta System

One of the earlier cellular systems in Canada was developed by the Westech Corporation of Alberta [10]. The system, known as AURORA, was used as a cellular mobile telephone system in Edmonton, Alberta. At the first stage, the system was designed only for 400-MHz applications. Later, to satisfy the overseas and other North American markets, Westech came up with its 800-MHz version of a cellular system.

These two systems are known as the AURORA 400 and the AURORA 800 systems. Both the 400 MHz and the 800-MHz system are modular in design and based on the principle of distributed control architecture. The AURORA systems are linked with the land telephone network at class 5 and toll-office levels, and require no modifications to the existing switching equipment. This also minimizes the data transmission requirement between different levels of system control, which means that low-speed data communication links can be used between controllers in the telephone hierarchy. Due to the modularity of the system, the expansion from low capacity to medium and even high-capacity systems is easily achieved. The architecture of the AURORA 400 and AURORA 800 systems is shown in Figure 4.14(a,b). The AURORA 400 system was supposed to cover an area of 1,920 km and provide service to an estimated 40,000 subscribers. A fully configured 120-cell system is intended to cover the whole province of Alberta. The basic system is a seven-cell cluster similar to AMPS in the United States. The system essentially has all those features and characteristics commonly required in cellular telephone system. The standard features are:

- Full-duplex transmission similar to land line telephone.
- Touch-tone dialing from the mobile.
- Speed dialing: up to 10 numbers can be stored in the memory and dialed out by recalling the memory.
- Calling compatibility with land telephone system, that is; the user can originate any type of call either by direct dial or operator assisted.
- On-hook dialing: mobile can originate a call when the handset is in the on-hook position. Ringback or busy state can be heard with an optional speaker attachment.
- Automatic tracking: so long as the mobile receiver is on, the system will track the whereabouts of the user without any action on the user's part.
- Out of range indication: a display on the mobile system will indicate that the user is out of range of the cellular coverage area.
- Electronic lock: used to prevent unauthorized usage, but the user can lock the system.

The optional features are:

- Call forwarding: to any land or AURORA mobile users.
- Queuing: an originating mobile can be placed in a service-waiting queue. The system will signal the mobile when the facilities are available for use.

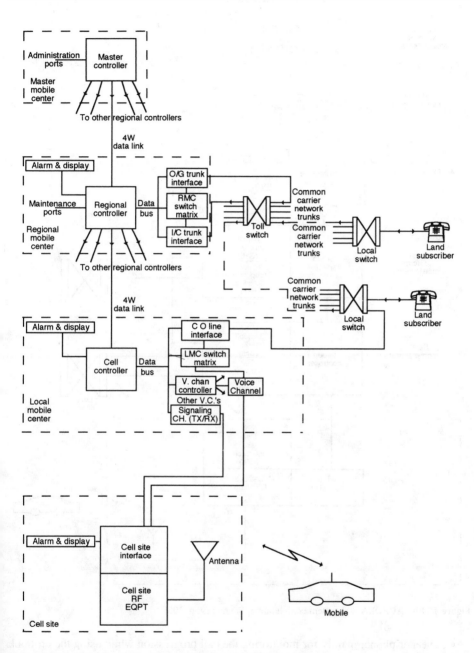

Figure 4.14(a) AURORA 400 system configuration. Source: [9, p. 303].

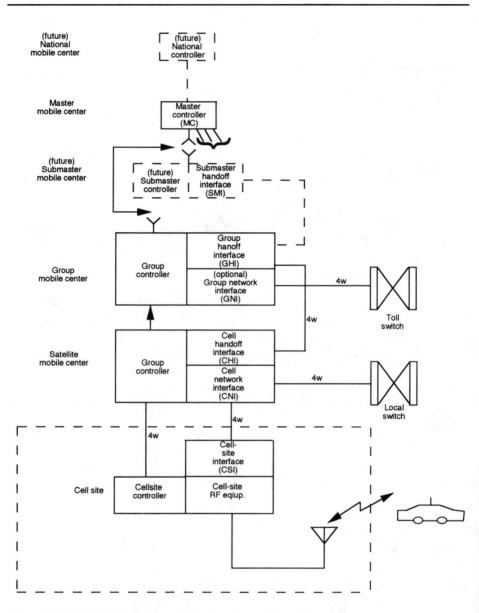

Figure 4.14(b) AURORA 800 system configuration. Source: [9, p. 305].

- Speaker phone: mainly for monitoring the call progression while using the on-hook dialing feature.
- 500-type telephone set: an optional 500 type telephone handset can be installed for access to outside computer facilities.

These features are available for both the AURORA 400 and AURORA 800 systems. Among the important standard features available to the users are: speed calling, on-hook dialing, automatic tracking, the out of range indicator, and electronic lock. Among the important optional features available to the subscribers are: call forwarding to land line or to another user of AURORA system, queuing, and speaker phone.

The architecture of the AURORA 400 system, as shown in Figure 4.14(a), interfaces with the land telephone system at both the local mobile center and the regional mobile center. The local mobile center controls several cell sites, while the regional center can only be a controlling point for a maximum of twenty-seven local mobile centers. A mobile call directed for a local land subscriber is first received by the cell-site RF equipment. After ensuring the validity and correctness of the called number, the message is sent through a cell-site interference unit to the receiving part of the signaling channel interface at the local mobile center. The cell controller, after proper credit verification, forwards the line through a central office line interface to the local switch and then to the land subscriber. After the land subscriber goes off-the-hook, a voice path is opened. The end-to-end connection is thus established through the mobile, to its cell-site RF equipment, to voice channel, to LMC switch matrix, to CO line interface, to the local switch, and finally to the called land party. The subscriber set cell-site interface unit ensures that the audio quality throughout the conversation is at least as good as that experienced during normal land-to-land calls. Calls originated by land subscribers in different geographical areas appear at the toll switch and are monitored by the regional controller. The regional controller, by going through its database, identifies the whereabouts of the mobile and requests the proper local mobile center to establish a path to the mobile. The conversational path as shown is through the calling party's local switch, to the toll switch, to the associated local switch, to the local mobile center, and finally to the called site where the called mobile is connected through a radio path.

A regional mobile center (RMC) is connected to the master mobile center through a four-wire data line. A maximum of six regional mobile centers can be connected to one master controller. A master mobile center maintains the master data. No direct links are required between regional mobile centers as the calling scheme makes full use of the existing switched telephone network. This eliminates the need for RMC-to-RMC communication in intertoll long-distance calls. As stated above, the AURORA 400 system has been modified to operate in a radio frequency band of 800 MHz. The AURORA 800 also makes the system compatible with the North American market. The system is FCC-compatible and is distributed in its control architecture. The AURORA 800 system, as shown in Figure 4.14(b), consists of: cell site control, cell controller, group controller, and master and submaster controller. The lowest level of control is at the cell site, where the controller provides on-site management of call handling and performs the tracking of cellular calls in progress for handoff purposes. The cell-site interface equipment provides line matching and interfacing between the cell site and the cell network interface (CNI) at the satellite mobile center (SMC). The CNI, as in the AURORA 400 system, interfaces to the local switch through which the connection to the land telephone network is established. The CNI also provides an interface to the cell site through a land line, through which a cellular mobile radio connection can be established. A call from the mobile can be routed to the local switch or rerouted to the same cell site under the control of the cell site controller. A cell handoff interface unit (CHI) houses a small four-wire handoff switch and is intended to route the call handoff between SMCs and the group mobile center (GMC). The cell site interface (CSI) coordinates the activities of CNI and communicates with the cell site controller and the group controller when call setup or handoff is necessary. The GMC maintains the database information for NNX or the mobile number groups on its service area. Additional database requirements are handled through a master mobile center (MMC). Each GMC is designed to control a multicell coverage area. The small switch at the group handoff interface (GHI) can provide a handoff of calls between SMCs within the group and also to any other multicell groups via the controlling GMCs. The optional group network interface (GNI) provides access to the toll switch for GMC interception of mobile-originated calls for extracting the billing data.

Since the mobile is identified at the local switch level, the interception of land calls to mobile is not done at GMC. Land-originated calls are terminated to the mobile at the "home" local switch of the called mobile, where the SMC takes over and directs the call to the mobile. When the called mobile has roamed to another area, the home SMC gets that information from the higher levels of the system control hierarchy and reroutes the call through the local switch via the network to the terminating local switch. The highest level of control in AURORA 800 system is at the master control level, which provides the primary database for all controllers under it and plays a critical part in message transfer and coordination between lower levels of controllers. The MMC collects and records data for such requirements as traffic and billing from lower levels of controllers.

4.5 THE U.K. SYSTEM: TOTAL ACCESS COMMUNICATION SYSTEMS (TACS)

In 1979, the World Administration Radio Conference (WARC) allocated the band from 890–960 MHz for mobile radio service. Following WARC's action, the conference of the European Post and Telecommunication (CEPT) recommended a frequency plan that made provisions for a mobile radio telephone service using a 890–915 MHz band (mobile to base) paired with a 935–960 MHz band (base to mobile). With 2×25 MHz bandwidth, there is a total of 1,000 channels. The first 600 channels were allocated equally to the two network operators. At present the top 400 channels are not available for cellular use, but are expected to be available in the near future. This 2×10 MHz of spectrum is held in reserve for the second generation cellular GSM system, which is now becoming operational.

4.5.1 Background

The U. K. system is much like the AMPS system of the U.S. except that the specified frequency spacing between channels is a standard 25-kHz CEPT channel spacing. The system allocates adjacent channels to alternate cells, and thus the full specification of CEPT for adjacent channel protection is not met. Other differences include the frequency deviation for voice and data signals. The voice frequency deviation is reduced from a ±12.0 kHz peak in AMPS to a ±9.5kHz peak. Data transmission is at the rate of 8 Kbps with a frequency deviation of ±6.4 kHz. A binary 1 is transmitted at $(f_c + 6.4)$ kHz, while a binary 0 is transmitted at $(f_c - 6.4)$ kHz, where f_c is the carrier frequency. The essential differences between AMPS and TACS are shown in Table 4.6. The decision to accept a version of AMPS, with minor modifications, was based on: a reduction in design time, a potential for exporting of the system, and commercial viability in terms of equipment costs.

The decision by the United Kingdom to adapt an advanced version of AMPS was based on considerations against contending European systems, such as NMT, MATS-E, C-900, and NAMTS. In the end a number of factors influenced the decision, but of particular importance were the advantages of the wider deviation of the AMPS system along with the results of a considerable number of field trials. In selecting a system based on the operational AMPS system of the U.S., the U.K. escaped the need for an extensive developmental test program. Because of this advantage, the government only allotted a two-year period during which the two nationwide systems had to be put into operation. One of these companies was a joint venture between British Telecom and Securicor, which implemented the initial phases of its cellular network, known as CELLNET (two operators per area), in January 1985. Motorola participated in the system planning and engineering with CELLNET

The initial coverage was limited to the London area. By mid March 1985, the network had a total of twenty-seven cell sites serving the London area and was initiating services in Birmingham, Manchester, and Liverpool. By August 1985, service was extended to ninety cells and served about 35,000 subscribers. About 82% of the total

Table 4.6
Major Differences Between the U.S. and U.K. Systems

Items	U.S. (AMPS)	U.K. (TACS)
Transmit frequency band:		
Base stations (MHz)	870–890	935–960
Mobile stations (MHz)	825–845	890–915
Channel spacing (kHz)	30	25
Voice signals:		
Modulation type	FM	FM
Peak deviation (kHz)	±12.0	±9.5
Control signals:		
Modulation type	FSK	FSK
Transmission rate (Kbps)	10.0	8.0
Peak deviation (kHz)	±8.0	±6.4
Coding	Manchester	Manchester

population was covered by 1986. Each of these two systems is supposed to offer coverage to at least 90% of the U.K. population by early 1990. The need to serve handheld portable radio telephones has been considered to be of particular importance in the CELLNET system. The CELLNET system has incorporated base receive diversity and has planned their system with small cell sizes in mind from the outset. Coverage in the center of London has been planned with, typically, 2-km radius cells in the core of the city and medium-sized cells which are typically 4 km in radius, in areas more remote from the center. Currently, this plan has made the London system the best operating system for portable coverage.

4.5.2 Characteristics and System Parameters

The basic requirements of the system are the same as those of the nationwide telephone network. This includes good voice quality, high reliability, low blocking, and low cost. Services are offered to vehicular mounted, transportable, and handheld equipment. Some of the characteristics provided to the subscriber, as shown in Table 4.7, are efficient use of spectrum, mobile registration techniques, the handoff process, preorigination dialing, charging information display, call completion, and growth. The requirements for some supplementary facilities provided by total access communication systems (TACS) are call forwarding, repertory dialing, message waiting, three-party conferencing, short-code dialing, and call holding. Charging information is transmitted from the cell site to the mobile on a real-time basis. This is particularly useful for passengers using phones in taxis. In addition to these basic network services, a range of value-added network services (VANS) are provided: information, secretarial, and personal answering and message-forwarding services.

Protection against theft and misuse of equipment is included by transmitting the equipment service number. This number is set at the beginning of equipment installation

Table 4.7
User and System Characteristics of TACS

Characteristic	Description
Efficient use of spectrum	Channels are repeated to get the best spectrum efficiency
Mobile registration	An enhanced scheme for registration allows automatic roaming, both nationally and internationally
Handoff	Standard feature
Preorigination dialing	Driver can store the called number before pressing the SEND key
Charging information	Transmitted during call to display charges; convenient for applications such as coin box and taxi phones
Call Completion	Allows successful completion of 99% of calls when initial two-way connection is established
Growth	Large cell site at the start followed by smaller cells as the system grows mix of large with small cells is possible

and cannot be altered. The major system parameters are described in Table 4.8. With a channel spacing of 25 kHz and a peak voice deviation of ±9.5 kHz. The carrier-to-interference ratio is about 3-dB lower than AMPS, which allows a shorter frequency reuse distance, and thus an efficient spectrum utilization. Signaling data on the dedicated channel is at a rate of 8 Kbps. Synchronization is chosen such that it does not appear in the data. Sufficient time for synchronization is given to the mobile at the beginning, when the mobile switches its power. In addition to data transmission from the base in the synchronous mode, the mobile transmits its data in the bursty asynchronous form. Forward error correction (from cell to mobile) is achieved by using a systematic linear block code (BCH) and repeating the data five times. At the receiver, a decision is made by performing majority voting. The objective is to achieve 95% accuracy in data transmission with the help of majority voting and improve its accuracy to 99.9% by using the error correcting code. A transfer of control data over the voice channel is accomplished by muting the audio. The absence of audio is for such a short duration that the mobile does not even recognize the muting. The data is repeated eleven times to combat the noisy surroundings

Table 4.8
System Parameters of TACS System

Category	*Parameters*
Number of channels*	600, 2 groups of 300 channels; each in two subbands
Number of signaling channels	21
Cell radius	2–20 km
Frequency range:	
Base-station transmit	935–950 MHz
Mobile-station transmit	890–905 MHz
Channel spacing	25 kHz
Frequency separation	45 MHz between transmit and receive channels
Maximum base station ERP	100W
Mode of transmission	Full duplex
Voice transmission	Phase modulation with peak frequency deviation of ±9.5 kHz
Speech-processing	2:1 syllabic compandor
Data transmission	Signaling-channel data is FSK modulated, Manchester encoding
Data-transmission rate	8 Kbps with peak frequency deviation of ±6.4 kHz
Error protection coding (data on voice):	Shortened, 63:51, BCH, repeated 5–11 times with bit-by-bit majority voting
Base-to-mobile	40:28, BCH, 5 repeats
Mobile-to-base	48:36, BCH, 5 repeats
Error detection	11–89 per 200 bits
Error correction	5 per 200 bits, minimum
Supervisory tones:	
Closed-loop identification	SAT frequencies 5,970; 6,000; and 6,030 Hz
On-hook condition	8-kHz tone sent with SAT simultaneously on audio channel

*In E-TACS, the number of channels is increased to 2,047; group-A signaling channels are from 23 to 43 group-B signaling channels are from 323 to 343.

of the mobile. In the reverse control and voice channel the data is repeated five times. The code structure is such that it provides a single error-correcting capability and four error-detecting capabilities.

Similar to AMPS in the U.S., the maximum ERP per channel from the base station is limited to 100W. An area where the TACS system has deviated from the AMPS specifications is on the power classes of mobile. Mobile nominal power is divided into four classes with a maximum of 10W assigned to class 1 stations, while the low-power mobiles belonging to class 4 have a nominal value of 0.6W. There are seven mobile attenuation codes, which allows the power to be reduced by about 30 dB below the maximum nominal power. Mobile power levels and the attenuation codes are shown in Tables 4.9 and 4.10.

4.6 THE JAPANESE SYSTEM

Although public mobile service has been offered in Japan for more than 30 years, the growing needs and demands were never met before the introduction of the cellular system because of limited channel availability. Potential users who were on a waiting list for a

Table 4.9
Mobile Power Levels

Class	Nominal ERP	Transmit Power (Watts)
1	Very high power mobile	10.0
2	High-power mobile	4.0
3	Mid-range hand-held portable	1.6
4	Low-power hand-held portable	0.6

Table 4.10
Mobile Attenuation Codes

Mobile Station Power Level	Mobile Attenuation Code	Nominal ERP (dBW) Power Class			
		1	2	3	4
0	000	10	6	2	−2
1	001	2	2	2	−2
2	010	−2	−2	−2	−2
3	011	−6	−6	−6	−6
4	100	−10	−10	−10	−10
5	101	−14	−14	−14	−14
6	110	−18	−18	−18	−18
7	111	−22	−22	−22	−22

long time turned out to be real users after the introduction of the cellular system. The Japanese system provides nationwide services to more than 450 cities, using nine mobile telephone switching centers to the common channel interoffice signaling network [13–22].

4.6.1 Background

A cellular telephone system at 800 MHz was introduced in the Tokyo metropolitan area towards the end of 1979. The initial system covered an area of about 600 km and was divided into 13 radio zones. Since that time both the service area and the total number of subscribers has steadily grown. By 1985 there were 51 radio zones with a total of 43,000 subscribers. Service was extended to 25 percent of the area and to 60 percent of the total population. As of July 1987, there were a total of 107,000 subscribers with the service area extended to about 70% of the whole country. Since that time the growth has been exponential. Figure 8.3 provides details of subscriber growth. Studies of traffic patterns in the Tokyo metropolitan traffic areas have revealed an approximate bell-shaped curve, which is proportional to $10^{-d/20}$ where d is the distance in km from the center of the city. The distribution with respect to time is geographically variable, as the busy hours of each radio zone are different. Design conditions for traffic behavior are shown in Table 4.11. The average number of calls/day per subscriber is 3.0 with a mobile originating-to-terminating call ratio of 2:1. The concentration ratio is 10% during the busy hour (10% of total calls per day arrive during peak one-hour period). The results of the test reveals that most of these design parameters are exceeded. All control channels are used simultaneously (simulcast) to identify the mobile. The dialing number for a land-to-mobile call is: 0 (trunk prefix) + 30 (mobile identification) + NN (mobile area code) + XXXXX (mobile telephone number). For calls from mobile-to-land, a subscriber dials: 0 (trunk prefix) + ABCDE (toll and office code) + XXXX (subscriber number). In all cases, the calling party is charged for the call. Speech quality, as discussed in Section 4.1, is measured in terms of the sound articulation index. An AI of 80% or more has to be maintained throughout the conversation. Radio zones are divided into two groups of 15 MHz each in 800–900 MHz region. Both group 1 channels and group 2 channels have 15 MHZ of

Table 4.11
Design Conditions for Traffic Behavior

	Conditions
Number of calls per subscriber	3 calls/day
Originating-to-terminating call ratio	2:1
Average holding time	120 sec/call
Busy-hour concentration	10%
Busy-hour traffic	0.01 erlang/subscriber
Handoff ratio	0.3–0.5 times/call

individual coverage with a 5-MHZ overlap between groups. These two subbands of coverage are shown in Figure 4.15. The channel separation between base station and mobile, and between mobile and base station, is 55 MHz.

4.6.2 Characteristics and System Parameters

The principal parameters and their descriptions are shown in Table 4.12, which is divided into four categories: radio system, exchange and control system, communication system, and mobile unit. As listed under the radio system parameters, the maximum number of radio channels is 1,000, divided into two groups of 500 channels each. Each band has a channel spacing of 25 kHz. The speech modulation is by narrowband FM with a maximum frequency deviation of ±5 kHz. This assures the median $C/N = 17$ dB at 90% of all locations on the boundary of cells. Data is FM-modulated at 300 bps, with a peak frequency deviation of ±4.5 kHz. For fast synchronization, data is Manchester encoded. Control channels are divided into paging and access channels. Data is continuously transmitted in the forward paging channel, and thus the bit synchronization is achieved easily without alternating the sequence. The forward paging channel is (43:31) BCH encoded.

The reverse paging channel (mobile to base) has the encoded data format of (11:7) and is the shortened form of a (15:11) BCH code word. The reverse paging channel data

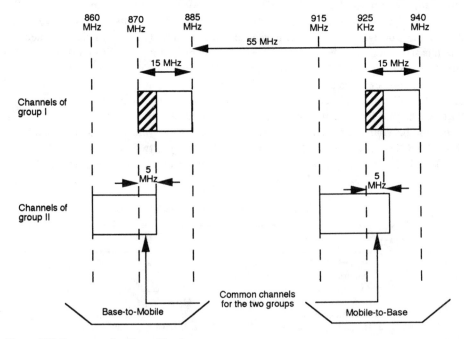

Figure 4.15 Frequency allocation subbands.

Table 4.12
Parameters of Japanese System

Category	Parameters
Radio system:	
Frequency	800-MHz band (860–885 and 915–940 MHz, 2 bands)
Number of radio channels	1,000 (500, 2 bands)
Channel spacing	25 kHz
Frequency separation	55 MHz between pair of transmit and receive channels
Modulation (speech)	FM (maximum frequency deviation: ±5 kHz)
Signal processing	2:1 syllabic compandor
Base-station transmitting power	25W
Mobile-station transmitting power	5W
Radius of urban coverage	5 km (small cell)
Radius of rural coverage	10 km (large)
Control signal modulation	FSK; peak deviation: ±4.5 kHz; Manchester coding; transmission rate: 300 bps
Base-to-mobile error protection	Forward paging/access channel: 43:31, BCH
Mobile-to-base error protection	Reverse paging/access channel: 11:7, BCH
Number of radio channels	128 per base station
Maximum number of base stations connectable to a mobile control station	32
Maximum number of mobile control stations connectable to a switching center	6
Exchange and control system:	
Tracking exchange	Speech channel switching initiated by S/N monitoring at base station
Regional calling	Determined by location registration
Radio-section signaling	Paging/access channel: digital; speech channel: digital/tone
Selective calling	On paging channel
Speech channel selection	Assigned by MCS through access channel
Communication System:	
Speech connection	Full duplex
Speech quality	Speech AI > 80%
Speech processing	2:1 syllabic compandor
Grade of service	3 out of 100 calls
Exchange	Fully automatic with push-button dialing
Mobile Unit:	
Transmitter output	5W
Spurious emission	60 dB or more below carrier at transmitter output
Receiver spurious response	More than 70 dB below the receiver sensitivity
Number of synthesizer channels	1,000
Channel switching time	Less than 50 ms

consists of 24 bits for bit synchronization, 8 bits for word synchronization, 7 bits of information, and 4 bits of parity for the data. A mobile station randomly originating a call is controlled by the access channel. To prevent collision by double seizure of the access channel, the base station, upon receiving the request from the mobile station, provides busy status on the access channel. The data frame composition for forward and reverse access channels and reverse paging channel is shown in Figure 4.16. The mobile unit location is detected at the base station by the field strength of the access signals received from the mobile units. Speech channels are assigned by the access channel, and a loop check signal is exchanged on the assigned channel between the mobile and the base station. For a handover connection, the base station monitors the S/N ratio on the speech channel and asks the control station for handover after detecting a deterioration in signal quality. The handover time is less than 50 ms. A full-duplex communication with a GOS better than 3% and a speech AI of 80% is achieved. Mobile transmitter output is limited to 5W of power, and its spurious emission to −60 dBc.

4.6.3 Overview of Japanese System

The present Japanese network and mobile system configuration are shown in Figures 4.17 and 4.18. The mobile network consists of an automobile switching center (AMC), a mobile control station (MCS), a mobile base station (MBS), a mobile subscriber station (MSS), and the transmission lines between the AMC, MCS, and MBS. The radio path is between the MBS and the MSS. The AMC is the interface between the land telephone system and the mobile telephone service. The AMC switches (D-10 ESS switch) calls between trunks to the land line and the access lines to the MBS. The central controller in the AMC controls the switch and the mobile data transmission equipment. The MCS controls the base station equipment and sets up the radio channels between mobile units and the base stations. The functions associated with various subassemblies of the MCS are:

- Data channel control equipment (DCE). Acts as a relay for control information between the AMC and other MCS equipment.
- Speech path equipment (SPE). Contains many speech path circuits and accommodates dial signal receivers. It is handled by the speech channel assignment equipment (SCAE).
- Paging channel control equipment (PCE). Handles the control of land-originated calls and is provided to each paging channel.
- Access channel control equipment (ACE). Handles the control of mobile-originated calls and is provided to each access channel.
- Handoff control equipment (CCE). Handles all the handoff controls.

The MBS contains the radio transmitters and receivers to communicate back and forth to the mobile. Regional centers, district centers, toll centers, and end offices are a part of

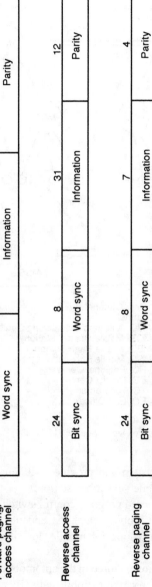

Figure 4.16 Data frame composition of NEC system.

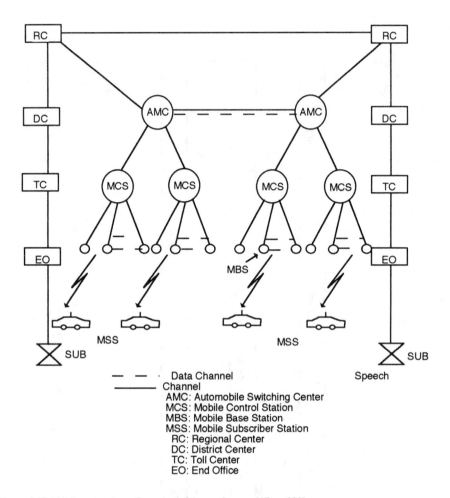

Figure 4.17 Mobile network configuration of Japan. Source: [17, p. 239].

the public switched telephone system and provide switching and control access to the fixed land subscribers to the mobile subscriber station.

The D-10 switching is the stored program control switch that performs the following functions: connection between a mobile telephone and the landline telephone system; charging the mobile subscribers; paging for mobile; location registration and switching for tracking exchange.

Mobile subscriber information is stored in the automobile switching center's memory and is accessible to other offices processing mobile calls. By a common channel signaling scheme, the AMC can access the memory and obtain information for the mobile from the other AMCs. An MCS controls the functions for setting up the radio path, handoff

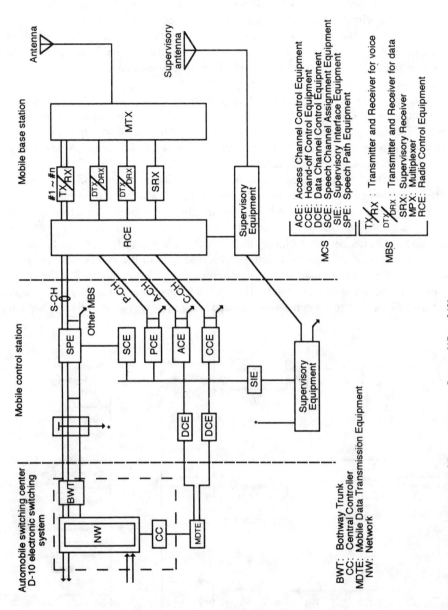

Figure 4.18 Equipment configuration of Japanese system. Source: [17, p. 243].

control, and location registration, as well as for supervising the mobile base station and the condition of radio circuits. The MBS transmits and receives signals from the mobile, and measures the circuit quality upon command from the MCS for handoff purposes. The station consists of transmitters and receivers for voice and supervisory data signals. A multiplex unit connects the receivers and transmitters to the antenna system. The multiplexor combines sixteen transmitters by high Q cavity resonators and a junction box. The insertion loss of the device is about 4 dB.

There are two types of standard antenna systems: a high-gain omnidirectional antenna mounted at the tower top, and a side-mounted antenna. A tower-top mounted antenna is an omnidirectional array antenna with an 11-dB gain over a halfwave dipole, which can multiplex up to four groups of a 16-channel multiplexed radio signal. A tower side-mounted antenna is a directional parabolic reflector.

The mobile unit (subscriber unit), which communicates to the MBS through radio channels, consists of an antenna and a duplexer through which the transmitter and the receiver access the antenna. A mobile control unit allows the subscriber's receiver to be tuned to hundreds of radio channels. The mobile receiver has a digital frequency synthesizer to provide tunable frequencies with very high stability. The transceiver unit for the mobile is shown in Figure 4.19. The analog and digital interface to the mobile control unit is shown on the right. The mobile antenna unit shown on the left is mounted at the center-top or on the side of the mobile vehicle. The control unit is a dial-in handset with a power switch and volume control switch, and is operated in the same way as an ordinary telephone set.

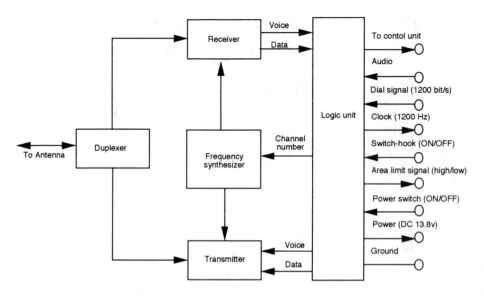

Figure 4.19 Mobile transceiver unit. Source: [17, p. 244].

4.7 THE NORDIC SYSTEM

The Nordic mobile telephone system (NMTS) was developed jointly by the telecommunications administrations of Denmark, Finland, Norway, and Sweden in order to establish a compatible automatic public mobile telephone system in the Nordic Countries. The system was put into operation during the end of 1981 with countrywide continuous coverage and full roaming capacity. In October 1985, the number of subscribers exceeded 200,000, which corresponds to about 1% of the population. The original system operated at around 450 MHz and was later updated to a 900-MHz operation in 1986 [21–22].

4.7.1 Background

It was considered essential that for calls to and from the mobile, the system should behave just like a landline telephone system. Based on this, the requirements for the system can be described as follows.

- It shall be possible to set up calls between the mobile stations and any fixed telephone subscriber or any other mobile telephone subscriber within the system, regardless of the country.
- The costs shall be charged to the calling station, regardless of whether it is located in the mobile system or in the fixed telephone network.
- The system shall provide for automatic roaming capability for the mobile subscribers within the Nordic countries.
- The introduction of the system shall not precipitate any significant changes in the fixed telephone networks.
- To subscribers, the system shall appear as a fixed landline telephone system.

Frequency bands of operation for the original system were 453–457.5 MHz for mobile to base, and 463–467.5 MHz from base station to mobile station. The upper 0.5 MHz in both bands are not used in Finland. Thus, with a 25-kHz frequency separation between channels, there are a total of 180 channels in all countries other than Finland, which only has 160 channels. The system now operates in the full-duplex mode with a frequency separation of 10 MHz.

4.7.2 Characteristic and System Parameters

As a result of the growing demand for service, new systems operating at 900 MHz were developed that had a full capacity of 1,999 channels. The capacity was enhanced by reducing the channel spacing to 12.5 kHz and increasing the bandwidth of operation to 25 MHz. The corresponding numbers for 450-MHz operation are 25 kHz and 4.5 MHz. The maximum base station ERP is 100W. The mobile in Stockholm area runs at the full power. Instead of reducing the mobile power, the sensitivity of the base station receiver is reduced. This method avoids the interference to the low-power mobile receiver by the high-power, more distant strong transmitter. The NMTS was designed originally with a hexagonal cell layout with omnidirectional antenna. In 1984, the cellular system in Stockholm became overloaded to the extent that 40% of the traffic was regularly blocked during the peak traffic hour. To overcome this situation, Swedish Telecom decided to use directional antennas to generate coverage areas in the form of sectors as shown in Figure 2.8. The original design was based on the realization that the traffic was highest at the center of the Stockholm system and decreased with distance from the center of the city. The conversion design realized uniform traffic throughout the area. The antenna used for sectorization was log periodic with side lobes 25-dB below the main beam and a front-to-back ratio of 25 dB. For data transmission, an efficient FFSK system was adopted. Here, binary 1 is represented by a full cycle of 1,200-Hz tones and a binary 0 is represented by one and a half cycles of 1,800 Hz tones. Unlike other systems of the world, the NMTS employed convolutional coding instead of blocked BCH codes for paging and access. For the handheld mobile, a low-power transmitter operating at a nominal power level of 1W was designed. The power level of the mobile at 450 MHz was set at 15W and 6W at 900 MHz. Thus there are striking differences between this system when compared to other international systems. The differences are in channel spacing, data modulation, and data coding. See Table 4.13.

4.7.3 Overview of NMT 450

The architecture of the system is shown in Figure 4.20. In this system, *base stations* (BS) are grouped into traffic areas (TA). The BS serves as an interface between the mobile

Table 4.13
Parameters of NORDIC 450 MHz and 900-MHz Systems

Parameters	NMT 450	NMT 900
Number of channels	180	1,000*/1,999
Cell radius (km)	1–40	0.5–20
Base-transmit frequency range (MHz)	463–467.5	935–960
Mobile-transmit frequency range (MHz)	453–457.5	890–915
Channel spacing (kHz)	25.0	25.0/12.5†
Frequency separation between transmit and receive channels	10.0	45.0
Maximum base ERP (W)	50.0	100.0
Nominal mobile transmitter power (W)	15.0	6
Mode of transmission	Full duplex	Full duplex
Voice-modulation transmission	PM	PM
Speech-processing transmission	—	2:1 syllabic compandor
Data transmission	FFSK‡	FFSK‡

*Corresponds to 25.0-kHz channel bandwidth.
†Both bandwidths are possible.
‡Peak frequency deviation: ±3.5 (kHz); NRZ encoded with a transmission rate of 1.2 Kbps.

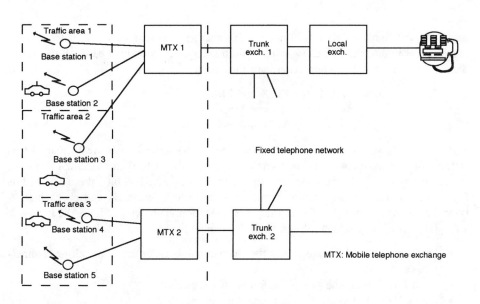

Figure 4.20 Architecture of NORDIC mobile telephone system.

radio and the land-based telephone system to perform switching of the speech path. The switching function is left to the mobile telephone exchange (MTX) which in turn controls one or more of the base stations. As an example, MTX2 controls both BS4 and BS5.

On every base station, one channel is used as a calling channel for signaling and is marked with a special identification signal. When free, one or more of the other channels are marked with a free traffic-channel identification signal. Standby mobile stations in an area under a base station are locked to the calling channel. It is, however, possible for the MTX to permit use of the calling channel for carrying conversation at times of high traffic. The call setup procedure is different in the two traffic directions, and thus different than other systems of the world. A mobile incoming call is sent over to the calling channel, which the mobile acknowledges by the return frequency on the outgoing calling channel. Information about the channel number is transmitted to the mobile station, which in turn switches to the traffic channel. Thus, the calling sequence for an incoming call to a mobile is same as in the AMPS. For calls initiated by the mobile, the mobile station looks for a free-marked traffic channel on which all signals are exchanged, including conversation. This is one of the major operational difference between this and other cellular systems. There are two parts to this medium: one a fixed 4W line circuit between the MTX and the BS, and the other a radio path between the BS and mobile. Since the fixed 4W circuit between the MTX and the BS carries digital signaling, group delay distortion is the controlling design parameter for the 900–2,100 Hz band. For a radio path between the BS and the mobile, cochannel interference between wanted and unwanted signals has to be kept below a certain threshold. Fading, as usual, plays an important part in this medium.

In the BS, the signal-to-noise ratio of an active channel is continuously evaluated and the result is reported to the MTX. Signaling within the system is divided between the MTX, the BS, and the mobile system. There are three signaling groups:

- Group 1: signaling between MTX and MS.
- Group 2: signaling between BS and MS.
- Group 3: signaling between MTX and BS.

Signals are transmitted on a 1,200-Hz binary signaling link. Signals are formatted into frames, the format being such that each frame contains 10–16 hexadecimal digits of information plus synchronization bits. In a frame containing 166 bits, there are 26 bits for frame and bit synchronization, while there are 140 information and check bits. A typical frame between the MTX and the mobile station is shown in Figure 4.21.

Different frame structures and what they represent are:

- Digits $N_1N_2N_3$ (12 bits) represent the channel numbers for traffic or calling channels in use.
- Digit P (4 bits) is the prefix for frame characterization.
- Digits Y_1Y_2 (8 bits) are the traffic area numbers.
- Digits $ZX_1X_2X_3X_4X_5X_6$ (28 bits) are the mobile subscriber numbers.
- When $Z = 15$, the information concerns BS digits $L_nL_nL_n$ (12 bits), which are telephone line signals for information.

Figure 4.22 shows the various kinds of signals exchanged between different components of the system. Between the MTX and a BS, the signaling consists of supervisory, maintenance, and alarm signals, as well as signals for individual remote control of each radio channel. These messages are sent at 1,200 bps and are designated as C in the figure. There is no signal exchange in the usual sense between a base station and a mobile station, but each established connection is accomplished by a continuous 4,000-Hz tone transmitted from the base station to the mobile station, where it is looped back to the base station. This is designated as B in the figure. In the base station, the signal-to-noise ratio is evaluated continuously and the result is reported to the MTX. In order to reduce MS complexity, the decision to switch a call in progress to another base station is left to the MTX, which makes a decision as to the field strength from the mobile in question on the

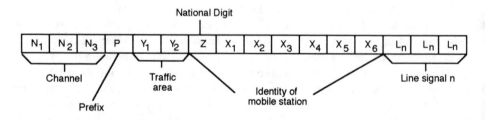

Figure 4.21 A Typical frame structure of the Nordic NMT 450.

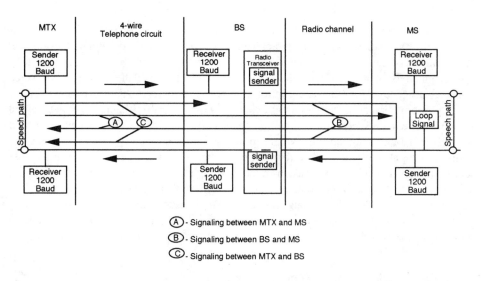

Figure 4.22 Signaling scheme of the Nordic NMT 450.

basis of information from the present BS, as well as information from any adjacent BS, about the signal-to-noise ratio. This information is designated as A in the figure.

4.8 MATS-E SYSTEM

The automatic mobile telephone system, MATS-E, was a joint-development venture of different Philips companies and the companies of the CIT-Alcatel group. This high-capacity system is designed for operation at both the 450 MHz and 900-MHz bands. However, its primary use today is at the 900-MHz frequency band. This system is designed for both rural and urban applications.

4.8.1 Background

Like other cellular systems, MATS-E is capable of accommodating a large number of subscribers with a grade of service comparable to a landline telephone system under the constraint of limited channel availability. This requires simultaneous usage of the same geographically separated frequencies and switching of calls in progress to cope with the mobility of the subscriber. By definition the system completes the switching function independent of the mobile location. The key functions of MATA-E include: optimized channel frequency reuse, efficient use of the control channel, off-air call setup (OACSU) and queuing of calls, an automatic nationwide roaming procedure, and an automatic handoff procedure between base stations.

As in other cellular systems, channel frequencies used at one cell are repeated at other cells that are separated geographically to minimize the cochannel interference. Efficient use of the control channel is accomplished by the mobile by randomly accessing the channel. Both OACSU and the queuing of calls improve the efficiency of voice channel usage by 10–30%. The control channel is used for OACSU, which uses the dialing information along with other relevant messages. A traffic channel is only assigned to the mobile upon successful completion of the call. Thus, the traffic channel is not tied up during the call setup and the waiting period for the called party to answer. Also, the calls arriving at the system when the traffic channel is busy are kept waiting in a queue until a traffic channel is available. In order to achieve the automatic roaming capability, the system tracks the whereabouts of an active mobile (mobile power is on) and the information is stored at the regional center. We shall describe this under the overview of MATS-E system. The automatic handover mechanism is similar to that of other cellular systems of the world.

4.8.2 Characteristics and System Parameters

The important system characteristics and parameters are listed in Table 4.14. The frequency range of operation, number of channels/system, channel spacing, and other modes of operation are similar to other systems of the world and require no further explanation. Up to three antennas can be installed at a base station and each antenna can have a

Table 4.14
Parameters of German MATS-E System

Category	*Parameter*
Number of channels	1,000
Cell radius	2–25 km
Typical number of traffic channels	1–100 per cell
Maximum number of traffic channels	192 per cell
Number of control channels per cell	1
Channel spacing	25 kHz
Base-transmit frequency range	935–960 MHz
Mobile-transmit frequency range	890–915 MHz
Frequency separation	45 MHz between transmit and receive channel
Mode of transmission	Full duplex
Voice-modulation transmission	FM
Data-modulation transmission:	
Control channel	FFSK
Control channel data rate	2.4 in-band Kbps (control channel)
Traffic channel data	Inaudible out-of-band signaling at 150 bps
Error protection coding (control channel)	Block code (16, 8; 5) with minimum Hamming distance of 5

maximum of 64 channels, with a combined total of 192 channels. The allocated frequency band 2×25 MHz provides 1,000 distinct radio channels. There are two types: traffic channels for voice communication and control channels for data transmission. At least one control channel must be used per cell. The possible cell radius ranges from 2–25 km. A cell radius of 2 km is the minimum for highly populated urban areas, while for rural or open areas the radius can be as large as 25 km. Adaptive power control at both BS and MS is employed depending on actual radio-transmission quality, which reduces the overall interference in the mobile system. As in other major systems of the world, frequency modulation is used for speech transmission. Data transmission over the control channel uses FFSK. Data is transmitted on both traffic and control channels. For data transmission over a traffic channel, out-of-band signaling of the amplitude-controlled PSK type at a lower rate of 150 bps is used. Traffic channel data is continuous at the rate of 2,400 bps. Error protection for the control channel is provided by block coding with a minimum Hamming distance of 5. This allows an automatic error correction of 2 bits.

4.8.3 Overview of MATS-E

The basic network architecture is shown in Figure 4.23. The MATS-E system is composed of two subsystems: the radio subsystem (RSS) and the switching and control subsystem (SCSS). The communication between these two subsystems is solely through digital messages. The RSS consists of the MS and the BS, with their transmitters, receivers, and antennas for radio transmission between them. The SCSS performs all necessary switching and management functions of the MATS-E provided by the MSC. For cases where the system is large, the MSC can be split into local centers (LC) and regional centers (RC).

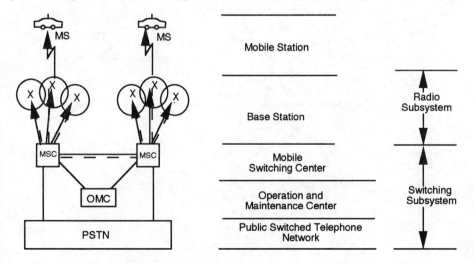

Figure 4.23 MATS-E network architecture. Source: [23, p. 31].

The LC registers the mobile and processes the call. The RC acts as a database for the system. The subscriber database contains information related to class of service and present location on each subscriber, which in turn helps in the handoff process. Administration and maintenance is centralized at the operation and maintenance center (OMC). The basic unit in the RSS is the cell, which is geographically bounded due to the signal sensitivity of the receiver. Several cells are grouped together and form a paging area where the MS can freely move and the call in progress can be switched automatically. The mobile can be tracked in a paging area. When the mobile is paged in a particular paging area, that page is received simultaneously in all the serving cells (simulcast). The number of cells per paging area is programmable and several paging areas are controlled by one MSC. While the mobile stations and the base stations are linked by a radio path, the BS and MSC are linked by a ground link, known as the approach link (AL). It can be implemented by 24-channel PCM links on T1 carriers, 30/32-channel PCM links, conventional analog lines, or by radio links. Figure 4.24 shows the mobile radio band for communication. As discussed in the last section, both the transmit and receive bands have a bandwidth of 25 MHz, split into 1,000 channels of 25 kHz each. Additionally, 999 interstitial channels

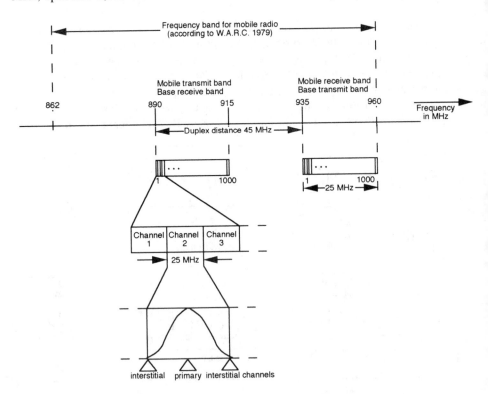

Figure 4.24 900-MHz frequency band allocation in MATS-E. Source: [23, p. 31].

(between two primary channels), with carrier frequencies shifted by 12.5 kHz, can be addressed. Most of the channels are used as TCH, which are available for speech or transparent data transmission; the others are CCH. The number of traffic channels assigned per cell depends on the actual traffic requirements. Usually one CCH per cell is assigned. However, in an urban environment with a high traffic demand, more than one control channel can also be used. In addition to the data transmission over CCH, a continuous data stream is exchanged on the TCH during a conversation, using out-of-band signaling with a subcarrier at 3.6 kHz. The data transmission, not audible to the subscriber, serves four functions: supervision of transmission quality; transmission of charging information to the mobile set; initiation of supplementary services; and significant reduction of speech-blanking time during handover procedure.

4.8.4 Message Formatting

As stated in the last section, there are two types of digital data signaling that take place in the MATS-E system: the signaling over dedicated CCH and signaling over the TCH in parallel with the speech transmission. Both these signaling types are continuous. The signaling over TCH is inaudible and is done by out-of-band transmission. On the other hand, signaling over CCH is in-band and utilizes the available channel bandwidth fully. As stated in Section 4.8.2, OBS consists of amplitude controlled phase shift keying (APSK) at a data rate of 150 bps. This modulation is relatively insensitive to the burst noise behavior of the channel. For transmissions at 900 MHZ on a radio path, both types of signaling are additionally modulated using FM. Both the IBS and OBS are optimized in order to fit into 25 kHz of channel spacing. The Gilbert-Elliott (binary symmetric channel) stochastic model was used for data transmission performance prediction. The model has two states: state G, a good state where the error rates (P_G) are low; and another state where error rate is high (P_L). The multipath fading leads to a high bit error rate. This also occurs when the mobile is at the boundary of the cell. Considering both the states, a mean bit error rate of 10^{-3} is accepted as a design objective. The effective data rate of 800 bps from MS to BS, and 1,000 bps from BS to MS, with the extra 200 bps for access control, is required. For the above throughput requirement, the actual data rate of 2.4 Kbps has been chosen as a reasonable design value. Message formatting in CCH is shown in Figure 4.25. CCH is organized in both directions into 80-ms time slots that hold 192 bits. The message format from BS to MS consists of 16 bits each of bit and frame synchronization. Information and redundancy for error control have 64 bits each. The access control information (ACI) field consists of 32 bits. The 128 bits of message, with 64 bits of redundancy, are organized into 8 code words of 16 bits each. A dotting sequence (101010) is used for bit synchronization. The format for messages from the mobile to the base station consists of a 16-bit period allowed for smooth carrier keying, a 32-bit dotting sequence for bit synchronization, 16 bits for the frame synchronization, and 128 bits of coded message divided into 8 code words of 16 bits each. Code words in both messages

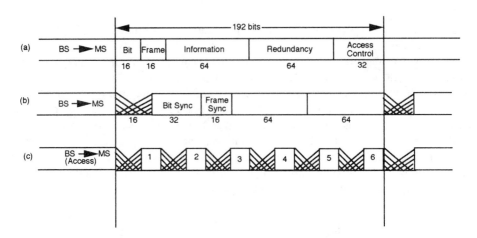

Figure 4.25 Message formats on control channels (CCH). Source: [23, p. 32].

from BS to MS and from MS to BS are interleaved in order to randomize the burst errors. A (16, 8) block error correcting code is used, which is derived from a cyclic (17, 9) code. This code can correct up to two errors. The access control message field from the base station to the mobile station asks certain mobile units to respond back to the base station. Those mobiles that are requested to respond in the frame from BS to MS respond in the message format shown in Figure 4.25(c). The active mobile that is requested to send a message does so after exactly one time-slot delay from the complete reception of the request message. A guard space of 16 bits is left between the message provided by different mobiles. This guard space is needed to accommodate the different propagation delays due to different locations of the mobile within a cell. The message from BS to MS is continuously sent and every active mobile within a cell is synchronized to this message frame. For synchronization, a preamble of the message consisting of 16 bits each of bit and frame synchronization are used.

4.9 THE C-450 GERMAN CELLULAR RADIOTELEPHONE SYSTEM

The C-450 system is the third public radiotelephone system in Germany, following network A and network B. Network A operated in the frequency range of 157–171 MHz and was discontinued in 1977. Network B started in 1972 and was discontinued in 1992. Network C, operating in the range of 450 MHz, started its operation in 1985 and has been designed by Siemens to meet the midterm demands for mobile until the 900-MHz digital system becomes operational in unified Europe.

4.9.1 Background

The configuration of the radiotelephone network C-450, with its interconnection with the telephone DDD network, is shown in Figure 4.26(a). The interconnection between the

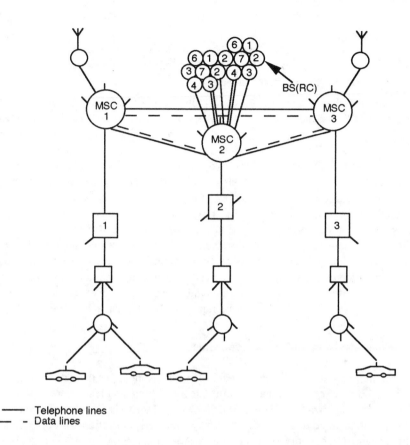

BS(RC)

Telephone lines
— - Data lines

Figure 4.26(a) Structure of C-450 network and its interconnection with the telephone DDD network. MSC = mobile switching center; RC = radio cell; BS = base station. Source: [31, p. 23].

mobile switching center and the base is with the standard CCITT signaling number seven. The network has the highest traffic in Hamburg and Munich. About 30% of all the radio channels and 40% of all traffic are concentrated in 7% of all radio cells. Channel loading is 30–50 subscribers. The basic cluster is a 7-cell configuration with about 175 cells in total. Once the mobile power is on, it is automatically registered in the active data file of the cell in use. If the subscriber is staying in his or her own MSC area, an "active" entry is made in the home data file of the MSC. If the subscriber's home base is in another MSC area, an active entry is stored in the external subscriber data file. The file also contains the user's home base MSC from his or her telephone number, and the through data channel causes an entry to be made in the radio subscriber's home data file. In addition to its "active" entry, the home MSC also contains the following individual user information:

- Subscriber number;
- Authorization entries (for example license for international call);
- Priority entries (for example, if the subscriber belongs to a safety group);
- Other entries (for example, suspension).

During conversation, a data message of 184 bits for supervision is continuously exchanged between the BS and the subscriber. This is achieved by inserting 4 bits of supervising bits every 12.5 ms by compressing speech. The gaps are of 1.136-ms duration and represent about 9.1% of a 12.5-ms speech slot. During this interval speech quality is checked, and if necessary the subscriber is handed over to another channel of either the same cell or another cell. The handover process lasts 0.3 seconds and is inaudible. The other use of this interval is for registration of charges, to be displayed at the subscriber unit; directives for mobile power adjustment; and switchover of the operating mode(data transmission from speech). On the receive side, the speech is decompressed. Data over voice channel is shown in Figure 4.26(b). Speech protection is provided against unwanted listening-in on a call by ordinary radio receivers by scrambling the phase-modulated analog speech. An *identification card* (IC) the size of a credit card switches the radiotelephone operation, and thus allows the flexible use of various radiotelephones. This is particularly useful for rental car service, where the car can be rented with a cellular telephone.

4.9.2 Characteristics and System Parameters

The basic requirements of the system are the same as those for the nationwide telephone system: good voice quality, high reliability, low blocking, and least cost to the subscriber. See Table 4.15. With 4.44-MHz transmit and 4.44-MHz receive bands and channel spacing of 20 kHz, there are a total of 222 full-duplex channels. Operation with both 25 kHz and 10-kHz channel bandwidth is also possible. Additionally, the system can operate in the interleaved channel spacing of 10 kHz with 12.5 kHz. The typical cell radius is 2–30 km. The lower radius of 2 km is achieved with a directional antenna at the cell site. As in other world systems, the maximum cell-site transmit power is limited to 100W. Nominal mobile power is 15W. Power control for both the base and the mobile is over the range of 35 dB and is achieved in eight steps. Speech is phase modulated with a peak frequency deviation of ± 4 kHz and is amplitude companded. Data modulation is FSK with a peak deviation of ± 2 kHz. The baseband data is NRZ-formatted and is transmitted at the rate of 5.28 Kbps with an effective data rate of 1.82 Kbps. The data-rate reduction is due to block coding and message repetition.

4.10 COMPARISONS, HIGHLIGHTS, AND SUMMARY

We have described seven major systems of the world. As the reader will note, there are lot of similarities among systems, but there are also areas where they sharply differ with each other. We provide here a summary of these differences. Major system characteristics

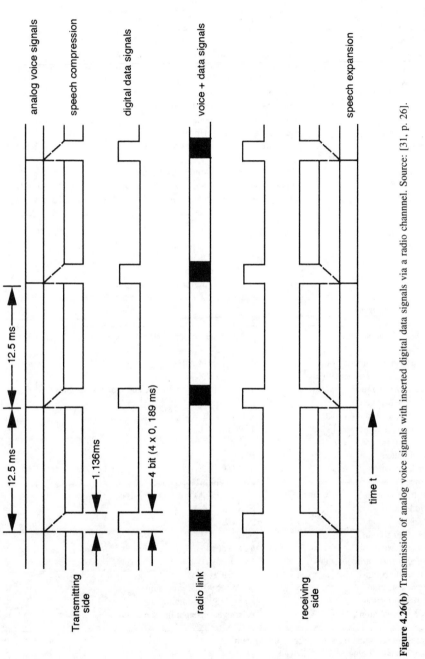

Figure 4.26(b) Transmission of analog voice signals with inserted digital data signals via a radio channnel. Source: [31, p. 26].

Table 4.15
User and System Characteristics of German C-450 System

Characteristics	Description
Efficient use of spectrum	Channels are repeated to get the maximum spectrum efficiency; total number of cells: 175; arranged in clusters of 7 cells
Mobile registration	Each active mobile user is registered in his or her MSC, which contains both home and external subscriber data file
Handoff	Position of mobile monitored continuously to make handoff easy
Preoriginate dialing	Off-air call setup is standard
Queuing operation	Standard when all channels are busy; applicable to incoming and outgoing calls
Speech protection	Scrambling the phase-modulated signal provides speech privacy
User identification	Allows user to make a call without subscribing to service; advantageous for car rental; car with cellular telephone can be rented

Table 4.16
System Parameters of German C-450 System

Parameters	Description
Number of channels	222 full duplex; up to 95 voice and 3 control channels per cell
Cell radius	2–30 km
Base-transmit frequency range	461.3–465.74 MHz
Mobile-transmit frequency range	451.3–455.74 MHz
Frequency separation	10 MHz between transmit and receive channels
Channel separation	20/10 kHz*
Maximum base station ERP	100W
Nominal mobile power	15W
Voice transmission:	
Modulation	PM
Peak deviation	±4 kHz
Processing	2:1 syllabic companding
Data transmission:	
Modulation	FSK
Peak deviation	±2 kHz
Base band coding	NRZ
Transmission rate	5.28 Kbps
Effective transmission rate	1.82 Kbps
Error protection	15:7, BCH
Error detection	40 per 150 bits, minimum
Error correction	20 per 150 bits, minimum
Message protection	Adaptive, repeated in case of error

Note: Siemens is the manufacturer of the system.
*Also 25 and 10 plus 12.5 interleaved.

have been summarized in Table 4.17. The RF frequency bands where cellular systems operate are at 450 MHz, 800 MHz, and 900 MHz. Advantages and disadvantages of 800/ 900 MHz bands over 450-MHz operational bands have been narrated in Chapter 1. The AMPS in the United States has a maximum of 816 channels, divided into two groups. Both groups have a channel frequency separation of 30 kHz and are compatible with each other in all respects. This is the only system in the world with a channel bandwidth of 30 kHz. Other systems of the world have a channel bandwidth of 25 kHz. The German C-450 MHz system can have both a 10 kHz and a 20-kHz bandwidth, while the NMT 900 system has a 12.5-kHz bandwidth. All world systems have two types of channels: traffic and control. Control channels are further divided into paging and access channels. The traffic channel is used for voice transmission, while the data is transmitted over the control channel. The NMT 450 and NMT 900 systems can also use traffic channels for data during the peak hours of traffic. Cell-site peak transmit power is uniformly 100W in most cases, except for the Japanese and NMT 450 MHz systems where the peak powers are limited to 25W and 50W, respectively. The maximum mobile power associated with the NMT 450-MHz system is 15W. The other world systems have power limited to less than 7W. Systems use either FSK or FFSK modulation for data with widely varying frequency deviations. The associated signaling data rate is a maximum of 10 Kbps with the AMPS, though the throughput is substantially less due to encoding and repeats of the messages. Baseband data is either Manchester or NRZ encoded. Different types of modulation are used to send data over the voice channel. All systems use the blank-and-burst technique for data transmission over traffic channels, except MATS-E and the German C-450 system, which uses out-of-band signaling and inserts data by compressing voice. In MATS-E signaling, data is amplitude-controlled PSK at 3.6-kHz subcarrier. All systems allocate the voice channel prior to calling and called parties go off-hook, except the MATS-E and C-450 systems. These two systems use the OACSU technique and assign the traffic channel only after both the parties answer. This technique increases the traffic carrying capacity of the system by more than 30% over other systems. Presently only the C-450 systems use speech scrambling to prevent eavesdropping.

Apart from the technical similarities and dissimilarities among systems, the other major issue is the compatibility among different systems of the world. In Europe, the only compatible systems between countries are: the Scandinavian countries of Denmark, Finland, Norway, and Sweden with NMT 450 and NMT-900 systems. Belgium, Luxembourg and Netherlands are also compatible due to the NMT-450 system. Compatibility exists with systems in Ireland and the United Kingdom. This means that if you are a subscriber in any one of these countries you may roam in a limited number of countries using your phone. In order to eliminate the problem of compatibility, the European PTTs have agreed to a Pan-European standard, which is being put into operation under the name *Groupe Speciale Mobile* (GSM) in 1991.

A summary of systems in different countries is shown in Table 4.18. The world cellular market by system type, and the top-ten cellular market penetration rates in the world are shown in Tables 4.19 and 4.20. These tables are by no means complete due to

Table 4.17

Summary of Different World Systems

	U.S. and Canada (AMPS)	U.K. (TACS)	Japan (NTT)	Nordic (NMT450)	Nordic (NMT900)	German (MATS-E)	German (C-450)
Number of channels	2 × 416	2 × 500	2 × 500	180	1999	1000	222
Cell radius (km)	2–20	2–20	2–20	1–40	.5–20	2–25	2–30
Cell repeat pattern (N)	7, 12	4, 7, 12, 21	9, 12	7, 12	9, 12	7	
Cell receiver frequency (MHz)	825–845	890–915	860–885	453–457.5	890–915	890–915	461.3–465.74
Cell transmitter frequency (MHz)	870–890	935–960	915–940	463–467.5	935–960	935–960	451.3–455.74
Frequency sep. between receiver and transmitter (MHz)	45	45	55	10	45	45	10
Channel spacing	30	25	25	25	12.5	25	20
Cell-site transmitter power (W)	100	100	25	50	100		100
Mobile transmitter power (W)	3	7	5	15	6	4	7
Voice:							
Modulation	FM	FM	FM	PM	PM	FM	PM
Frequency deviation (kHz)	±12	±9.5	±5	±5	±5	±8	±4
Signalling:							
Modulation	FSK	FSK	FSK	FFSK	FFSK	MSK	FSK
Formatting	Bi-φ	Bi-φ	Bi-φ	NRZ	NRZ		NRZ
Frequency deviation (kHz)	±8.0	±6.4	±4.5	±3.5	±3.5		±2
Bit rate (Kbps)	10	8	.3	1.2	1.2	2.4	5.28

Table 4.18
Cellular Systems in Different Countries

System	Countries Where Used
AMPS	Anguilla, Antigua, Argentina, Australia, Bahamas, Barbados, Barbuda, Bermuda, Bolivia, Brazil, Brunei, Canada, Cayman Islands, Chile, Costa Rica, Curacao, Dominican Republic, El Salvador, Guatemala, Hong Kong, Indonesia, Israel, Jakarta, Korea, Laos, Malaysia, Mexico, Netherlands, New Zealand, Pakistan, Paraguay, Peru, Philippines, Puerto Rico, Samoa, Singapore, St. Kitts/Nevis, St. Lucia/St. Vincent/Grenadines, St. Maarten, St. Martin/Bartholemy, South Korea, Taiwan, Thailand, Trinidad/Tabago, United States, Venezuela, Virgin Islands, Zaire
TACS	Austria, Bahrain, China, Egypt, Ghana, Hong Kong, Hungary, India, Ireland, Italy, Kenya, Kuwait, Macao, Malaysia, Malta, Mauritius, New Zealand, Nigeria, Singapore, Spain, Sri Lanka, and United Kingdom
NTT	Columbia, Hong Kong, Japan, Jordan, Kuwait, Singapore
NMT-450	Andorra, Austria, Belgium, China, Czechoslovakia, Denmark, Estonia, Faeroe Island, Finland, France, Hungary, Iceland, Indonesia, Latvia, Lithuania, Luxembourg, Malaysia, Morocco, Netherlands, Norway, Oman, Poland, Russia, Saudi Arabia, Spain, Sweden, Thailand, Tunisia, Turkey, Yugoslavia
NMT-900	Algeria, Cyprus, Denmark, Finland, Netherlands, Norway, Sweden, Switzerland, and Thailand
MATS-E	France, Kuwait
C-450	Germany, South Africa, Portugal

Source: [26].

Table 4.19
World Cellular Market by System

System	Number of Subscribers	Total Subscribers (%)	Number of Countries
AMPS	8,174,448	59.4	42
TACS	2,214,860	16.1	21
NMT-900	906,650	6.6	9
NMT-450	892,600	6.5	24
Others (450, 900 MHz)	1,572,150	11.4	11
Total	13,760,708	100.0	108

Source: [26].

the rapid adaptation of systems in different countries. Presently there are 113 countries who have adopted, or are in the process of adopting, cellular technology. This number is growing every day. However, in countries where cellular systems are either currently in operation or will be in service in the near future, those systems are most probably based on one of the one or more of the systems described here.

Table 4.20
Top-Ten Cellular Penetration Rates of the World

Country	Number of Subscribers	Population (%)	Years of Service
Sweden	564,060	6.7	11
Finland	276,110	5.8	10
Norway	227,240	5.4	10.5
Iceland	12,240	5.1	6.0
Denmark	168,900	3.2	10.5
Hong Kong	168,420	3.0	8.5
Switzerland	164,080	2.9	4.5
Faeroe Islands	1,330	2.8	3.5
United States	7,557,148	2.6	8.5
Singapore	66,000	2.4	4.0

Source: [26].

REFERENCES

[1] Fisher R. E. "A Description of the Bell System 850 MHZ High Capacity Mobile Telecommunication System," ICC 1976.

[2] Callendaer, M. H. "An Integrated Voice and Data Mobile Service," *IEEE Vehicular Technology Conference*, 1981.

[3] Huff, D. L., and J. T. Kennedy. "The Chicago Developmental Cellular System."

[4] Huff, D. L. "The Developmental System," *The Bell System Technical Journal*, Vol. 58, No. 1, January 1979, pp. 249–269.

[5] Fisher, R. E. "Portable Telephone for 850 MHz Cellular System," *IEEE Vehicular Technology Conference*, 1980.

[6] Wells, J. D. "The Evaluation of Cellular System Design," *IEEE Vehicular Technology Conference*, 1984.

[7] MacDonald, V. H. "The Cellular Concept," *The Bell System Technical Journal*, Vol. 58, No. 1, January 1979, pp. 15–41.

[8] TC27-004A-204. "An Overview of Canadian telephone Service," *Datapro Research Corporation*, April 1986.

[9] Ali, M. "The Aurora Cellular Mobile Telephone Systems," *IEEE Vehicular Technology Conference*, 1983.

[10] Mikulski, J. J. "DynaTAC Cellular Portable Radiotelephone System Experience in the U. S. and U.K.," *IEEE Communication Magazine*, Vol. 24, No. 2, February 1986, pp. 40–46.

[11] Watanabe, K., et. al. "High Capacity Land Mobile Communications Systems in NTT," *IEEE Vehicular Technology Conference*, 1987.

[12] Kikuchi, T., and T Nishigaichi. "Economical Radio and Switching Equipment for Nationwide Land Mobile Telephone Service," *IEEE Vehicular Technology Conference*,

[13] Kikuchi, T., et. al. "Improved Digital Signaling on Land Mobile Communication System," *IEEE Vehicular Technology Conference*, 1982.

[14] Toya, M., et. al. "800 MHz Mobile Telephone for Cellular System," *IEEE Vehicular Technology Conference*, 1982.

[15] Ueda, J., et. al. "Design and Technology for Cellular Portable Telephone Equipment," *IEEE Vehicular Technology Conference*, 1980, pp. 1–4.

[16] Kobayashi, K., et al. "Detachable Mobile Radio Units for the 800 MHz Land Mobile Radio System," *IEEE Vehicular Technology Conference*, 1984.

[17] Kubota, H., and T. Kikuchi. "High Capacity Automobile Telephone System," *IEEE Vehicular Technology Conference*, March, 1979.

[18] Ito, S., and Y. Matsuzaka "800 MHz Band Land Mobile Telephone-Overall View," *Review of the Electrical Communication Laboratories*,

[19] Seki, S., N. Kanmuri, and A Sasaki. "Detachable Unit Service in 800 MHz-Band Cellular Radio Telephone System," *IEEE Communications Magazine*, Vol. 24, No. 2, 1986, pp. 47–51.

[20] Suramoto, M., and M. Shinji. "Second Generation Mobile Radio Telephone System in Japan," *IEEE Communications Magazine*, Vol. 24, No. 2, 1986, pp. 16–21.

[21] NMT DOC 4. Nordic Mobile Telephone Group, "Preliminary Specification for the Base Station," 1977.

[22] NMT DOC 1, 1977, "NORDIC Mobile Telephone-System Description."

[23] Preller, H. G, and W. Koch. "MATS-E, An Advanced 900 MHz Cellular Radio Telephone System: Description, Performance, Evaluation and Field Measurements," *IEEE Communications Magazine*, Vol. 24, No. 2, February, 1986, pp. 30–39.

[24] Schmidt, W. "Protocol Design of the EC 900 Cellular Radio System," *IEEE Vehicular Technology Conference*, 1984.

[25] Mabey, P. J. "Predicting the Range and Throughput of a Mobile Data System," *IEEE Vehicular Technology Conference*, 1982.

[26] Steward, S. P. " The World Report," *Cellular Business*, May, 1992.

[27] Hennen, H. A., and K. Mulford, et al. "Land Mobile Voice Privacy Communications," *IEEE Vehicular Technology Society IEEE*, 1980, p. 3.

[28] Ehrlich, N. "The Advanced Mobile Phone Service," *IEEE Communications Magazine*, March 1979, p. 11.

[29] Fluhr, Z. C., and P. T. Porter. "Control Architecture," *Bell System Technical Journal*, Jan. 1979.

[30] Chadha, K. J. S., et al. "Mobile Telephone Switching Office," *Bell System Technical Journal*, Jan. 1979.

[31] Spindler, K. "The German Cellular Radiotelephone System C," *IEEE Communication Magazine*, Jan. 1986, Vol. 24, No. 2, p. 26.

Chapter 5
Cell Enhancer

5.1 INTRODUCTION

A properly designed cellular system can provide almost unlimited coverage within a cosmopolitan area. This is especially true when the service area consists of regular terrain and there are no obstacles such as tunnels, hills, tall buildings, topographical depressions, and so forth. However, as cellular systems mature beyond the initial push to provide basic service in an area, more attention is needed to provide optimization and improvement on the basic service. This is rather important as the cellular providers compete based on their quality of coverage. Virtually every cellular system provider has coverage areas within its territory where it would be profitable to enhance or supplement coverage. These areas fall into the following categories:

- *Holes* are relatively small dead spots and the zones of marginal operation within the area of cellular operation. Within these holes, the problems can range from poor voice quality or frequently dropped calls to total lack of service.
- *Fringe areas* are edges of the existing service area where the service is marginal and a large customer potential exists. Only better coverage will attract more customers.
- *Corridors* are long, narrow zones of high traffic area extending outward from the main service area or tying together two adjacent systems. One example is a busy highway passing through a region of low population density where continuity of service is desirable from the customer's point of view.

These problems can easily cause customer dissatisfaction and damage the carrier's reputation as a reliable service provider, even though coverage in most highly populated areas is excellent. This will slow down the growth of the subscriber base. Also, if another service provider covers the area better, the customer may switch systems.

Some possible solutions to these coverage problems are: relocation of existing cell sites, addition of new cell sites, or installation of cellular repeaters. Relocating cell sites may solve one set of problems but could cause others that are just as serious. Addition of new cell sites is a costly solution, which leaves the third choice: the installation of cellular repeaters. The economics are in favor of repeaters. A cellular repeater that is capable of providing 30–40 channels at 250 mW of ERP per channel (24 dBm/channel) can be installed for less than $40,000, which is no more than 20% of the cost of a regular cell site. Cellular repeaters have become an accepted technology to use to provide service in shadow areas caused by terrain or buildings and to provide service in convention centers, tunnels, and so forth. In addition to providing coverage, the enhanced service also provides an opportunity for cellular operators to increase incremental revenue with a low investment. The air time usage is also increased by providing cellular service in areas that were previously inaccessible.

The cellular repeater in its most elementary form is a nontranslative RF signal booster consisting of two antennas, two linear broadband RF amplifiers, and two diplexers of appropriate bandpass characteristics, as shown in Figure 5.1. Here, the repeater is situated such that its donor antenna receives a strong forward RF signal from the donor cell. Usually, the line-of-sight path, as shown in Figure 5.2, is desired. The signal is boosted by the forward amplifier and transmitted by the repeater's reradiation antenna toward the area where coverage is required. A similar reverse path is provided for signals received by the mobile or portable and transmitted towards the donor cell site. Depending on the application, the booster beams the amplified signals into a hole and outward, into the fringe area of coverage or the low-population density area of a long highway. Simultaneously, it picks up the signals from mobiles in the boosted coverage area, amplifies them, and transmits them back to donor cell site.

In view of the above, we first establish the design requirements of the booster in Section 5.2, followed by several practical design examples in Section 5.3. In Section 5.4,

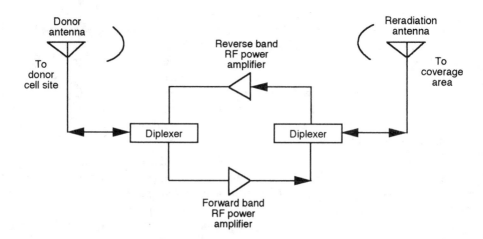

Figure 5.1 Simple bidirectional and nontranslative repeater (enhancer).

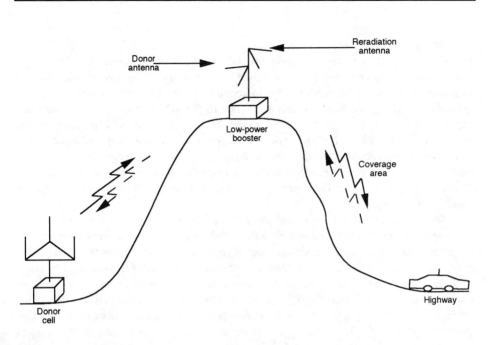

Figure 5.2 Typical cellular (enhancer) repeater location.

we examine the three different equipment configurations and discuss their advantages and disadvantages; we also provide a typical specification of a booster. In Section 5.5, we provide a set of enhancer applications within United States. In Section 5.6 we draw conclusions.

5.2 DESIGN CRITERIA

The proper design of a repeater requires careful considerations of feedback problems between donor and reradiation antennas; transmitter intermodulation; enhancer or booster siting; and delay profile.

The possibility of RF feedback, leading to RF oscillations, exists if the donor and the reradiation antenna are not properly isolated. The same problem may also be encountered if the shielding of the RF amplifier, diplexer filters, and coaxial cabling is not adequate. Isolation between the two booster antennas must exceed the net RF gain of the booster plus some margin, typically about 15 dB, to ensure long-term stable operation. The practical value of the achievable isolation is about 75 dB unless there is a building or other structure isolating these antennas. Considering this value of isolation, the desired margin, and the antenna front-to-back ratio, the stable booster operation permits a maximum booster gain of 70 dB. A somewhat lower gain is adequate for most of the applications.

Transmitters with multichannel capability are likely to generate *intermodulation* (IM) if the amplifiers are not properly backed off. To prevent interference with other systems, boosters are engineered to keep the IM level well below acceptable levels. The configuration shown in Figure 5.1 helps in keeping the level of interference low, as the forward channel signals from the donor cell site to the mobile pass through a different amplifier than the signals from the mobile to the donor cell site. The separate path reduces the number of signals by half and essentially eliminates the possibility of the forward path channels interacting with the reverse path channels, which helps in reducing IM interference. Since the number of channels, the amplifier gain, and the input signal level all affect the generation of IM, practical power output has been limited to 24 dBm, as stated in the introduction.

The power output from a cell enhancer is appreciably less than that from a conventional cell site, which mandates careful siting of the repeater so that proper coverage is obtained. Areas served by approximately equal signal strengths from the cell directly and by the cell repeater (enhancer) can cause a severe delay problem for signaling and voice band data, and careful consideration is required to solve this problem.* An equal, or nearly equal, level of signals from these two sources can cause multipath fading, which definitely affects the signaling channels. The effect of delay is not serious on voice. This problem is known as "black hole" and generally occurs on the fringes of the boosted coverage area, as shown in Figure 5.3. Proper siting of the booster can minimize this problem, but complete elimination may not even be feasible.

5.2.1 Detailed Design Considerations

As discussed above, the design of a cell enhancer requires many technical and application engineering considerations. These detailed technical considerations are: output power level per channel, linearity of a broadband power amplifier, maximum gain of a repeater, noise figure, repeater selectivity, and signal delay. Among the application engineering considerations are proper siting of the repeater, antenna selection, and handoff. We discuss each of these below.

5.2.2 Technical Considerations

5.2.2.1 Output Power Level per Channel

For proper design of the amplifier, the technical requirements stated above have to be met. The output power level is an important design criterion, based on which other parameters can be established. Initial developmental authorizations by the FCC set a

---

*The words "repeater," "enhancer," and "booster" are used interchangeably.

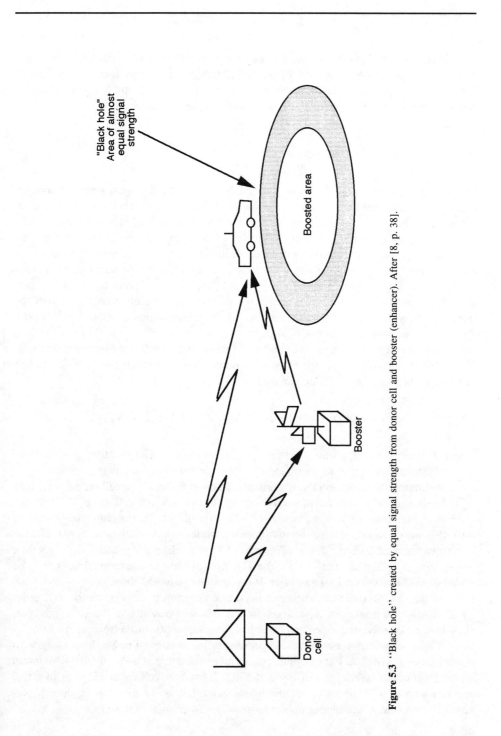

Figure 5.3 ''Black hole'' created by equal signal strength from donor cell and booster (enhancer). After [8, p. 38].

maximum forward channel effective radiated power (ERP) level of 250 mW or +24 dBm. Assuming a line-of-sight coverage of the cell enhancer, this power level is sufficient to meet (properly designed) UHF receiver selectivity for up to 5 miles of operation. A cellular repeater is generally intended to operate within this distance.

5.2.2.2 Linearity of the Power Amplifier

Linearity of the power amplifier in the forward direction (from donor antenna to reradiation antenna) is controlled by accounting for the number of channels reradiated at a level of +24 dBm without generating excessive intermodulation products. In the reverse direction, the required power-handling capacity of the amplifier is generally low as the low-powered mobiles and portables are at varying distances from the repeater site. The level requirement for intermodulation products as imposed by the FCC is −13.0 dBm or below for broadband devices. Equation (5.1) relates the third-order intercept (TOI) point to the power-per-channel output, S, of the amplifier, and the number of channels passing through the amplifier, N, which satisfies the FCC requirement of intermodulation products. Obviously, the formula assumes that the output signals from undesired cells are well below those of the desired donor cell. In case other undesired inputs are significant (greater than 10 dB below the donor cell level), their resulting output must be taken into account before arriving at the capacity of the power amplifier (PA) [3].

$$I = 1.5(S) - 0.409 + 24.75(\log_{10}N) - 1.437[(\log_{10}N)^2] \qquad (5.1)$$

where I denotes the amplifier third-order intercept point (TOI) in dBm; S is the per-channel transmitter output in dBm; and N is the number of equal power channels.

Assuming the total number of available channels to be 333, a cell repeat pattern of 7, and uniform traffic, the maximum number of channels, N, per cell site is 47. With the individual ERP level set to a maximum of +24 dBm, the reradiation antenna gain set to +15 dB, and the transmission line loss between diplexer and antenna to be +1 dB, the computed value of transmit power (T_x), from (5.1) is 10-dBm per channel (24 − 15 + 1). Substituting $S = 10$ dBm and $N = 47$, the TOI from the above equation comes out to be +51.8 dBm. Thus the desired amplifier TOI requirement is +52 dBm.

In the reverse direction, the most important criterion in arriving at the TOI point of the PA is the maximum input level from a mobile operating in the coverage area. Based on experience, the repeater power-handling capability in the reverse direction is arrived at by considering two mobiles operating at the maximum power level as close to the enhancer as possible. For example, two mobiles operating at +4W of ERP (maximum power level of the mobile) at a distance of 500 feet from the cell enhancer, with an active repeater gain of +53 dB and a reradiation antenna gain of 15 dB, will produce a per-channel power amplifier output of approximately +28 dBm at 900 MHz:

$$36 - l + 15 - 1 + 53 = 36 - [37 + 20 \log_{10}(900) + 20 \log_{10}(500/5280)] + 67$$
$$= 27.5 \approx 28 \text{ dBm}$$

where l is the free space loss for 500 feet. Here the cable loss between the receiver antenna and the repeater amplifier is also 1 dB, which is the same as in the forward direction. Now, applying (5.1), the value of TOI is $1.5(28) - 0.409 + 24.75[\log_{10}(2)]\text{TOI} - 1.437[\log_{10}(2)]^2 \approx 49.0$ dBm. Plots for TOIs of 52 dBm and 49 dBm for different numbers of equal power channels are shown in Figure 5.4. Monitoring of the mobile power by the donor cell can further reduce the power-handling requirement of the reverse power amplifier. Thus, the burden on the reverse PA caused by a close-in mobile can substantially be reduced.

5.2.2.3 Maximum Repeater Gain

The maximum active gain of the repeater is limited not so much by the requirements of a particular application, but in actuality by the desired isolation between the donor and the reradiation antennas. Antenna isolation depends on three factors: antenna front-to-back ratio, free-space attenuation between antennas, and the shielding provided by terrain or structures between antennas. In other words, the maximum allowable repeater gain must satisfy

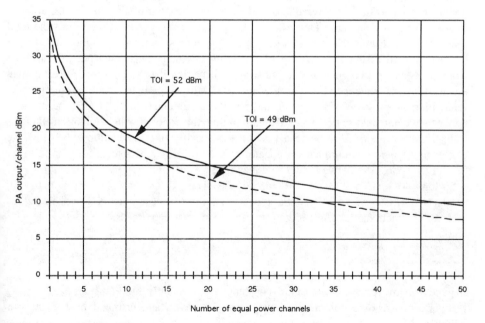

Figure 5.4 PA output per channel versus number of equal power channels.

$$(G_d - G_{d-r}) + (G_r - G_{r-d}) + G_{RP} + M \leq L_{FS} + L_{SH} \qquad (5.2a)$$

or

$$G_{RP} \leq G_{d-r} + G_{r-d} + L_{FS} + L_{SH} - G_d - G_r - M \qquad (5.2b)$$

where

G_{d-r} = radiation suppression for the donor antenna in the direction of the reradiation antenna;

G_{r-d} = radiation suppression for the reradiation antenna in the direction of donor antenna;

L_{FS} = free-space loss between donor and reradiation antennas;

L_{SH} = attenuation due to shielding between two antennas;

G_d = gain of the donor antennas;

G_{RP} = maximum allowable repeater gain;

G_r = gain of the reradiation antenna; and

M = desired gain margin (typical value = 15 dB).

In (5.2a), the parenthetical quantities $(G_d - G_{d-r})$ and $(G_r - G_{r-d})$ represent the gain of the donor antenna towards the reradiation antenna and the gain of the reradiation antenna towards the donor antenna. Thus, the left-hand side of this equation represents the total loop gain, which must be less than total loop loss represented by the right-hand side of the equation to assure the stability of the loop. One can conclude from this that the isolation between two antenna sets must be at least M-dB (15 dB typical) greater than the gain of the enhancer to minimize the possibility of RF feedback. Assuming the reradiation suppression of the donor and reradiation antennas to be 20 dB, the distance between two antennas to be 40 feet, the gain of both the donor and the reradiation antennas to be 10, the value of M to be 15 dB, and the other shielding attenuation to be 0 dB, we can compute the maximum active repeater gain at 900 MHz by using (5.2a) as follows.

Free space loss between two antennas is: $37 + 20 \log_{10}(d_{miles}) + 20 \log_{10}(f_{MHz}) = 37 + 20 \log_{10}(40/5280) + 20 \log_{10}(900) = 53.67$ dB. Thus, $G_{RP} < 20 + 20 + 53.67 + 0 - 10 - 10 - 15$ or, $G_{RP} < 58.67$. Thus, the repeater gain in this case must be less than 59 dB. In most applications the repeater gain is limited to less than 70 dB.

5.2.2.4 Noise Figure

In general, a low receiver noise figure (NF) plays a significantly important part in improving the performance of a communication system. However, as illustrated by the analysis below, an NF as high as 10 dB can be tolerated in a typical repeater application. Assuming the maximum repeater gain to be 70 dB, the reradiation antenna gain to be 15 dB, and the minimum ERP per channel to be 10 dBm towards mobile, these values will limit the transmitter PA input to −75 dBm (10−15 −70). Assuming the minimum sensitivity of the repeater to be −120 dBm provides a margin of 45 dB. A noise figure of 10 dB will still provide a sufficient margin of 35 dB. In the reverse direction, a mobile operating in line of sight and located at a maximum distance of 5 miles will encounter a free-space path loss of 110 dB. If the mobile ERP is 0 dBW, the receive power at the reradiation antenna input will be 30−110 = −80 dBm. Assuming a reradiation antenna gain of 10 dB, the reverse bandpower amplifier input is −80 + 10 = −70 dBm. Once again, a −120 dBm amplifier sensitivity provides a margin of 50 dB. Assuming a worst case fading margin of 40 dB still provides a repeater NF of 10 dB. Considering both the forward and the reverse band, a minimum NF of 10 dB is acceptable. One should note that here the discussion is based on single-channel input to the amplifier as this provides the lowest signal input where the sensitivity is important.

5.2.2.5 Repeater Selectivity

The problem of system selectivity is related to the rejection of the nearest channel of the competitor by the desired repeater. Let us assume that system B is the desired repeater and system A is the repeater of the competitor operating in the same geographical area. As was explained in Chapter 2, an FCC ruling mandates that two systems must cover the same area. Let us further assume that these two repeaters have a geographical separation of 500 feet and that they both have equal power per channel at their respective cell sites. The physical separation will provide an added attenuation of about 76 dB to the nearest interfering channel at the donor antenna of system B. Thus, if the desired signal output of the channel at the reradiation antenna of repeater B is +24 dBm, then the undesired signal will be at the level of −52 dBm. If we now arbitrarily impose the requirement that the interference channel level be at −100 dBm at the repeater output towards the coverage area, the receiver preselector of system B must provide an added attenuation of 48 dB to the nearest undesired channel. This is shown in Figure 5.5.

Here we are assuming that the total number of channels is 832, divided equally between system A and system B. In this case the desired selectivity of 48 dB must be met between the first control channel of system B (channel 334) and the last control channel of system A (channel 333), which has a frequency separation of 30 kHz (channel 333 operates at f_c = 879.99, channel 334 operates at f_c = 880.02). The desired selectivity figure of 48 dB is virtually impossible to achieve by conventional filtering as it requires a filter quality factor, Q of 29,334 (880.02/0.03). As suggested in [8], a stagger-tuned SAW filter operating at an IF of around 45 MHz can be designed to provide a passband width of around 10 MHz and attenuation skirts that reach 50 dB, within 360 kHz of the 3-dB frequency. Of course, a 360-KHz skirt represents a loss of 12 control channels (30 × 12 kHz), leaving only nine channels for use. Thus the control channel assigned to system A must be 360 kHz away from the 3-dB response of the control channel assigned

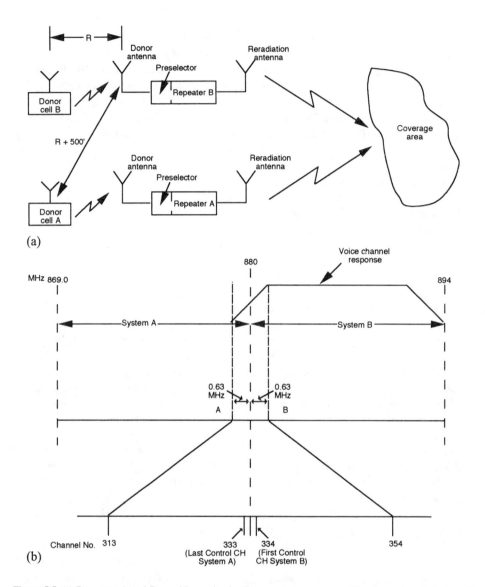

Figure 5.5 (a) Repeaters A and B provide service in the same coverage area. (b) voice and control channel response requirement of a selective repeater.

to B. This restriction is not a serious problem and can be lived with. Thus, a stagger-tuned SAW preselector filter can be designed to limit the adjacent control channel level to −100 dBm.

An alternative approach to achieve this selectivity is to use a modified system, as shown in Figure 5.6, that consists of wideband and a narrowband subsystems in both the

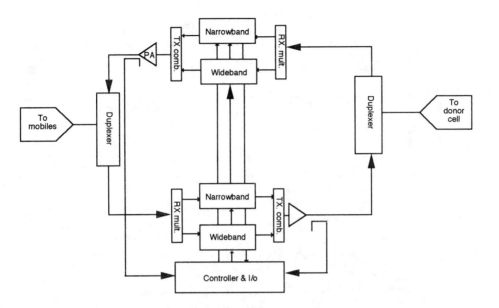

Figure 5.6 Low-power booster with separate paths for forward and reverse signals. After [8, p. 40].

transmit and the receive part of the booster. Here the narrowband part of the system contains a single-channel (30-kHz passband) amplifier with dual synthesizer to control both the input and the output sides of the amplifier. The narrowband part of the circuitry repeats and translates the frequency of the control channel transmitted from the donor cell site. The wideband part of the transmitter and receiver subsystem in this case only amplifies the voice channels, and thus the required bandwidth is reduced to 9.4 MHz ($\approx$ 10.0 MHz–0.6 MHz, total bandwidth available for speech). It should be noted that no frequency translation takes place in the wideband boards. The modified design can easily provide about 50 dB of attenuation at 600 kHz from the edge of the passband, and will also provide sufficient gain to all the passband voice channels. Here, the two IF filter amplifier/boards in the forward and the reverse signals paths are the key to the system. One is a programmable wideband IF board for voice channels and the other is the narrowband IF board with a single-channel high selectivity curve, as shown in Figure 5.7. Thus, the interference from system A voice channels is controlled and meets the attenuation requirement of 50 dB at the assigned control channel of system B. Also, the interference from the assigned control channel of system A to the assigned controlled channel of system B is easily minimized due to the narrowband response of these subsystems.

5.2.2.6 Signal Delay

Delay plays a major role in the performance of the system when the signal levels directly from the donor cell and the level through repeater are same at the mobile. If the delay

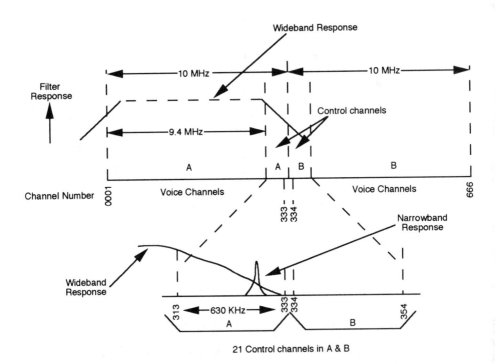

Figure 5.7 Narrow and wideband response of the enchancer. After [8, p. 48].

through the repeater is excessive (i.e., a significant portion of the baseband data bit), severe intersymbol interference may occur degrading the bit error rate (BER) performance of control channel data streams. Test results show that the maximum tolerable delay for the cell site and mobile radios are on the order of 10% of the bit duration for lower order modulations such as FSK. Assuming a 2.4-Kbps data rate, the maximum permissible delay is 41.6 ms for equal-level signals from the cell site and the repeater. Based on 5 miles of coverage by the booster, it is easy to see that the repeater should not cause a delay value of greater than 15 ms $(41.6 - (5 \times 5.4) \approx 15.0)$. This is based on a free-space delay nearly equal to 5.4 ms per mile. As the data rate is increased, the specification for booster delay has to be made even tighter. The operation of two equal-level RF signals arriving at different times at the mobile receiver is viewed as a two-ray multipath interference system. This will cause fading and has to be viewed critically. However, with a 10% limit on the total multipath delay, the performance of the enhancer is not any worse than the cellular system without a repeater.

5.2.3 Application Engineering

In addition to the technical factors controlling the performance of the repeater, there are other important considerations that balance many variables associated with potential applications. We discuss these in the following sections.

5.2.3.1 Site Selection

Repeater placing must assure line-of-sight operation both from the cell site to the donor antenna and from the reradiation antenna to the mobile. If the line of sight is maintained, the distance from the donor cell site is not critical if a high-gain donor antenna is used. Increasing the donor antenna gain can have a negative impact if the donor antenna is not at line of sight, with respect to the cell site, due to incident angle diffusion of the incoming signal. For direct path, a workable range of level at the donor antenna input is −50 dBm to −70 dBm.

5.2.3.2 Antenna Selection

For line-of-sight applications, the choice of a directional high-gain antenna will reduce the undesired signals from other cell sites. This will reduce the power level of other cells that might otherwise place a burden on the repeater's forward power amplifier linearity, resulting in higher IM products. Where line of sight is not possible, the donor antenna should be placed such that the desired signal level from the donor cell is maximized. The placement should account for diffusion of the incident angle. As with the choice of donor antenna, a high-gain reradiation antenna will also reduce the linearity requirement of the power amplifier.

5.2.3.3 Handoff Considerations

With equal repeater gain in both the forward reverse directions, the reciprocal relationship used in developing handoff parameters is maintained. Thus, if the coverage area of the donor cell is increased, adjacent cell coverage area has to be decreased.

5.3 Application Examples

In this section we shall discuss and give examples of several applications for cell enhancers.

Example 1

The enhancer shown in Figure 5.8(a) is located at a mountain top 5 miles from the fixed cell site. Serving area to the mobile is from 100 feet to 1.5 miles on the other side of the mountain. The cell site and mobile ERPs are 165 dBu and 150 dBu, respectively. The cell site, donor, and reradiation antenna gains are 15 dB, 10 dB, and 10 dB, respectively. Amplifier gain in both the forward and reverse directions is 60 dB. The transmission line losses from the donor antenna to the forward amplifier and from the forward amplifier to the reradiation antennas are 1 dB each. Assuming the frequency of operation to be 900

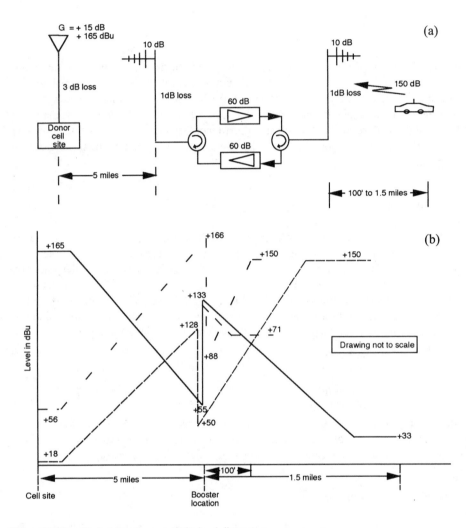

Figure 5.8 Example 1: (a) system setup; (b) level diagram.

MHz and the transmission line loss at the donor cell site to be 3 dB, find the signal levels at the mobile and at the cell site.

Free-space loss (FSL) for 5 miles = 37 + 20 log$_{10}$(5) + 20 log$_{10}$(900) = 110 dB. Similarly, FSL for 100 feet and 1.5 miles is 62 dB, and 100 dB respectively. Computed levels at the mobile and the cell site when the mobile is at 1.5 miles from the repeater are:

$$P_{RM} = P_{T,\,cell} - FSL_{5\,mile} + G_E - FSL_{1.5\,mile}$$
$$= 165 - 110 + (10 + 60 + 10 - 2) - 100 = 33 \text{ dBu}$$

$$P_{RC} = P_{T, \text{mobile}} - FSL_{1.5 \text{ mile}} + G_E - FSL_{5 \text{ mile}}$$
$$= 150 - 100 + (10 + 60 + 10 - 2) - 110 = 18 \text{ dBu}$$

Similarly, levels at the mobile and at the cell site when mobile is at 100′ from booster station are: $P_{RM} = 71$ dBu, $P_{RC} = 56$ dBu

A level diagram for this example is shown in Figure 5.8(b). Assuming the mobile to have a dipole antenna of gain 2.1 dB, the mobile receiver input power levels will become 35.1 dBu and 73.1 dBu, respectively. Similarly, the cell site receiver inputs with a specified gain of 15 dB are 33 dBu and 71 dBu. Thus both sides of the link behave about the same.

Example 2

An enhancer with amplifier gain of 70 dB in both directions, as shown in Figure 5.9(a), is located on a mountain top 8 miles from the fixed cell site. The cell site and the mobile are transmitting at 50 dBm and 36 dBm, respectively. Assuming both the donor and the reradiation antenna gains to be 15 dB, find the mobile and cell site received power levels when the mobiles are located at 1, 2, and 3 miles away from the repeater site. Assume the cabling loss from the donor antenna to the repeater and from the repeater to the reradiation antenna to be 1 dB. For 8 miles,

$$FSL = 37 + 20 \log_{10}(d_{\text{miles}}) + 20 \log_{10}(f_{\text{MHz}})$$
$$= 37 + 20 \log_{10}(8) + 20 \log_{10}(900) = 114.2 \text{ dB}$$

Similarly, free space loss for mobile at 1, 2, and 3 miles is 96, 102, and 105.6 dB respectively. The received signal to the mobile at 1 mile is

$$P_{RM,d=1 \text{ mile}} = P_{T, \text{cell}} - FSL_{8 \text{ mile}} + G_1 - L_1 + GE - L_2 + G_2 - FSL_{1 \text{ mile}}$$
$$= 50.0 - 114.2 + 15 - 1 + 70 - 1 + 15 - 96 = -62.2 \text{ dBm}$$

Similarly, $P_{RM,d=2 \text{ mile}} = -68.2$ dBm and $P_{RM,d=3 \text{ mile}} = -71.8$ dBm. The received signal at the cell site when mobile is located at 1 mile from the booster is

$$P_{RC,d=1 \text{ mile}} = P_{T, \text{mobile}} - FSL_{1 \text{ mile}} + G_2 - L_2 + G_E - L_1 + G_1 - FSL_{8 \text{ mile}}$$
$$= 36 - 96 + 15 - 1 + 70 - 1 + 15 - 114.2 = -76.2 \text{ dBm}$$

Similarly, $P_{RC, d=2 \text{ mile}} = -82.2$ dBm and $P_{RC,d=3 \text{ mile}} = -85.8$ dBm. Level diagrams for cell site and mobile transmitting are shown in Figures 5.9(b,c).

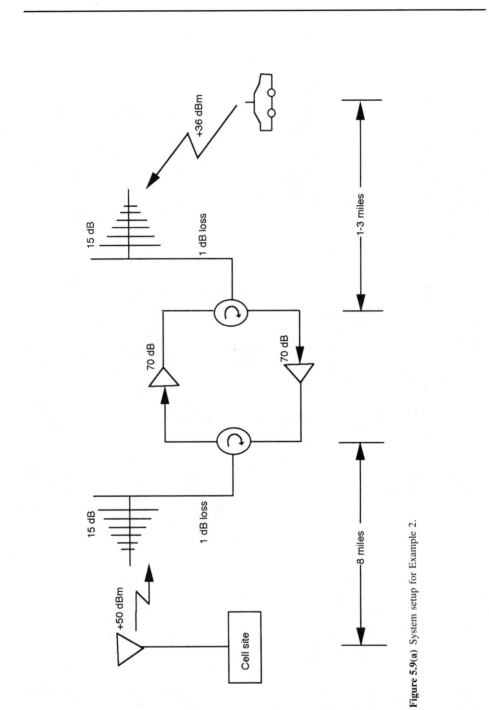

Figure 5.9(a) System setup for Example 2.

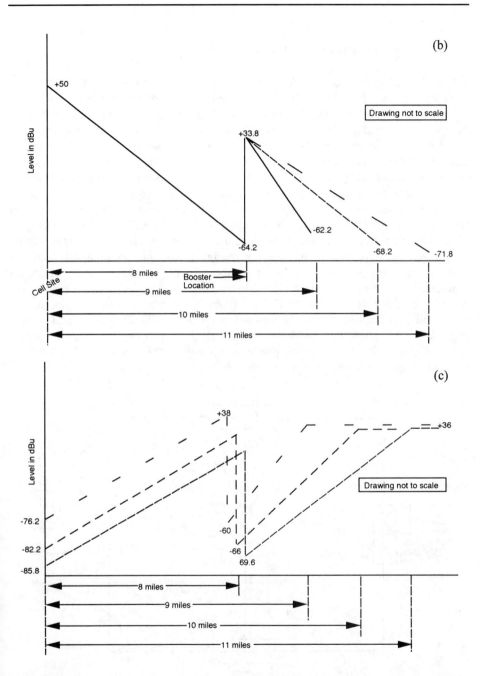

Figure 5.9(cont.) Level diagram from: (b) cell to mobile for Example 2; (c) mobile to cell for Example 2.

5.4 Equipment configuration and design specifications

In this section, we discuss three suggested configurations of the booster [6,8]. The first two configurations are shown in Figures 5.1 and 5.6. The booster configuration shown in Figure 5.1 is a simple configuration and has advantages with respect to controlling the intermodulation products. We discussed this in Section 5.1.1. In order to meet the sensitivity requirement of the repeater, the interference from the competitive channels must be minimized. We suggest a booster configuration, as shown in Figure 5.6, with separate forward and return amplifiers. Here, each path has both narrow and wideband amplifiers. This will make it possible to meet the speech and data quality requirements at the mobile. Of course, this realization is far more complex than the previous configuration shown in Figure 5.1. The third configuration, with a single wideband amplifier, is shown in Figure 5.10. This configuration consists of two circulators, one for receiving and transmitting signals from the cell site and the other for receiving and transmitting signals from the mobile. In order for the amplifier to be used by the WCC or RCC, the minimum bandwidth requirement for the system is 12.5 MHz. The exact filter placing is based on whether the system is used for RCC or for WCC. For proper diagnostics, couplers are placed both before and after amplifier. Here, the total adjustable gain of 70 dB is achieved with two stages of amplification. The first-stage preamplifier gain is 30 dB, followed by a power amplifier gain of 40 dB. A variable potentiometer adjusts the power amplifier input for different applications. Irrespective of which realization we choose for our design, the

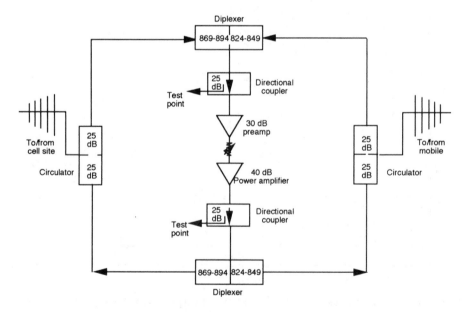

Figure 5.10 Single wideband amplifier configuration.

overall specifications can be arrived at based on our discussion in Section 5.2. The relevant parameters are shown in Table 5.1. The frequency bands of operation are the basic transmit and receive bands for the cellular radio specified by FCC in the U.S. An adjustable gain of 30–70 dB will satisfy most of the requirements. With a +24 dBm per channel power requirement, the design for 100 channels will require an average output power of +34 dBm or 2.5W. Third-order IM products must be less than −13 dBm when two frequencies at +26 dBm are reradiated from the repeater. A third-order intercept point (TOI) in the range of 45 dBm or more should be acceptable. The noise figure, as we discussed in Section 5.1.1, is not critical and can be as high as +10 dB. Donor and reradiation antenna gains can be in the range of 10–15 dB. Front-to-back ratios of donor and reradiation antennas in the range of 10–13 dB are an acceptable choice.

5.5 CELL ENHANCER APPLICATIONS IN THE UNITED STATES

Cell enhancers have been used in different parts of the United States for providing or improving the coverage on freeways and golf courses, and in tunnels and convention centers. We provide below some examples of these actual applications based on [3,6–7].

5.5.1 Specific U.S. Applications

5.5.1.1 Freeway Coverage between Olympia and Tacoma, Washington

On a heavily traveled freeway, even small shadow areas may be intolerable if they cause calls to drop and customers to complain. The repeater application described here concerns a stretch of Interstate 5 between Olympia and Tacoma, Washington. Prior to installing the repeater, there was a quarter-mile stretch of the freeway at Nisqually where calls

Table 5.1
Design Specifications of a Conceptual Booster

	Parameter
Frequency band, upward	824–849 MHz
Frequency band, downward	869–894 MHz
Booster gain	30–70 dB
Average power output	
for 100-channel booster	34 dBm
Linearity	< −13 dBm*
Noise figure	10 dB
Antenna gains (donor and reradiation)	10–15 dB
Front-to-back ratio	10–13 dB

*Third-order intermodulation products must be less than −13 dBm when two frequencies are reradiated simultaneously.

dropped out due to loss of coverage. With the repeater, there is now continuous coverage through this area.

The Peninsula Engineering Group placed CMRF-800 cellular repeater antennas on a 70′ wood pole while the repeater itself was placed at the base of the pole, in a driveway of a private residence. The antenna used for receiving the signal from the donor cell is a Scala PRBB-850, 17-dB gain, 12-deg directional parabola. The same kind of antenna is used towards the coverage area, mounted at the 30-foot level on the pole. The donor cell is a Northern Telecom system, loaded with 11 channels located 6.5 miles from the repeater site. A 10-dB omni no-tilt antenna is used at the donor cell. The coverage area, as stated above, is a quarter-mile stretch of I-5, 2 lanes north and south located 2.5 miles away from the repeater site. The cell site ERP is 100W and the signal level at the top of the pole is −50 dBm. The repeater gain is padded down from a maximum of 70 dB to 45 dB. Accounting for transmission line loss within the repeater the ERP per channel is +23 dBm. Repeater gain in this case is adjusted to 45 dB after accounting for available antenna decoupling, number of channels (11) passing through the repeater, and the maximum power authorized by the FCC.

5.5.1.2 Golf Course Coverage at Phoenix, Arizona and Kent, Washington

Golf courses in general are interesting sites for repeater applications. Coverage is almost mandated due to a strong desire on the part of relatively affluent golfers to be able to use their phones on the course. We describe here two applications, one in the desert surroundings of Phoenix, Arizona, with no trees and the other in Kent, Washington, with wet and forested surroundings. The coverage area in Kent is a 0.5 by 1 mile 18-hole golf course; in Phoenix it was desired to provide coverage along a commuter route as well, so the coverage area is 2 miles by 2 miles with a quarter-mile minimum distance to the repeater. In both cases the donor cell was part of a Northern Telecom system located 6–6.5 miles from the repeater and providing 100W of ERP with an omni antenna. In both places the repeaters were placed at a private residence.

In Phoenix, the PEGI CMRF-800 repeater was mounted in a closet with one Celwave PD10085, 10 dB, 65-deg panel antenna-mounted on the front of the house pointing towards the donor cell and another panel antenna-mounted on a fence at the back of the house pointing towards the coverage area. The panel antennas were selected for their "low profile" in a residential neighborhood. The signal at the repeater input was at −59 dBm; with 10-dB antenna gain on both sides, 65-dB repeater gain, and 2 dB of total transmission line loss. The output towards the mobiles is 250 mW (+24 dBm) per channel.

In Kent, the repeater and antennas were installed on a 60-foot tower at a private residence. Though the repeater is a little closer to the donor cell than in the Phoenix installation, the incoming signal level is down to −80 dBm due to excessive absorption of RF energy in trees. In this case a 17-dB gain Scala parabola was chosen towards the cell site, and a 10-dB Celwave PD 10085 panel towards the coverage area. With full 70-

dB gain and a transmission loss within the repeater of 6 dB, output toward the coverage area is +11 dBm. In both these cases, the repeater provided the coverage in an area where there was no coverage before.

5.5.1.3 Tunnel Coverage in Pittsburgh

Tunnels are difficult to provide coverage in with conventional technology; RF does not propagate into the mouth of the tunnel very well. While it is annoying not to provide coverage inside the tunnels, in most cases it would not be economically justifiable to put in a cell site to cover the tunnel. Coverage was provided in three tunnels in Pittsburgh at the Squirrel Hill, Fort Pitt, and Liberty tunnels. These tunnels range in length from 3600 feet to 5,000 feet. Prior to repeater installation, in all cases there was no coverage in the main sections of the tunnels. After repeater installation, tunnel coverage is virtually complete in all cases except for occasional problems at the far end of the Fort Pitt tunnel. At Liberty tunnel, the donor cell has a 17-dB Scala parabola, while the Fort Pitt donor cell antenna has a gain of 26 dB. With a 65-dB repeater gain and either a 9 or 12-dB Yagi antenna, the repeater ERP per channel in the tunnel is +22 or +24 dBm. In all cases the overall system dropped-call rate improved by 25% after the repeater installation.

5.5.1.4 Convention-Center Coverage

Most convention centers are built partly underground or with construction techniques that do not allow good coverage inside from an outside signal. Moscone Convention Center in San Francisco is a concrete structure that is largely underground. Without the repeater, it is not possible to provide coverage. Cellular operators can generate substantial revenue by being able to provide coverage inside convention centers during trade shows and conferences. However, as in the case of tunnels, it is difficult to justify the cost of installation of a cell site. We describe here the use of a booster in the Moscone Convention Center in San Francisco.

Two PEGI CMRF-800 cellular repeaters were used. Two repeaters were necessary because there are two halves of the convention center, with no conduit available to run a coaxial line and an antenna from one side to the other. The two repeaters are fed from two different donor cells, both with 45 channels and three sectors. One donor is 2 miles from the convention center; the other is three-quarters of a mile. The two donor cells have an ERP of 91W and 87W. Both these signals are at a level of −75 dBm at the repeater input. The repeaters were installed in the utility closets. The first repeater is a CMRF-800 with 65-dB gain, providing output to three Katraine panel antennas (KAT7402127) having a 105-deg beamwidth and a 9-dB gain. The output power per channel is 10 dBm. The other repeater has Celwave PD10040 panel sectored 60-deg antennas having gains of 8 dB. The signal level at the output is set to about 10 dBm per channel.

5.6 CONCLUSIONS

For optimum coverage of a CGSA, cellular systems often makes use of a cell enhancer. This allows the cellular operator to adequately provide service in areas that are otherwise difficult to cover. In this chapter, we have outlined the design requirements, equipment configuration, and applications of such cell enhancers. In the next chapter, spectral efficiency of cellular system will be defined. Proper justification for changing over to digital cellular from analog cellular will be provided.

PROBLEMS

5.1 Other then holes, fringe areas, and highway corridors, name other applications for a booster.

5.2 Justify the statement: ''Increase of front-to-back ratios of the donor and the reradiation antennas enhances the stability of the booster.''

5.3 For the configuration in Figure 5.11 find the value of X from the repeater to the mobile such that the signal level due to the donor cell and booster are the same. What problems will this create? Suggest some means of overcoming the problems. Assume the total distance $X + Y$ to be 10 miles. Cell enhancer gain can be assumed to be 60 dB.

5.4 The repeater layout for tunnel application is shown in Figure 5.12 where the cell site is assumed to be at a distance of 5 miles. Cell site and mobile transmit power are 165 dBu and 150 dBu, respectively. Let the placing of the second repeater (CE) inside the tunnel be 1,000 feet from the booster at the entrance. and the location of the mobile from the second repeater be 1,050 feet. Find the signal level at the cell site and at the mobile assuming the attenuation of the cable to be 0.03 dB/feet and the second repeater gain to be 30 dB. Overall gain of the first repeater is 71 dB.

5.5 For the configuration shown below, draw the level diagrams from cell to mobile and from mobile to cell. Assume that the mobile transmits at + 3W. Cell site transmitter output is +47 dBm, as shown in Figure 5.13.

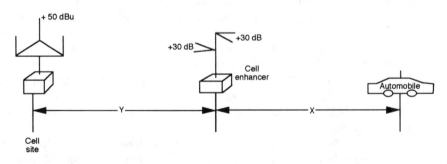

Figure 5.11 System layout.

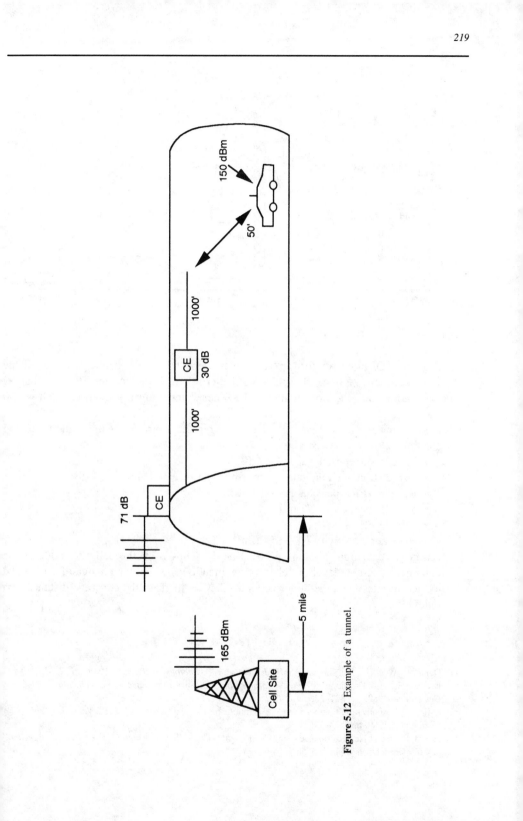

Figure 5.12 Example of a tunnel.

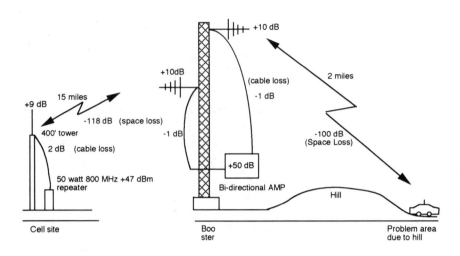

Figure 5.13 System layout for Problem 5.

5.6 Find the TOI point for forward and reverse boosters when there are 395 available voice channels. Assume the cell repeat pattern to be 7, the booster ERP per channel towards mobile is +24 dBm, and the ERP per channel towards the donor cell is +30 dBm.

5.7 Find the maximum permissible value of the repeater noise figure under the following conditions:
- Maximum repeater gain of 60 dB.
- Reradiation antenna gain for transmit and receive bands is 13 dB.
- Repeater ERP per channel is10 dBm.
- Mobile location is 1/2 mile.
- Mobile ERP is 3W.
- Repeater sensitivity is −115 dBm.

5.8 Assuming data transmission speed to be 10 Kbps from the cell site, allocate the maximum tolerable level of delay for a typical booster serving an area of 1 mile and located 6 miles from the donor cell site. Can we increase the coverage distance to 5 miles?

REFERENCES

[1] Quinn, E. W. "The Cell Enhancer," *IEEE Vehicular Technology Conference,* 1986, pp.77–83.

[2] Drucker, E. H. "Development and Application of a Cellular Repeater," *IEEE Vehicular Technology Conference,* 1988, pp. 321–325.

[3] Leff, B. J. "Application Engineering Considerations for Cellular Repeaters," *IEEE Vehicular Technology Conference,* 1989, pp. 532–534.

[4] Howat, F. "Cell Like Performance Using The Remotely Controlled Transmitter," *IEEE Vehicular Technology Conference,* 1989, pp. 535–541.

[5] Isberg, R. A., et al. ''Performance Tests of a Low Power Cellular Enhancer in a Parking Garage,'' *IEEE Vehicular Technology Conference,* 1989, pp. 542–546.

[6] Cell Expander, ''Preliminary Technical Description,'' *Astronet Corporation,* 1984.

[7] Coperich, F. ''FCC Requirements for Type Accepting Signal Boosters and Licensing Distributed Antenna Systems,'' IEEE, 1982

[8] Leslie, S., et al. ''Low-Power Cellular Boosters,'' *Cellular Business,* September 1988.

Chapter 6
Spectral Efficiency

6.1 INTRODUCTION

The radio spectrum is a finite resource, and it is important that it be exploited efficiently by all users. The predicted growth of cellular in Europe and the United States is such that the bands allocated to mobile radio will become congested in the near future unless steps are taken to deploy modulation and multiple access techniques that improve spectrum utilization. As the existing frequency allocations for mobile radio become saturated, there are at least three possible approaches that can be considered for its solution.

The first technique is to explore higher frequency bands for cellular as the technology and the economic conditions permit, bearing in mind the worsening propagation conditions as the frequency is increased. The second technique is to move fixed services to higher microwave and millimeter wavebands with the justification that fixed systems can better tolerate the adverse propagation characteristics at the higher frequencies. Spectrum relieved from these services can then be allocated to cellular radio. However, this may not be an acceptable solution due to complex political and economic factors.

The third option, which is most favored by regulatory agencies, is to make more efficient use of the current spectrum allocation. This approach is also justified, since the new and higher spectrum cannot be demanded if the present allocation is not even utilized efficiently. Conversion of analog cellular to digital falls in this category, we shall verify that this conversion achieves the higher capacity. Two terms have come into use for evaluating the efficiency of cellular systems: channel efficiency and spectral efficiency. For conventional communication systems where the spectrum is not reused, these two efficiencies are proportional to each other. In cellular, however, where channels are repeated, these two efficiencies are different. Here, the channel efficiency is defined as the maximum number of channels that can be provided for a given spectrum allocation. Spectral efficiency, on the other hand, is defined as the maximum number of calls that can be served in a given area. Another direct measure of spectral efficiency is load-per-unit spectrum, or erlangs per MHz. It is a valid figure of merit in most communication systems, but it fails to account for the geographic density aspect of the cellular radio. For example, if the service is provided in two cities with varying populations and the same Erlangs-per-unit spectrum, the spectral efficiency will be higher for the smaller city than for the large city, which is more spread out geographically. The composite unit "load per unit spectrum per unit area," or erlangs/MHz/km^2, has more meaning for cellular radio. An alternative definition of spectral efficiency used in cellular land-mobile radio systems is voice channels/MHz/km^2. The above two measures of spectral efficiency are directly related. It can be shown that the conversion from channels/MHz/km^2 to erlangs/MHz/km^2 can easily be done given the desired blocking probability and the formula used for traffic calculation.

In view of the above, this chapter is divided into the following sections. Section 6.2 opens the discussion of spectrum efficiency by defining mathematically the modulation and multiple access efficiencies. Section 6.3 evaluates and compares the modulation efficiency of some analog and digital systems and suggestes some approaches for improving the spectral efficiency of the present analog systems. Here we also take up the important study of the single sideband systems (SSB) and provide reasons for not choosing this for mobile communication. We also discuss the multiple access efficiency of FDMA and TDMA systems. Section 6.4 attempts to provide spectral efficiency for future CDMA systems. Lastly, Section 6.5 covers the conclusions that can be drawn from the above information.

6.2 TECHNIQUES FOR INCREASING SPECTRUM EFFICIENCY

The efficient use of the frequency spectrum is the most important problem in mobile communication. In order to realize the efficient use of spectrum, a variety of techniques have been proposed or are already implemented in cellular systems [1,4,6,7,16]. The main techniques involve narrowing the channel bandwidth, the channel assignment algorithm, delay connection, off-air call setup, variable bit rate control, and information compression. Efficiency also depends upon the choice of multiple access scheme. These techniques are listed in Table 6.1.

An increase in the number of channels can easily be obtained by reducing the channel bandwidth. For example, if the channel bandwidth is reduced from 30 kHz to 15 kHz, the number of channels can be doubled and thus the capacity can be increased more than two times. The reason for an expected capacity increase by more than two times is due to the nonlinear relationship between the number of channels and the associated traffic. This will be discussed in detail in Section 6.3. For the time being, let us understand the basic principle of this technique. Here, we will vary the channel bandwidth with distance—when the mobile is close to the cell site, the reduced bandwidth is allocated to the mobile; when the mobile is far away from the cell site, a higher bandwidth channel

Table 6.1
Techniques for Spectrum Utilization

Objective	Techniques
Increase the number of channels	Reduce the channel bandwidth
Improve spatial and timewise frequency spectrum reuse	Timed prechannel assignment
	Dynamic and hybrid channel assignment schemes
	Delay connection service
	TASI
Reducing invalid frequency spectrum use	Variable bit rate control
	Off-air call setup
	Reduction of uncompleted call
Traffic reduction without losing information	Information compression
Low data rate transmission	Low bit rate voice coding
	Lower symbol rate
	Lower overhead
Improved cochannel interference	Lower C/I requirement for digital cellular

is allocated. Unfortunately, in analog FM, as the channel bandwidth is reduced the noise susceptibility is increased, which in turn forces a need to raise the carrier-to-interference (C/I) ratio. The increase in C/I ratio effectively offsets the capacity increase as channels can then only be repeated with more geographical separation. Improvement in spatial and timewise frequency reuse spectrum can be obtained by increasing or decreasing the number of radio channels in a cell according to the traffic per unit time. The traffic pattern in a given area can provide the channel requirements. Dynamic and hybrid channel assignment schemes provide additional capacity to a cell under a high traffic load condition, as discussed in Chapter 3. Timewise frequency spectrum reuse can be obtained by assigning the channel to other users at the time of silence. Since, on the average, half of a conversation is spent listening, transmitting when no information is being passed is spectrally inefficient. The technique has been adopted by GSM in Europe and is being pursued by Hugues in United States [15–19]. The basic principle, termed discontinuous transmission (DTX), is to switch the transmitter on only for those periods when there is active speech to transmit. In this way, the average interference on the air will be reduced, thus allowing a smaller frequency reuse cluster size. It has been found that given an average voice activity of 50%, the spectrum efficiency could be doubled under certain idealized conditions. The GSM system implements it by using an adaptive threshold voice activity detection (VAD) algorithm. The equipment determines when a speaker is not talking and interrupts transmission. The receiver detects the DTX and fills the empty frames with "comfort noise." The net effect of this is to increase the number of channels.

Assignment of the channel according to service class (speech or data) or according to the desired speech quality can optimize the spectrum usage. Here, we can vary the channel bandwidth with data rate—for a higher data rate, allocate the wider channel bandwidth. In the GSM system, provision has been made for data channels with adaptation rates of 12 Kbps (for 9,600-bps data), 6 Kbps (for 4,800-bps data) and 3.6 Kbps (for 2,400 bps-or-less data) on full or half-rate channels. Thus, the operational channel bandwidth is data dependent. The other technique for increasing capacity is to allow for the delay connection (Erlang C), which increases the number of erlangs of traffic per channel and thus achieves more efficient spectrum utilization. The off-air call setup technique assigns the traffic channel only after both the parties answer. A reduction in the number of uncompleted calls by using recording and call transfer services can also increase spectrum utilization. This technique is being used in MATS-E and C-450 analog systems as discussed in Chapter 4.

Data compression, if it can be achieved without sacrificing the quality of speech unacceptably, can be translated into a reduction in transmission time or bandwidth. The technique has a high potential in digital cellular systems. System capacity can easily be increased by using lower bit rate voice coding. The American and European digital systems are projected to be using 16 Kbps presently. In the near future it is expected that the same quality of voice can be achieved at 8 Kbps. This will mean a doubling of the capacity of the system. Even 4.8 Kbps speech encoding is expected within 5 years, which will allow a quadrupling of the capacity of the proposed digital system. Thus, a reduction in speech encoding means a direct increase in system capacity. If we choose to express spectral capacity in terms of conventional communication systems as bps/Hz, then it is clear that for the same channel bandwidth, by reducing the bit rate, spectral efficiency can be improved. A lower symbol rate means less channel encoding and less complex equalization. As an example, more channel encoding will be required to combat the effect of a 3-ms delay spread at the 135 Ks/sec symbol rate than if the symbol rate is 16 Ks/sec. The limit of 3-ms delay spread overlaps about 40% of the adjacent symbol at the 135 Ks/sec rate, but only overlaps about 5% at the 16 Ks/sec rate. Thus, fewer parity bits are added for error detection and correction in a fading environment for the 16 Ks/sec rate than at the 135 Ks/sec rate. This will effectively increase the throughput and thus the capacity, or spectral efficiency, of the system. Cellular systems, whether FDMA or TDMA, require certain overhead in terms of additional signaling or supervisory bits. Additionally, both systems require separate signaling channels. Both these requirements reduce the system capacity. Since the number of signaling channels in AMPS is 21 out of 333, the capacity of the system is nearly equal to 94%. Thus, 6% of the capacity is wasted due to signaling channels. In a TDMA system, this reduction in capacity is far less. About 1% of the total channel capacity is allocated to signaling channels. On the other hand, the overhead requirements of the TDMA system are greater than an FDMA system due to critical synchronization requirements. In a TDMA system, the subscriber transmits and receives in a burst mode; thus, a certain number of bits are required at the beginning of each slot for bit synchronization. On the other hand, the requirement is not so critical for an FDMA system due to its continuous transmission. Additionally, in a TDMA system guard bits are required between two adjacent slots so that the overlapping of two subscriber's bits due to their operation at varying distances from the cell site can be prevented. In FDMA this requires 2–3% of the total system capacity, while a digital cellular system like GSM may require as much as 30% of the total capacity in the overhead part. The digital system of the United States only requires 16% of the total capacity[19].

Lastly, an improved cochannel interference performance (lower interference) can be attained in a digital system due to its having a lower required C/I ratio than an analog FDMA system. As discussed in subsequent sections of this chapter, a C/I ratio of 13 dB is adequate for digital transmission than a C/I ratio of 18 dB is for analog FM. Due to lower C/I ratio requirements, digital cellular radios can operate under higher cochannel interference than analog FM for the same voice quality requirement. Due to a 5-dB reduction in the C/I ratio for digital systems, the cochannel interference can be three times greater than for an analog system for the same voice quality service. Before we leave this section let us compare the spectral efficiency of major digital cellular systems of the world, expressed interns of bps/Hz.

Example 6.1

Data rates per channel and the channel bandwidth for American, European, and Japanese digital systems are shown in Table 6.2. Find the spectral efficiency of these systems in terms of bps/Hz. Assuming R_d to be the data rate per channel and the channel bandwidth to be B Hz, spectral efficiency on per channel basis can be defined as $\eta_c = R_c/B$ bps/Hz.

Table 6.2
Spectral Efficiency of Digital Cellular Systems of the World

System	Data Rate, R_d, per Channel (Kbps)	Channel Bandwidth B (kHz)	Spectral Efficiency (bps/Hz)
American digital cellular, TIA 45.3	48.6	30	1.6
European digital cellular, GSM	270.83	200	1.35
Japanese digital cellular, JDC	42.0	25	1.68

6.3 DEFINITION OF SPECTRAL EFFICIENCY

In order to judge the efficiency of spectrum usage for different modulation and multiple access systems, it is first necessary to precisely agree upon a definition of spectral efficiency [1]. This standard definition can then be used to resolve the conflicting claims regarding the relative efficiency of existing and proposed future systems. A precise measure of spectral efficiency will also allow one to estimate the capacity of various existing and proposed cellular land-mobile radio systems as well as setting up a minimum standard as a reference of measure.

With the present rate of growth of subscribers, analog cellular systems will soon be saturated. To ensure that cellular systems can grow to meet future demand, more efficient digital systems are being proposed. The reason why there is higher efficiency in digital systems is due to their lower susceptibility to cochannel interference, which leads in turn to a lower cochannel reuse distance. In the analog domain, lower bandwidth systems (which, unfortunately, have limitations and problems) have been proposed for higher efficiency. As we shall show in this and the next section, lower bandwidth analog systems will require higher C/I ratios, which in turn will partially or fully compensate for the improvement due to bandwidth reduction. This leads us to search for other approaches, namely, digital technology.

6.3.1 Measure of Modulation Efficiency

We can estimate the overall efficiency by knowing the modulation and the multiple access efficiencies separately. Factors covering the radio communication task include information transfer, channel transparency, channel sounding, and interference immunity. An improved quality of modulation acts favorably for these factors. As discussed in the introduction, the measure of spectral efficiency with respect to modulation [1,3–9] can be defined as:

$$\eta = \frac{\text{total number of channels available within the system}}{(\text{total available bandwidth}) \cdot (\text{cluster area})} \tag{6.1a}$$

$$\eta_m = \frac{B_t/B_c}{B_t(NA)} = \frac{1}{B_c \cdot N \cdot A} \tag{6.1b}$$

where η_m is the modulation efficiency in channels/MHz/km^2; B_t is the total bandwidth available to the system in MHz; B_c is the voice channel bandwidth or channel spacing in MHz; N is the number of cells per cluster(cluster size); and A is the area of a cell in km^2. From (6.1b), we observe that the spectral efficiency of modulation is independent of B_t and only depends upon channel bandwidth B_c and cluster area NA. We call this the modulation efficiency as the channel bandwidth B_c is a function of the modulation system. By decreasing the channel bandwidth, the modulation efficiency of the system can be increased provided it does not force an increase in the cluster area NA. Assuming that the cell area of the system is controlled by geographical constraints (propagation related) rather than the modulation, then the efficiency can be expressed as

$$\eta_m \propto \frac{1}{NB_c} \tag{6.2}$$

From the above definition it is clear that the spectral efficiency can be increased by reducing the channel bandwidth and cluster size, N. As discussed in Chapter 2, the cluster size N in turn depends (directly proportional) upon the C/I ratio. Thus the lower the value of C/I ratio, the higher the spectral efficiency. One of the questions that can now be asked is—Why not increase the spectral efficiency by reducing the bandwidth of the present analog FM system? We shall provide the answer to this important question in detail below but for the time being let us just note that by reducing the channel bandwidth, the C/I ratio is increased and thus it partially or fully compensates for the potential spectral improvement due to bandwidth reduction.

For two different modulation systems, x and y, the relative efficiency of system x with respect to system y can be written as

$$\eta_r = \frac{(\eta_m)_x}{(\eta_m)_y} \tag{6.3a}$$

or

$$\eta_r = \frac{(B_c)_y \cdot N_y}{(B_c)_x \cdot N_x} \tag{6.3b}$$

Dropping the subscript c, for convenience,

$$\eta_r = \frac{B_y \cdot N_y}{B_x \cdot N_x} \tag{6.3c}$$

The number of cells in a cluster, N, depends upon the tolerance of a given modulation format to interference from the nearby cells reusing the same channel. Assuming that the channel ambient noise is insignificant compared to cochannel interference, then N can be expressed in terms of D/R, or, C/I, as follows:

$$\frac{C}{I} = \frac{1}{6}\left(\frac{D}{R}\right)^\alpha = \frac{1}{6}(3N)^{\alpha/2} \tag{6.4}$$

Here, only the first tier of six cochannel interferers are considered, which adequately represents the practical systems. Assuming the fourth-power propagation law in the urban environment, (6.3c) can be expressed in terms of the C/I requirements of the two systems as follows:

$$\eta_r = \frac{\left(\frac{C}{I}\right)_y^{1/2} B_y}{\left(\frac{C}{I}\right)_x^{1/2} B_x} \tag{6.5}$$

This equation demonstrates the interesting property that the relative system efficiency decreases as the square root of the increase in C/I, but directly in proportion to the reduction in channel bandwidth. Thus, a modulation scheme that offers a reduction in bandwidth will have to raise the C/I ratio more than proportionately in order to maintain the same efficiency. In other words, the signal power has to be raised by a factor of four if we reduce the bandwidth to half. Example 2 applies this concept to a digital system.

Example 2

For digital cellular radio, the carrier power C can be expressed as

$$C = E_b R_b = E_s R_s \tag{6.6a}$$

where E_b and R_b are the bit energy and the baseband data rate while E_s and R_s are the symbol energy and channel data rate. Since the interference level is dependent on the cochannel power level and the system configuration (cluster size), we can assume this to be the same for both the modulations. Assuming, the relative efficiency η_r to be unity, we obtain from (6.5);

$$\frac{(E_s R_s)_x}{(E_s R_s)_y} = \left(\frac{B_y}{B_x}\right)^2 \tag{6.6b}$$

Denoting $(E_s R_s)_x = (E_s)_x (R_{sx})$ (and similarly for system y), we rewrite (6.6b) as

$$\frac{(E_s)_x R_{sx}}{(E_s)_y R_{sy}} = \left(\frac{B_y}{B_x}\right)^2 \tag{6.6c}$$

For linear modulations, the channel bandwidth can be related to the channel data rate, that is, $kB_x = R_{sx}$ and $kB_y = R_{sy}$. Substituting this in (6.6c), we obtain,

$$\frac{(E_s)_x kB_x}{(E_s)_y kB_y} = \left(\frac{B_y}{B_x}\right)^2 \tag{6.6d}$$

or

$$\frac{(E_s)_x}{(E_s)_y} = \left(\frac{B_y}{B_x}\right)^3 \tag{6.6e}$$

This relationship is rather important as it clearly shows that as the channel bandwidth is reduced by a factor of two, the energy per symbol has to be increased by a factor of eight. Thus, the imposed power penalty is 9 dB for reducing the channel bandwidth to one-half. For digitized voice, a comparative assessment of two systems can easily be done by measuring the BER of two systems. In other words, symbol energy is higher for the system providing improved performance provided the interference density is the same for both the systems. Before we leave this section, let us discuss a few more examples based on the above equations.

Example 3

Here we make use of (6.5) to compare the capacity of the present U.S. analog cellular system versus the forthcoming digital cellular system. Since the channel bandwidth for both these systems, at 30 kHz, is the same, and the performance dependence on system bandwidth is eliminated, we can compare the capacity improvement as a function of C/I ratio. Assuming, $(C/I)_y = 18$ dB for present analog FM system, and with $(C/I)_x$ of 13 dB, 12 dB, 10 dB, and 8 dB for digital systems we obtain (relative efficiency):

- $\eta_{r1} = 1.78$ for $(C/I)_x = 13$ dB;
- $\eta_{r2} = 2.51$ for $(C/I)_x = 10$ dB;
- $\eta_{r3} = 3.16$ for $(C/I)_x = 8$ dB.

Subjective tests on speech quality have shown that the performance of digital modulation will be same as that of analog FM with about 13-dB carrier-to-noise ratio. Due to a reduced C/I ratio, the cochannel reuse distance is also less. Thus, there are more channels available in a unit area and hence the relative efficiency is increased to a level nearly equal to 1.78 compared to analog FM.

Example 4

Assuming the required carrier-to-interference ratio of 12 dB for a 30-kHz digital system and about 18 dB for a 10-kHz digital system, one can easily show that the capacity of the 10-kHz system are increased in spite of the reduction in bandwidth. Applying (6.5), the relative spectral efficiency η is

$$\frac{30\sqrt{10^{1.2}}}{10\sqrt{10^{1.8}}} = 1.5$$

Thus in spite of the increase in the C/I ratio for 10-kHz bandwidth, the spectral efficiency of a 10-kHz system is greater than that of a 30-kHz digital system. This is not, however, true in case of analog FM systems, where the capacity is actually decreased when the channel bandwidth is reduced from 30 kHz to 10 kHz. This provides another advantage of digital system. To gain further insight into this problem, the reader is referred to problem 4 at the end of the chapter.

6.3.2 Alternate Measures of Modulation Efficiency

As explained in the introduction, one measure of spectral efficiency is channels/MHz/km^2 and the other measure is erlangs/MHz/km^2 [4,9–14]. We have already discussed the first measure. Let us now discuss the second measure and establish the relationship between the two. At the end of this section some adjustments to this definition will be made with assumptions. Following the definition of an erlang as the quantity of traffic on a voice channel or a group of channels per unit time, one can relate the two definitions as shown in the following equation. From equation (6.1a),

$$\eta_m \text{ erlangs/MHz/km}^2 = \frac{\text{traffic offered by } (B_t/B_c) \text{ channels}}{B_t \cdot (NA)} \tag{6.7}$$

We can also express the above equation in terms of traffic per cell by a slight rearrangement of the above equation,

$$\eta_m \text{ erlangs/MHz/km}^2 = \frac{\text{traffic offered by } [B_t/(B_c/N)] \text{ channels}}{B_t \cdot A} \tag{6.8}$$

The trunking efficiency factor can also be included in the above equation to represent the total amount of carried traffic through the system, as follows.

$$\eta \text{ erlangs/MHz/km}^2 = \eta_t x \eta_m = \frac{\text{traffic carried by } [(B_t/B_c)/(\eta_t/N)] \text{ channels}}{B_t \cdot A} \tag{6.9}$$

where η_t is the trunking efficiency, which provides a measure of how many calls are carried through the system out of the total number of calls received by the system. Essentially, there are two traffic formulas, Erlang B, and Erlang C, as discussed in Chapter 3. Erlang B is a "pure loss" or blocking formula in which blocked calls are cleared. That is to say, if a call arrives when all channels are busy, the call is immediately cleared from the system. Thus, the holding time in the system (waiting time) for this is zero. On the other hand, in Erlang C a call arriving when all the trunks are busy will wait within the system indefinitely until a channel becomes free. It gives a probability of delay greater than t seconds in terms of the number of channels available to the system. The following observations can be made about the above equations.

- The voice quality depends upon the cluster size N, which is a function of Carrier-to-interference ratio of the modulation technique used in the system.
- The relationship between the total available bandwidth of the system and traffic carried is nonlinear; that is, for a given percentage increase in B_t, the amount of additional traffic carried is more than the increase in B_t.
- Knowing the average traffic per user (erlangs/user) during the peak hours and the erlangs/MHz/km^2, the capacity in terms of users/MHz/km^2 can be derived.

- The spectral efficiency in erlangs/MHz/km^2 obviously depends on the blocking probability or the waiting time (Erlang C).

We shall illustrate the use of these equations with an example.

Example 5

A 333-channel cellular system with 312 voice channels and a 7-cell repeat pattern will have 44 channels per cell. This assumes uniform traffic throughout the system. Calculate the spectrum efficiency of the system in terms of erlangs/MHz/Km^2 using a maximum and minimum radius cell size of 4.2 miles and 1.05 miles. Assume the coverage area to be 5,000 square miles, the total available bandwidth for the transmit and receive sides to be 10 MHz each, and the blocking probability to be 2%. Cells are assumed to be hexagonal in shape. Cell area with 4.2 mile radius = 2.6× 17.64 = 45.9 mi^2. Cell area with 1.05 mile radius = 2.6 × 1.1025 = 2.87 mi^2. The total number of cells in 5,000 mi^2 area of coverage with larger and smaller radius cells is 5,000/45.9 = 109 and 5,000/2.87 = 1743. The total carried traffic per cell by 44 channels with 2% blocking probability = 34.7 × 0.98 = 34 erlangs (using the Erlang B table from the appendix in Chapter 3). Thus the spectral efficiency for larger and smaller cell size systems is (34 × 109)/(10 × 5,000) = 0.074 erlangs/MHz/mi^2 and (34 × 1,743)/(10 × 5,000) = 1.18 erlangs/MHz/mi^2.

It should be noted that as the cell size decreases, the cell capacity grows. The ultimate capacity of a land mobile radio system using a particular modulation system is achieved by employing a minimum cell size, which can be achieved in a matured system through cell splitting. As discussed in Chapter 2, the minimum system size the designer can attain is limited by some practical considerations, such as handoff rate, cell site tolerance, acceptable level of cochannel interference, and power control. Cell size as small as that nearly equal to a 1-km radius is possible in a digital cellular system. Smaller cell radius is also the concept used in high-capacity microcellular systems, where the minimum cell radius can be of the order of few hundred meters, compared to few kilometers for analog cellular radio.

We have now completed the two most popular definitions of spectrum efficiency, namely, channels/MHz/km^2 and erlangs/MHz/km^2 and their relationship. Once the number of channels and desired blocking probability are known, traffic intensity in Erlangs can be found and thus the conversion from channels/MHz/km^2 to erlangs/MHz/km^2 is complete. If the total bandwidth of the system is fixed, one can equivalently express the spectrum efficiency in terms of channels/km^2. In other words, so long as the context of the problem is known, we can find the spectral efficiency in terms of erlangs/km^2 or channels/km^2. Based on this concept, one can express spectral efficiency in the following different forms [4, 9–10,16]. If the specified blocking probability of the system is P_B, then the erlangs/cell can be found by using Erlang B, Erlang C, or the Poisson formula.

$$\eta_1 \text{ erlangs/cell} = \text{traffic offered by } [(B_t/B_c)/N] \text{ channels} \qquad (6.10)$$

Here, B_t/B_c is the total number of channels in the system divided among a cluster of N cells. Thus it is equal to: traffic offered by m channels/cell. Once again, the carried traffic can be found by multiplying the offered traffic by the trunking efficiency, $(1 - P_B) = \eta_t$.

Equation (6.10) can be functionally expressed as

$$\eta_1 = f(m, P_B) \qquad \text{erlangs/cell} \tag{6.11a}$$

where the notation $f(x, y)$ represents a function of (x, y).

Accounting for the cell area A_1, traffic η_1 can be expressed in terms of erlangs/km^2:

$$\eta_2 = f(\eta_1, A_1) = f(m, P_B, A_1) \tag{6.11b}$$

If the average holding time of the call is T, then the spectrum efficiency can be expressed in terms of calls/hr/km^2:

$$\eta_3 = f(\eta_2, T) = f(m, P_B, A_1, T) \tag{6.11c}$$

If we assume that on the average subscribers make k calls during the busy hour of the day, then the spectral efficiency can be expressed in terms of number of users/km^2:

$$\eta_4 = f(\eta_3, k) = f(m, P_B, A_1, T, k) \tag{6.11d}$$

If the area of coverage is A_t km^2, then the total number of users served by the system is

$$\eta_5 = f(m_4, A_t) = f(m, P_B, A_1, T, k, A_t) \tag{6.11e}$$

Spectral efficiency can also be expressed in terms of η_6 (calls/hr/cell), η_7 (number of users/cell during peak traffic hour), and η_8 (number of users/channel during peak traffic hour):

$$\eta_6 = f(m, P_B, T) \tag{6.11f}$$

$$\eta_7 = f(m, P_B, T, k) \tag{6.11g}$$

$$\eta_8 = \eta_7/m \tag{6.11h}$$

Let us use the above equations in the following an example.

Example 6

In the U.S. cellular system, the total allocated bandwidth is 12.5 MHz and the channels have a bandwidth of 30 kHz. Compute the spectral efficiency by the above alternate definitions and assume the following parameters: $A_1 = 10$ km^2 (cell size), T = average holding time of a call = 120 sec, k = average number of calls per user during peak traffic hour = 0.8, and the total area of system coverage = 5,000 km^2. Carry out these definitions for both omnidirectional and 120-deg sectored cell.

With 30 kHz of channel separation, the total number of channels in 12.5 MHz is 416 (half duplex). Out of these, 21 channels are allocated for signaling. Thus the number of voice channels is 395. With a 7-cell cluster size, the number of channels per cell is 56 (some cells can have 57 channels). Using Erlang B with $P_B = 0.02$ for omnidirectional cases, the total erlang offered is 45.9. Thus,

- η_1 = 45.9 erlangs/cell;
- η_2 = 45.9/10 = 4.59 erlangs/km^2;
- With T = 120 secs, η_3 = 4.59 × 3600/120 = 137.7 calls/hour/km^2;
- η_4 = 137.7/0.8 = 172.1 number of users/km^2;
- η_5 = 860,500 users in the system;
- In addition, η_6 = 45.9 × 3600/120 = 1377.0;
- η_7 = 1377/0.8 = 1721.3 users/hour/cell;
- η_8 = 1721.3/56 = 30.7 users/channel/hour.

For 120-deg sectored cases, the number of channels/sector is 56/3 = 19. Some sectors will have 18. Assuming 19 channels/sector, the offered value of traffic with $P_B = 0.02$ is.

- 12.3 × 3 = 36.9 erlangs/cell = η_1;
- η_2 = 36.9/10 = 3.69 erlang/km^2;
- η_3 = 3.69 × 3600/120 = 110.7 calls/hour/km^2;
- η_4 = 110.7/0.8 = 138.4 users/km^2;
- η_5 = 138.4 × 5000 = 691,875 users/system;
- η_6 = 36.9 × 30 = 1107 calls/hour/cell;
- η_7 = 1107/0.8 = 1383.8 users/hour/cell; and
- η_8 = 1660.5/57 = 29.13 users/channel/hour.

6.3.3 Modulation Efficiency Improvement of Analog Systems

In this section, we discuss methods of improving the spectral efficiency of analog systems. We also address the hardware and software modifications necessary to achieve these improvements. If the existing hardware and software are kept, major cost saving can occur. The basis of these discussions is the capacity increase while retaining the present FM modulation. We shall see that, realistically, the capacity of the present FM system can be increased by about 60%, beyond which the capacity increase is not practical. The following schemes have been proposed in the literature [7,4,12]:

- Scheme 1: multiple channel bandwidth system;
- Scheme 2: overlay/underlay approach;
- Scheme 3: hybrid scheme—combination of schemes 1 and 2;
- Scheme 4: channel offset approach.

The multiple channel bandwidth approach is based on reducing the modulation deviation and thus reducing the spectral occupancy of the FM signal [9]. With $f_m = 3.1$ kHz and a peak deviation of ±12 kHz, the modulation index is 3.8 in the present 30-kHz U.S. system.

If the channel bandwidth is reduced to 15 kHz with the same f_m, the deviation can be reduced, the number of channels can be doubled, and the capacity can be increased more than two times. Unfortunately, as the deviation is reduced, the susceptibility to noise is increased. Subjective tests on voice quality have revealed that for toll-quality voice the C/I ratio in a 15-kHz channel is about 24 dB, while for a 30-kHz channel the accepted value is 18 dB. As discussed in Chapters 2, and 3, the increase in C/I ratio effectively offsets the capacity increase, since the channels cannot be geographically used as often as in the 30-kHz case. Thus, by reducing the deviation ratio only a slight improvement in channel efficiency can be achieved. The second scheme, the underlay/overlay scheme, does not reduce the bandwidth of the channel but rather creates a smaller cell within a larger cell, as discussed in Chapter 2. Channels allocated to the smaller cell, closer to the cell site, operate at the same power level but are only used when the mobile is closer to the cell site. As the inner cell radius is reduced the cochannel separation distance can also be reduced, which makes it possible to have the same D/R ratio for both the inner and the (larger) outer cell. The scheme effectively increases the capacity to 1.6, as discussed here. The hybrid scheme suggested by Lee [3] adjusts the power and makes use of underlay/overlay technique suggested by Helpern [20]. The scheme provides improvements over standard one-cell systems, but the increment in capacity is lower than in the second scheme. The channel offset approach, as claimed by Lee [3], reduces the cochannel interference as the interference energy from near by cells is reduced and the crosstalk, in the form of intelligible speech, is practically eliminated due to the FM capture effect. Unfortunately, the scheme does not account for the interference introduced due to other shifted groups of frequencies, as discussed below. We now elaborate on each of these four schemes.

6.3.3.1 Scheme 1: Multiple Channel Bandwidth System

Here, two configuration are considered: a combination of 30 kHz and 15-kHz systems, as shown in Figure 6.1(a) and a three-channel combination of 30 kHz, 15 kHz, and 7.5-kHz systems, as shown in Figure 6.1(b). Assuming C/I requirements of 18.0 dB, 24.0 dB, and 30 dB for 30-kHz, 15-kHz, and 7.5-kHz channel bandwidth, we can find the radius of the smaller cells in terms of the larger cells, as shown below. Here, the smaller cells are assigned with lower BW channels and consequently will operate at a higher transmit power due to an increased C/I requirement. Limiting the cochannel interference to the first tier of six cochannel cells, we write

$$C/I = \frac{1}{6}\left(\frac{D}{R}\right)^{\alpha}$$

(6.12a)

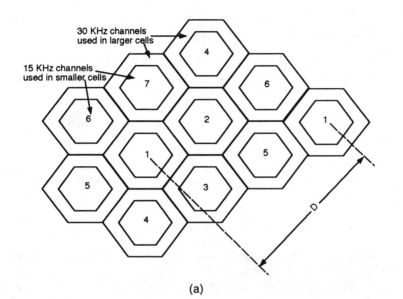

(a)

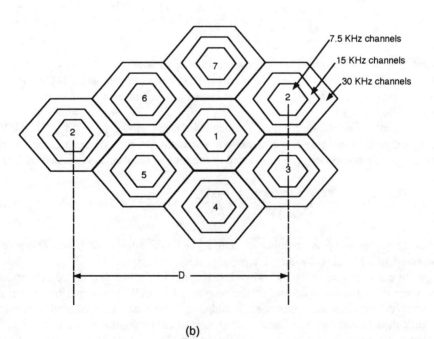

(b)

Figure 6.1 Multiple channel bandwidth systems: (a) a two-channel bandwidth system and (b) a three-channel bandwidth system.

Since the cochannel separation D is same for both 30 kHz and 15-kHz channels, we can rewrite (6.12a) as

$$\frac{(D/R_{15\text{kHz}})^4}{(D/R_{30\text{kHz}})^4} = 10^{(24-18)/10} \tag{6.12b}$$

Here, we assume the exponent α to be 4, or

$$R_{15}/R_{30} = 0.7079 \tag{6.12c}$$

Thus,

$$A_{15}/A_{30} = 0.5 \tag{6.12d}$$

From the above equation, it is seen that the area of the inner cell is approximately equal to the area of the outer ring. Assuming the total available bandwidth of the system to be 10 MHz and the number of channels to be the same in both the inner cell and the outer ring (assumes the same amount of traffic), the channel allocation can be done as follows:

$$\left(\frac{2}{3}\right)\frac{10,000}{30} + \left(\frac{1}{3}\right)\frac{10,000}{15} = 222 + 222 = 444 \tag{6.12e}$$

With an $N = 7$ reuse pattern, the original number of available channels per cell is 47 (333/7). In the modified scheme, the number of channels per cell is 63 (444/7). Thus, the capacity increase in terms of the number of channels over standard one-cell size configuration is 63/47 or 1.34.

A similar calculation for the three-bandwidth case shows $A_{30}:A_{15}:A_{7.5} = 1:2:4$. If we assume that the inner ring has a higher traffic density than the outer ring, then a three-channel bandwidth system can be divided in the following ratio, as suggested by Lee [9]:

$$\frac{1}{3}\frac{10,000}{30} + \frac{1}{3}\frac{10,000}{15} + \frac{1}{3}\frac{10,000}{7.5} = 111 + 222 + 444 = 777 \tag{6.12f}$$

Once again, assuming the number of cells to be 7, the number of channels per cell is increased to 111. This should be compared with the original cluster of 47. Thus the capacity can be increased to 111/47 = 2.4. It should be noted that as the cell radius of the inner cells is decreased and as the number of subcells increases, the problems in handing off between cells increase. In addition, the change in bandwidth necessitates modification of each mobile transceiver. The increased complexity due to handoff may forbid using three-channel bandwidth systems. Thus, a practical gain in capacity increase is limited to the two-channel scheme and is 1.34, as shown above.

6.3.3.2 Scheme 2: Overlay/Underlay Approach

This scheme, suggested by S. W. Halpern, again creates a subcell within an original large cell [20]. However, in this case both the larger and smaller inner subcell have the same channel bandwidth and thus power is kept the same in both groups of channels. Channels allocated to the inner cells are only used when the mobile is substantially closer to the base station. Figure 6.2 shows the cell configuration. Since R is decreased, the reuse distance D can also be decreased in order to keep the same D/R ratio for both the outer and the inner groups of channels. Since the D/R ratio for both groups of channels is same, from (6.4) it can be seen that the C/I ratios are same. This scheme provides an increase in capacity by a factor of 1.6, as shown below.

For a cluster size N of seven cells, the *D/R* ratio is 4.6, and therefore,

$$D_1/R_1 = D_o/R_o = 4.6 \tag{6.13a}$$

For smaller cell the cluster size is 3, thus $D_1/R_o = 3$. Therefore,

$$\frac{D_1/R_1}{D_1/R_o} = \frac{4.6}{3} \tag{6.13b}$$

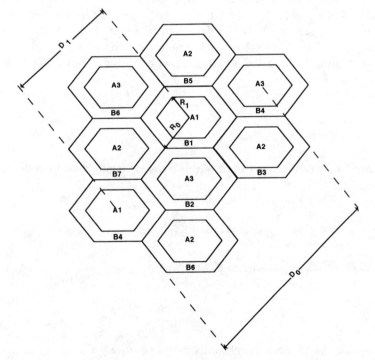

Figure 6.2 Cellular grids partitioned into two groups with $N_A = 3$ and $N_B = 7$.

or

$$\frac{R_o}{R_1} = \frac{4.6}{3} \qquad (6.13c)$$

Thus, $R_o = 1.5333R_1$, or $A_1 = 0.43A_o$

Assuming the total number of voice channels to be 395 (the total allocated spectrum = 12.5 MHz). The inner hexagon can have $0.43 \times 395 = 169$ channels and the balance of 226 channels can be allocated to the outer ring. Thus, the number of channels per cell is $169/3 + 226/7 = 88$. The number of channels can now be compared with the original allocation of $395/7 = 56$ channels per cell. This assumes uniform capacity in all cells. Thus, the capacity of the system is increased by a factor of 1.6. Since the bandwidth of the two groups of channels is the same, the channel allocation can be software controlled and no hardware changes are required to the mobile and base transceivers. Also, it imposes virtually no constraints on the operation of the system.

6.3.3.3 Scheme 3: Hybrid Scheme

This is the combination of Halpren's underlay/overlay scheme and Lee's method of two-bandwidth allocation. As shown below, this scheme really does not provide any improvement over scheme 2. The analysis here assumes the number of channels per cluster, N, equals 7 and the total available bandwidth to be 10 MHz, with a C/I ratio of 24 dB and 18 dB for 15 kHz and 30-kHz channel bandwidth. The ratio D_1/R_1 is assumed to be 6.2, as shown below.

$$\frac{C}{I} = \frac{1}{6}\left(\frac{D_1}{R_1}\right)^4 \qquad (6.14a)$$

The factor 6 in the denominator of (6.14a) is based on interference from the first tier of six cells.

$$\frac{D_1/R_1}{D_1/R_o} = \frac{6.2}{3.0} = 2.1 \qquad (6.14c)$$

Keeping $D_1/R_o = 3.0$, as per scheme 2, we obtain

$$\frac{R_1}{R_o} = 0.48 \qquad (6.14c)$$

Thus, $A_1 = 0.23 A_o$. Dividing the channels in this ratio we get

$$\frac{K\dfrac{10,000}{15}}{K\dfrac{10,000}{15} + (1 - K)\dfrac{10,000}{30}} = 0.23 \qquad (6.14d)$$

or $K \approx 0.13$. Thus, the channels are divided among two bandwidths as:

$$\frac{0.13 \times 666}{3} + \frac{0.87 \times 333}{7} = 28 + 41 = 69$$

Since the original cluster size is 7, the number of channels per cell with a 30-kHz channel bandwidth is 47. Therefore, the capacity is increased by $69/47 = 1.5$, which is slightly lower than the increase in capacity in scheme 2 by Halpern. This is shown in Figure 6.3.

6.3.3.4 Scheme 4: Channel Offset Scheme

Figure 6.4 shows the one-third channel offset scheme using a 7-cell reuse pattern. The closest interferers to the center cell 1 are due to three 1/3-offset cells $1'$ and three 2/3-offset cells $1''$. The C/I formula can be developed based on the desired carrier energy and the inserted interference energy, or

$$\frac{C}{I} = \frac{C}{\displaystyle\sum_{i=1}^{3}(\alpha_i' + \alpha_i'')} = \geq 18 \text{ dB} \qquad (6.15a)$$

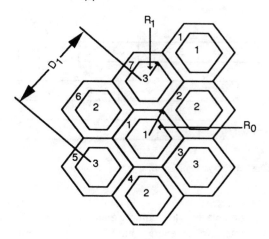

Figure 6.3 Hybrid scheme.

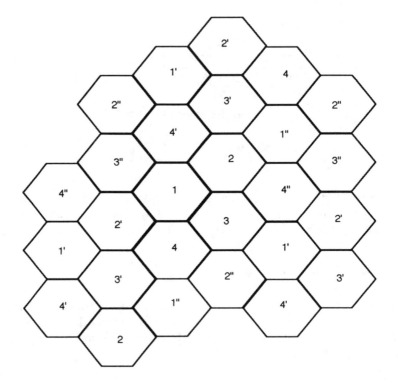

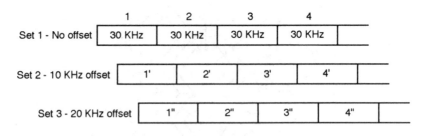

Figure 6.4 One-third channel offset system ($N_c = 4$).

where α_i' and α_i'' are the interference components of 1/3 and 2/3 offset interferers, respectively. Since the interference energy introduced is proportional to the spectral overlap, the above equation can be simplified as

$$\frac{C}{3(0.666 + 0.333)I_i} \geq 18 \text{ dB} \qquad (6.15b)$$

where $0.666I_i$ and $0.333I_i$ are the interference energy introduced by 1/3 offset (2/3 overlapping) and 2/3-offset channels (1/3 overlapping). Thus, $C/3I_i \geq 63$, or $C/I_i \geq 189$. Since, $C/I_i = (D/R)^n = 189$, with $n = 4$, $D = 3.71R$. Since $D/R = \sqrt{(3N)} = 3.71$, $N = 4.6$.

From the above one can conclude that a 4-cell cluster, as shown in Figure 6.4(b), can be selected. By choosing sectored implementation, additional cochannel interference suppression can also be obtained. The capacity improvement obtained by this method is almost a factor of two since the same channels distributed originally over seven cells are now distributed over four cells. Unfortunately, the technique suggested above does not account for the fact that the frequency offset has also introduced interference from the other shifted group of frequencies. For example, both 2 and 1′ and 2 and 1″ are now also interfering in addition to 2 with 2′ and 2″. However, a strong force is in favor of the FM modulation, which may suppress this crosstalk due to its capture effect. A summary of these four techniques is provided in Table 6.3.

6.3.3.5 Spectral Efficiency of SSB Systems [2,7–11]

As proved in the previous section, the spectral efficiency of a cellular system is a function of channel bandwidth and the carrier-to-interference ratio. In other communication systems where the frequencies are not repeated, bandwidth is the sole parameter of spectral efficiency and does not depend on carrier-to-interference ratio. Advocates of single sideband systems compare the 5-kHz channel bandwidth with 30 kHz FM and claim that the SSB system is six times more spectrally efficient than FM systems, ignoring the second

Table 6.3
Techniques of Spectral Efficiency Improvement for FM Systems

Scheme	Description	Bandwidth (KHz)	Capacity (x Times a Standard System)	Comments
Scheme 1(a)	Two-channel BW	30, 15	1.34	
Scheme 1(b)	Three-channel BW	30, 15, 7, 5	2.33	Difficult to implement
Scheme 2	Channel BW same	30	1.6	Maximum capacity increase (best scheme)
Scheme 3	Hybrid of schemes 1 and 2	30, 15	1.5	Capacity less than scheme 2
Scheme 4	1/3 channel offset		≈2*	Questionable approach

*Claimed.

factor, carrier-to-interference ratio. In this section we shall find the spectral efficiency of 3-kHz, 5-kHz, and 7.5-kHz SSB systems and compare these findings with a 30-kHz FM system. Both 5 kHz and 7.5 kHz were suggested for cellular applications; 3 kHz is used here only for comparative purposes and has not been recommended for commercial applications.

As a first step towards finding the spectral efficiency we have to establish the required value of carrier-to-noise ratio so that FM speech of comparable quality is reproduced at the system output. We shall assume that the total available bandwidth is 12.5 MHz and express the spectral efficiency in terms of erlangs/km^2. From [6], the required baseband signal-to-noise ratio for FM in multipath surrounding with preemphasis/deemphasis and two branch diversity combiner is 38.23 dB. For comparable quality of voice, the SSB system must equal this S/N ratio. Since the SSB system is a linear modulation, the carrier-to-noise ratio is the same as the baseband signal-to-noise ratio. Based on the work of Lee [10], this carrier-to-noise ratio requirement can be arrived at as follows:

$$(S/N)_{2BR-FM} = [(C/N) - 3] + G_{preemphasis/deemphasis} + G_{2BR-FM} \qquad (6.16)$$

Here, the factor -3 dB is the loss encountered in the FM system without preemphasis/deemphasis and diversity. Thus, in this case the baseband signal-to-noise ratio is 3 dB degraded from the carrier-to-interference ratio. The second factor, preemphasis/deemphasis gain, is given by:

$$\frac{(f_2/f_1)}{3} \qquad (6.17)$$

where, f_2 and f_1 are the highest and the lowest passband frequencies, with $f_2 = 3000$ Hz and $f_2 = 300$ Hz.

$$G_{preemphasis/deemphasis} = \frac{(3000 \text{ Hz}/300 \text{ Hz})^2}{3} = 33.3 \Rightarrow 15.2 \text{ dB} \qquad (6.18)$$

The third factor is 8 dB, the advantage in signal-to-noise ratio due to a two-branch diversity combiner over a single FM receiver. The factor that is ignored here is the companding gain, which is assumed to effect both FM and SSB systems equally. The function of the compandor is to reduce the noise power in the channel bandwidth. From (6.16) through (6.18), we can write:

$$(C/I)_{SSB} = (S/N)_{SSB} = (S/N)_{2BR-FM} = (18 - 3) + 15.2 + 8.0 = 38.23 \text{ (dB)} \qquad (6.19)$$

The above equation simply states that, for equivalent quality of FM voice, the required carrier-to-interference ratio for SSB is 38.23 dB and this compares with an 18-dB carrier-to-interference ratio for FM under Rayleigh fading.

With $C/I = 18$ dB for FM and $C/I = 38.25$ dB for SSB systems, the cluster size N for these systems, from (2.8) and (2.10) are

$$N = \sqrt{\frac{2}{3}10^{1.8}} = 6.5 \approx 7 \qquad (6.20)$$

$$N = \sqrt{\frac{2}{3}10^{3.823}} = 66.6 \approx 66 \qquad (6.21)$$

With these values of N, the distance to the cochannel cells are $N = 7$, $D = 4.6R$, and, for $N = 66$, $D = 14.1R$.

With 12.5 MHz of total bandwidth, the number of channels per cell is shown in Table 6.4. It is clear from Table 6.4 that, for SSB systems with 5 kHz and 7.5 kHz channel bandwidths, the capacity is lower than in an FM system with a 30-kHz BW under Rayleigh fading. SSB systems with a 3-kHz BW can have a capacity (channels/cell) that is slightly higher than in FM systems. However, as stated above, a 3-kHz channel is not recommended for toll-quality service for cellular applications under Rayleigh fading.

As a second step in this analysis, let us find the cell size for SSB systems with different bandwidths, assuming the cell size for the FM system to be 10 km. Once the cell size for an SSB system is known, the spectrum efficiency in terms of erlangs/MHz/km^2 can be found. Since the bandwidth for SSB systems is not 30 kHz, the reference noise level for these systems has to be adjusted linearly according to their bandwidths. The relative (with respect to 30-kHz FM) noise level for a 3-kHz SSB system is $10 \log(3 \text{ kHz}/30 \text{ kHz}) = -10$ dB. Similarly, the noise level for 5-kHz and 7.5-kHz SSB systems are -7.8 dB and -6.0 dB, respectively. Thus, the additional C/I (over 30-kHz FM) required for SSB systems with 3-kHz bandwidth is

$$(C/I)_{SSB} - (C/I)_{FM} = 38.2 - 18.0 - 10.0 = 10.2 \qquad (6.22a)$$

Similarly, the additional C/I ratios for 5-kHz and 7.5-kHz SSB systems are

Table 6.4
Channels per Cell for FM and SSB Systems With Rayleigh Fading

System	Channel BW (kHz)	Cluster Size, N	Channels per Cell
FM	30	7	59
SSB	3	66	63
SSB	5	66	37
SSB	7.5	66	25

$$(C/I)_{SSB} - (S/I)_{FM} = 12.43 \qquad (6.22b)$$
$$= 14.23$$

respectively. This adjustment of C/I power levels for SSB systems is shown in Figure 6.5.

Assuming the propagation exponent to be 4.0, the cell size radius for SSB systems with a 3-kHz bandwidth is $-40 \log_{10}(R_{3\,kHz}/10) = 10.23$ or $R_{3\,kHz} = 5.5$ km, where the denominator 10 represents cell radius of the FM systems. With a 5-kHz bandwidth, the radius is $R_{5\,kHz} = 4.9$ km and, with a 7.5-kHz bandwidth, it is $R_{7.5\,kHz} = 4.4$ km. From these, one can easily observe that the higher the C/I (over 30 kHz FM) for the SSB system, the smaller the cell size. In other words, due to a higher required C/I ratio, the mobile cell boundary has to be nearer the cell site. For the same voice quality at the cell boundary, a 30-kHz FM system with 10-km radius is equivalent to a 5.5 km, 3-kHz SSB radius cell. Similar conclusions can be drawn for 4.9-km and 4.4-km SSB cells. Thus, an FM system permits a larger cell size, while the SSB cells are smaller for the same voice quality.

The distance to the cochannel cells for these cases is shown in Table 6.5. It is clear from Table 6.5 that the cell size for FM systems is large and the distance to the cochannel cell is lower. The reverse is true for SSB systems (small cell size and large distance to the cochannel cell). Since the cochannel cells are nearer in an FM system, the capacity for an FM system is higher. Since the cell sizes for these systems are known, spectral density in terms of channels/km^2 or erlangs/MHz/km^2 can be found, as shown in Table 6.6. Here, the blocking probability, $P_B = 0.02$, and Erlangs/MHz/km^2 are the carried traffic. Thus, in a Rayleigh fading environment, a 30-kHz FM system is more spectrally efficient than both 5 kHz and 7.5-kHz SSB systems. Also, for the same signal strength at the cell boundary, more cells are required for the SSB system than the FM system.

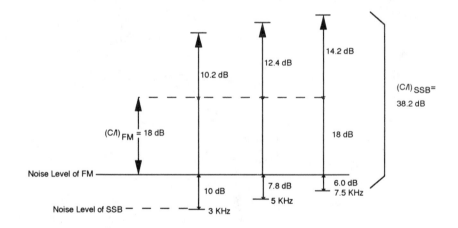

Figure 6.5 Additional power requirements for SSB systems.

Table 6.5
Cell Radius and Cochannel Distance for Different Systems

System	Cell Radius (km)	Distance to Cochannel Cell (km)
FM, 30 kHz	10	46
SSB, 3 kHz	5.5	77.6
SSB, 5 kHz	4.9	69.1
SSB, 7.5 kHz	4.4	62.0

Table 6.6
Cell Size for Different Systems

System	Cell Radius (km)	Number of Cells	Channels per Cell	Erlang/km^2	Erlang/ MHz/km^2
FM, 30 kHz	10	1	59	0.16	0.013
SSB, 3 kHz	5.5	3.3	63	0.17	0.014
SSB, 5 kHz	4.9	4.2	37	0.09	0.007
SSB, 7.5 kHz	4.4	5.2	25	0.005	0.005

6.3.4 Multiple Access Efficiency of FDMA and TDMA Systems

The objective of the multiple access techniques is to combine signals from different sources onto a common transmission medium in such a way that, at the destinations, the different channels can be separated without mutual interference. In other words, multiple access systems permit many users to share a common medium in the most efficient manner. There are three basic types of multiple access techniques:

- Frequency division multiple access (FDMA),
- Time division multiple access (TDMA), and
- Code division multiple access (CDMA).

In FDMA, users share the radio spectrum in the frequency domain. The user is allocated a part of the frequency band, which is used throughout the conversation. In TDMA, the users share the radio spectrum in time domain. An individual user is allocated a time slot during which, he or she has access to the whole frequency band allocated for the system (wideband TDMA) or only part of the band (narrowband TDMA). The CDMA technique is a combination of FDMA and TDMA techniques. With CDMA based on spread spectrum, each user is assigned a unique pseudorandom user code and thus can access the frequency time domain uniquely. Spread spectrum and frequency hopping are two examples of CDMA. All three multiple access techniques should have an efficiency of unity provided the signals transmitted by users are orthogonal to each other. However, this is difficult to attain. In FDMA, the multiple access for speech efficiency is reduced due to the necessity of having guard bands between channels in order to reduce the filter roll-off requirements and also due to signaling channels. In TDMA, the efficiency is reduced due to inclusion of guard time and synchronization sequence. Similarly, the efficiency of CDMA system is reduced due to the nonorthogonality of the codes. In this section we shall confine ourselves to the discussion of FDMA and TDMA. In the next section we shall take up the study of CDMA separately.

The multiple access efficiency is defined in our context as the ratio of the total time-frequency domain dedicated for voice transmission to the total time-frequency domain available to the system. Obviously, high efficiency is a function of design and in most cases can be attained with higher complexity. From this definition, the multiple access efficiency factor is dimensionless and has an upper limit of unity. Mathematically, the multiple access (MA) efficiency factor η is defined as the MA efficiency in time domain multiplied by the MA efficiency in frequency domain or $\eta_t \cdot \eta_f$. We evaluate η for different multiple access systems below.

FDMA

For FDMA,

$$\eta_{MA,FDMA} = \eta_f \cdot \eta_t = \eta_f = \frac{B_c M_a}{B_t} \leq 1 \qquad (6.23a)$$

where M_a is the total number of voice channels available to the system; B_c is the voice channel bandwidth or channel spacing; and B_t is the total bandwidth available to the system in MHz. Obviously, its efficiency in time domain is unity, since an individual user occupies the channel for one-hundred percent of the time. Alternately, (6.23a) can be expressed as:

$$\eta_{MA,FDMA} = \frac{(\text{voice channel BW in terms of bps})M_a}{(\text{total BW in terms of bps})} \qquad (6.23b)$$

Figure 6.6 represents the layout a typical FDMA system. In general, the efficiency is evaluated by considering the guard band. A guard band is required in most communication systems. However, in cellular systems a guard band between channels is not included as the adjacent channel is never allocated to the same cell. Let us consider the U.S. FDMA system and evaluate its efficiency.

Example 7

The total number of voice channels available in a U.S. system is 395, with 21 channels for signaling. The total allocated bandwidth is 12.5 MHz. Thus, the FDMA efficiency of the system from (6.23b) is:

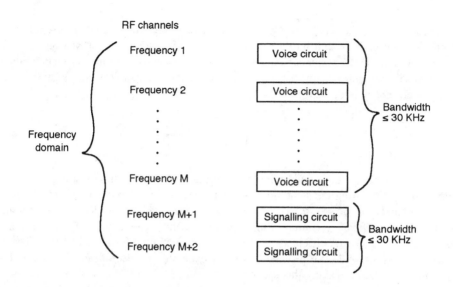

Figure 6.6 Layout of a typical FDMA system.

$$\eta_{FDMA} = \frac{B_c M_a}{B_t} = \frac{30 \times 395}{12,500} \approx 0.95$$

Assuming further a 7-cell configuration, the total number of voice channels per cell is $395/7 \approx 56$. Assuming a single signaling channel per cell, the spectral efficiency on a per cell basis is

$$\eta_{FDMA,cell} = \frac{30 \times 56}{12,500/7} = 0.94$$

TDMA

We define the TDMA efficiency for wide- and narrowband separately. The efficiency of the wideband TDMA is

$$\eta_{MA,TDMA} = \frac{\tau \cdot M_t}{T} \leq 1 \qquad (6.24a)$$

where τ is the time slot duration for speech or data transmission in seconds; T is the frame duration in seconds; and M_t is the number of time slots for voice transmission in a frame. Obviously, its efficiency in the frequency domain is unity. It is assumed that the total available band is shared by all users. Normally, in a wideband TDMA system the number of slots or users is ≥ 10. Cellular systems use the narrowband TDMA schemes where the total band is split into a number of subbands, each using TDMA techniques. For narrowband TDMA systems, frequency domain efficiency is not unity as the individual user channel does not occupy the whole frequency band available to the system. Narrowband efficiency is defined as

$$\eta_{\text{MA,TDMA}} = \eta_t \eta_f = \left(\frac{\tau \cdot M_t}{T} \right) \cdot \left(\frac{B_u \cdot M_u}{B_t} \right) \leq 1 \qquad (6.24\text{b})$$

where B_u is the bandwidth of an individual user during his or her time slot and M_u is the number of users sharing the same time slot in the system, but having access to different frequency bands. In (6.24b), the first factor is the wideband TDMA efficiency assuming that the FDMA efficiency is unity. Similarly, the second factor is the FDMA efficiency assuming that the TDMA efficiency is unity. The configuration for a narrowband TDMA is shown in Figure 6.7. We provide below an example of a European digital cellular system (GSM) before leaving this section.

Example 8

The GSM has an individual user data rate of 33.85 Kbps in which the speech with error protection has a rate of 22.8 Kbps. The data rate of the slow associated control channel

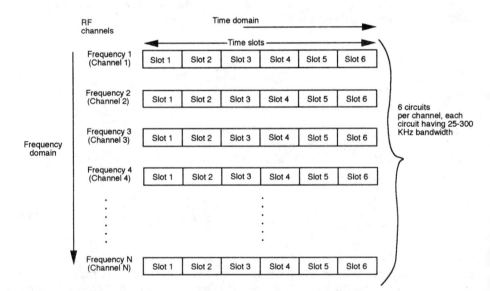

Figure 6.7 Narrowband TDMA system.

(SACCH) is 0.95 Kbps. Find the overhead data rate and the TDMA efficiency assuming that the system contains 125 channels of 200-kHz bandwidth each and with a total system bandwidth of 25 MHz. The frame duration is 4.62 ms divided equally between eight slots. Overhead data rate = (33.85 − 22.8 − 0.95) = 10.1 Kbps. The slot duration allocated to speech is

$$\tau = \frac{(33.85 - 10.1)}{33.85} \frac{4.62}{8} = 0.405 \text{ msec}$$

For the frame duration, $T = 4.62$ ms; $M_t = 8$; $B_u = 200$; $M_u = 125$; and $B_t = 25$ MHz. Substituting these values in (15.24b), the efficiency of GSM is

$$\left(\frac{0.405 \times 8}{4.62} \right) \left(\frac{200 \times 125}{25,000} \right) = 70.1\%$$

Thus, the overhead portion of the frame is $(1 - \eta_{\text{TDMA}})$ is 29.9%.

6.3.5 Overall Spectral Efficiency of FDMA and TDMA Systems

The overall efficiency of a cellular system considering both the modulation and the multiple access efficiencies is the product of these two efficiencies and is given by:

$$\eta = \eta_m \cdot \eta_{\text{MA}} \tag{6.25}$$

6.4 SPECTRAL EFFICIENCY OF THE CDMA SYSTEM

Among the many attributes of a CDMA system is the virtually unlimited capacity of the system. Unlike FDMA and TDMA schemes where the capacity is limited by the number of channels, in CDMA the capacity is interference limited. The reduction of interference directly translates to a capacity increase. Since many users share the same channel, interference can be induced from one user to another if there is not enough code isolation between the users. Since the voice signals are intermittent with a duty factor of roughly 40%, this interference is reduced naturally by this factor and the capacity can be increased by an amount inversely proportional to this factor. In other words, the capacity is increased roughly by a factor of 2.5. The other factor that enhances the capacity is the sectorization of the cell site. Assuming 120-deg cell sectorization, a threefold capacity increase is anticipated due to the reduction of interferers by one-third. In reality the increase in capacity will be some what less than three due to imperfect isolation among the three antennas. These two factors, voice activity and spatial isolation, are sufficient to theoretically increase the capacity of CDMA by a factor of 7.5 over FDMA. We illustrate the basics of CDMA from a capacity point-of-view by considering the simplest case of a single cell reverse link (from mobile to cell site) and illustrate basically on what factors the capacity depends and thus how it can be increased.

6.4.1 Single Cell Reverse Link Configuration

Let us assume that a single channel occupying the whole available bandwidth of the system is allocated to all users in one cell configuration. Let us also assume that the power control is exercised by the mobile such that the received powers of all mobiles at the cell site are the same. In this case the cell site receiver processes a composite signal of all mobiles containing the desired signal power of P and $(N - 1)$ interfering signals, each having the power of P. Thus the C/I ratio is

$$CIR = \frac{C}{I} = \frac{P}{(N-1)P} = \frac{1}{(N-1)} \tag{6.26a}$$

Since

$$CIR = \frac{C}{I} = \left(\frac{E_b}{I_o}\right)\left(\frac{R}{W}\right) \tag{6.26b}$$

Here, R is the information bit rate of the desired user and W is the total spread bandwidth of the system. From (6.26a,b), the ratio of the bit energy to interference power density ratio is

$$\frac{E_b}{I_o} = \frac{C/I}{R/W} = \frac{W/R}{(N-1)} \tag{6.26c}$$

The ratio W/R is the processing gain of the spread spectrum system, which is directly proportional to the number of users it can serve. On the other hand, the equation shows an inverse proportionality of the number of users to the bit energy-to-interference density ratio, E_b/I_o. Thus, the number of users can be increased as E_b/I_o is reduced. For digital voice transmission, E_b/I_o is the required value for a BER of 10^{-3} or better. For dual antenna diversity at the cell site and with a 1/3-rate convolutional code having the constraint length of 9, the required value of E_b/I_o has been projected to be 7 dB for quaternary phase shift keying (QPSK) modulation. The above equation does not include the effect of background thermal and spurious noise, μ (normalized value μ/P), in the spread bandwidth W. Including this as an additive degradation term in the denominator results in a bit energy-to-interference ratio of

$$\frac{E_b}{I_o} = \frac{W/R}{(N-1) + \mu/P} \tag{6.26d}$$

Rearranging (6.26d) in terms of the number of users, N, we get

$$N = 1 + \frac{W/R}{E_b/I_o} - \frac{\mu}{P} \qquad (6.26e)$$

Thus, the capacity of the system is reduced by μ/P, which is the ratio of background thermal noise plus spurious noise to power level.

As discussed above, these equations assume that the power levels of all the mobiles are the same. A mobile adjusts its power level based on the pilot signal it receives in the forward link. The pilot signal in the forward link, from the base to the mobile, provides a coherent reference signal to the mobile and allows the mobile to adjust its output power in an inverse proportion to the signal level it receives. The pilot signal is effective in the multipath fading environment of the cellular system, as both the pilot and the desired signal fade equally. The transmitter configuration at the cell site and the subscriber terminal is shown in Figure 6.8(a,b). The cell site transmitter adds the individual spread signals linearly and the phase randomness is assured as the spread codes of the individual subscribers are different. Additionally, the codes in the I and Q channels of the individual modulators are also different. The weighting factors λ_1, λ_2, ..., λ_n are adjusted in multicell systems to provide power controls. The subscriber transmitter is the simplified block diagram, which includes a speech vocoder, a forward error correcting coder, a QPSK modulator, and the direct sequence spreader. Due to coherent demodulation of the pilot carrier by the subscribers, the performance of the forward link in a single cell configuration is better than the reverse link from the subscriber to the cell site. For multiple cell systems, interference from other cells will tend to equalize the performance in the two directions.

For a fixed W/R ratio, one way to increase the capacity is to reduce the E_b/I_o, which is a function of modulation and coding. By providing powerful coding this ratio can be decreased, but the complexity of the system will also increase. Also, the E_b/I_o cannot be reduced indefinitely. The only other technique for increasing the capacity of the system is to reduce the interference (μ/P). Two techniques have been mentioned above, one based on natural behavior of speech and second based on antenna sectorization. Studies have shown that on the average a speaker only talks for about 40% of the time. Thus, for the remaining period of the time the interference induced by the speaker is eliminated. Since the channel is shared between all users, noise induced in the desired channel is reduced due to the silent interval of other interfering channels. Assuming a 120-deg sectored antenna at the cell site, the interference sources seen by any antenna are roughly one-third of those seen by an omnidirectional cell site antenna. Accounting for both speech activity and antenna sectorization, (6.26d) can be modified as:

$$\frac{E_b}{N_o} = \frac{W/R}{(N_s - 1)\,\alpha + \mu/P} \qquad (6.26f)$$

where, $3N_s = N$, and α is the speech activity factor, which usually has an approximate value of 0.4. With both of these factors accounted for, the capacity is increased roughly

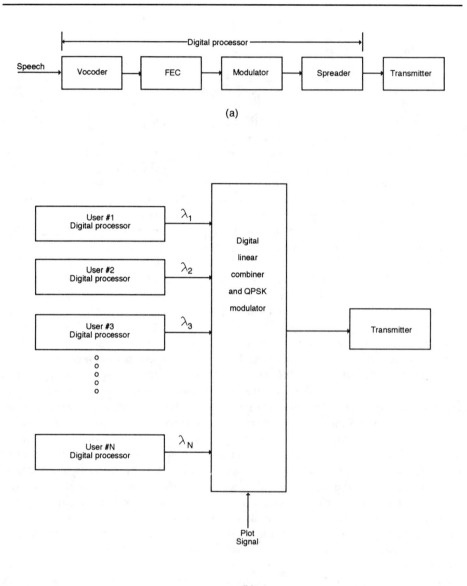

(a)

(b)

Figure 6.8 (a) Reverse link subscriber transmitter; (b) forward link cell site processor/transmitter.

by about 7.5 times (3×2.5). One can easily see this by substituting $\alpha = 0.4$ and $N_s = N/3$. Before we leave this section, let us estimate the capacity of a CDMA system based on (6.26e).

Example 9

The U.S. digital cellular system has an allocated bandwidth of 12.5 MHz. Assuming 8-Kbps digitized speech, estimate the capacity of the CDMA system assuming a voice activity factor of 2.5. E_b/I_o can be assumed to be 7 dB. Neglect other sources of noise. The number of subscribers, without accounting for the speech activity factor, is:

$$N = \frac{(12.5 \times 10^6)/(8.0 \times 10^3)}{10^{0.7}} + 1 = 312 \text{ users}$$

Since the speech activity factor is 2.5, the number of subscribers simultaneously served in each cell is 780. Assuming a 7-cell configuration, and with uniform traffic distribution, the maximum number of simultaneous users per cell using FDMA is 56. Thus the increase in capacity is nearly equal to 14. It should be noted here that we have allocated the total bandwidth to a single channel. In reality, there will be several spread-spectrum channels. In other words, the total bandwidth of 12.5 MHz may be divided equally between, say, half a dozen channels. We shall demonstrate this division process below by considering the forward link from the cell site to the mobile.

6.4.2 Capacity Based on Forward Link Consideration

Considering a 12-cell layout and assuming that the mobile is located at the boundary of a desired cell, as shown in Figure 6.9, the received C/I ratio can be written as:

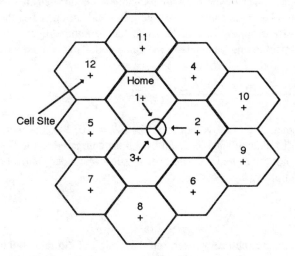

Figure 6.9 CDMA system and layout.

$$C/I = \frac{kR^{-4}}{k(M-1)R^{-4} + k(2M)\ R^{-4} + k(3M)\ (2R)^{-4} + k(6M)(2.6R)^{-4}}$$

$$= \frac{1}{3.312M - 1}$$

where, the total number of spread spectrum channels is M, k is a constant, and R is the radius of the cell. The first term in the denominator is due to undesired users in the home cell 1; the second term is the contribution of undesired users from adjoining cells 2 and 3; the third term is due to cells 4, 5, and 6; and the last term is due to cells 7 through 12, which are 2.6R away from the worst case location considered here. Assuming E_b/I_o = 7 dB, as above, the spread bandwidth to be 1.25 MHz, and the coded speech data rate to be 8 Kbps,

$$\frac{C}{I} = \left(\frac{E_b}{I_o}\right)\frac{R}{W} = \frac{10^{0.7} \times 8}{1.25 \times 10^3} = 0.03207 \tag{6.27b}$$

Thus,

$$M = \frac{(1/0.03207) + 1}{3.3123} = 9.715 \tag{6.27c}$$

Thus, the total number of permissible channels per system is 9.715. Assuming all channels can be allocated to all cells, the capacity per cell is 9.715 channels. This is the case when the power control scheme is not exercised. Let us take up the case when the interference is controlled or the power of the individual channel is controlled in such a way that the interference from other cells can be completely ignored. In this case, the above equation can be rewritten by dropping out all terms in the denominator except the first, which refers to interference among channels in the same cell. Thus, the equation is reduced to:

$$\frac{C}{I} = \frac{R^{-4}}{(M-1)R^{-4}}$$

Once again, evaluating the above equation for $C/I = 0.03207$, the value of M is 32.18. Thus the number of channels is increased by 3.3, as compared to the first case where the interference was not controlled. The number of subscribers per channel can be arrived at as shown in Example 9.

6.5 CONCLUSIONS

This is the basic chapter before we start our discussion of digital cellular radio. We have established the fact that the capacity of the present FM signal can only be increased by about 60% without any major change in the system. The potential use of single sideband systems in mobile communication is limited due to the increased value of the C/I ratio as the bandwidth is reduced. Though the number of channels is increased, SSB is less spectrally efficient because the cochannel distances are increased and the cell sizes are reduced. Both these factors work against the spectral efficiency increase. Ignoring the capacity increase due to C/I reduction, the digital TDMA system in the U.S. will increase the spectral efficiency by at least a factor of three, as there are three users per 30-kHz channel bandwidth. The potential increase in capacity of the CDMA system is enormous as the systems are only interference limited. This is unlike the case of FDMA and TDMA, where the capacity is limited due to the finite number of channels. The voice activity factor also helps in improving the spectral efficiency of the CDMA system. In the next chapter we shall discuss the European and the American TDMA systems and the proposed CDMA system in the United States.

PROBLEMS

6.1 A 70-channel radio system with a 7-cell repeat pattern will have 10 channels per cell, assuming the traffic to be uniform in each cell. Calculate the spectrum efficiency of the system in terms of erlangs/MHz/mi^2 using the maximum and the minimum radius of the cell to be 4 miles and 1 mile and assuming the total full-duplex BW = 3.5 MHz.

6.2 A typical cellular system in the U.S. covers an area of 3,000 sq. miles and uses a 7-cell repeat pattern. Assuming the cell radius to be 5 miles, find the spectral efficiency in terms of circuits/MHz/mi^2. Now, if the radius of the cell is reduced to 1 mile, find the new spectral efficiency. How does the spectral efficiency change when the reuse pattern, N, is 3 instead of 7 as in the original plan?

6.3 Assuming a digital cellular system with the required C/I ratio of 12 dB and using a 3-cell repeat pattern, find the relative efficiency of this system with respect to a standard analog cellular system using a 7-cell repeat pattern.

6.4 Find the relative spectral efficiency of a 7.5-kHz channel over a 15-kHz channel using analog FM system, assuming that the C/I requirement for these are 30 dB and 24 dB, respectively.

6.5 The European GSM system carries 8 users per carrier, having a bandwidth of 200 kHz. The required C/I ratio with and without frequency hopping is 9 dB and 11 dB, respectively. Compare the capacity of this system with an analog FM system having a bandwidth of 25 kHz and the required C/I of 17 dB.

6.6 The Japanese system carries 3 users per carrier, having a bandwidth of 25 kHz. The required C/I ratio with and without antenna diversity is 13 dB and 17 dB, respectively. Compare the capacity of this system with an analog FM system having a bandwidth of 25 kHz and the required C/I of 17 dB.

6.7 Assuming a TDMA system with 30-kHz channel bandwidth shared by 3 users, 1.25 MHz of total available bandwidth, and a cell reuse pattern of 4, find the capacity in terms of number of channels per cell. Find also the capacity in the same units assuming this to be the FDMA system with a cell reuse pattern of 7.

6.8 The American digital TDMA system has a total user data rate of 16.2 Kbps divided as follows: speech codec output is 7.95 Kbps; error protection on speech is 5.05 Kbps; SACCH (slow associated control channel) data rate is 0.6 Kbps; guard time, ramp time, synchronization, and so on are 2.6 Kbps. Find the frame efficiency and the overhead portion of the frame.

6.9 The minimum required *C/I* ratio for the Japanese digital system with and without antenna diversity is 13 dB and 17 dB, respectively. The pessimistic and optimistic number of channels per site are 429 and 750, respectively. Find the capacity in terms of Erlang/km^2 assuming the cell radius to be 1 km and the 120-deg sectored antennas.

6.10 The digital European cordless telephone system (DECTS) has a channel bandwidth of 1,728 kHz and a bit rate of 1,152 Kbps. Find the spectral efficiency in terms of bps/Hz. Repeat this for the English system, CT-2, having a channel bandwidth of 30 kHz and a data rate of 48.6 Kbps.

6.11 The North American digital cellular system carries 3 users per carrier, having a bandwidth of 30 kHz. The *C/I* ratio for the system with and without antenna diversity is 12 dB and 16 dB, respectively. Compare the capacity of this system with an analog FM system having a channel bandwidth of 30 kHz and the required carrier-to-interference ratio of 18 dB.

6.12 Prove that by reducing the channel bandwidth from 30 kHz to 15 kHz you do not increase the spectral efficiency of an analog FM system.

REFERENCES

[1] Batman, A., et al. "Cochannel Measurements for Amplitude Companded SSB Voice Communication," *IEEE Vehicular Technology Conference*, 1987, pp. 505–511.

[2] Ogose, S., and K. Daikoku. "SINAD Performance and Spectrum Efficiency for RZ SSB," *IEEE Vehicular Technology Conference*, 1986 pp. 342–349.

[3] Lee, W.C.Y. "Comparison of Spectrum Efficiency Between FDMA and TDMA," *IEEE Vehicular Technology Conference*, 1989, pp.165–168.

[4] McGeehan, H. H., et al. "Spectral Efficiency of Cellular Land Mobile Radio Systems," *IEEE Vehicular Technology Conference*, 1988, pp. 616–622.

[5] Lee, W.C.Y. "Spectrum Efficiency and Digital Cellular," *IEEE Vehicular Technology Conference*, 1988, pp. 643–646.

[6] Lee, W.C.Y. "New Concept Redefines Spectrum Efficiency of Cellular Mobile Systems," *Telephony*, November 11 1985.

[7] Lee, W.C.Y. "Narrowbanding in Cellular Mobile Systems," *Telephony*, December 1 1986, pp. 44–46.

[8] Boucher, L., and H. M. Hafez. "Evaluation of VHF FM, SSB, and ACSSB Radio Systems in the Interference Context of the Land Mobile Bands," *IEEE Vehicular Technology Conference*, 1990, pp. 244–246.

[9] Lee, W.C.Y. " New Cellular Schemes for Spectral Efficiency," *IEEE Transactions on Vehicular Technology*, 1987, pp.188–192.

[10] Lee, W.C.Y. "Spectrum Efficiency in Cellular," *IEEE Transactions on Vehicular Technology*, 1989, page 69–75.

[11] Nagata, Y., et al. "Analysis for Spectrum Efficiency in Single Cell Trunked and Cellular Mobile Radio," *IEEE Transactions on Vehicular Technology*, 1987, pp. 100–112.

[12] Sekiguchi, H., et al. "Techniques for Increasing Frequency Spectrum Utilization In Mobile Radio Communication System," *IEEE Vehicular Technology Conference*, 1985, pp. 26–31.

[13] Calhoun, G. *Digital Cellular Radio*, Chapter 15 on Choosing the Future: Evaluating the Alternatives, Norwood, MA: Artech House, Inc., 1988.

[14] Murota, K., et al. "Spectrum Efficiency of GMSK Land Mobile Radio," *IEEE Vehicular Technology Conference*, 1981,

[15] Schilling, D. L., et al. "Spread Spectrum for Personal Communications," *Microwave Journal*, September 1991.

[16] Lee, W.C.Y. "Overview of Cellular CDMA," *IEEE Vehicular Technology Conference*, May 1991, pp. 291–302.

[17] Gilhousen, K. S., et al. "On the Capacity of a Cellular CDMA System," *IEEE Vehicular Technology Conference*, May 1991, pp. 303–312.

[18] Pickholtz, R. L., et al. "Spread Spectrum for Mobile Communication," *IEEE Vehicular Technology Conference*, May 1991, pp. 313–322.

[19] Raith, K. "Capacity of Digital Cellular Systems," *IEEE Vehicular Technology Conference*, May 1991, pp. 323–332.

[20] Halpern, S. W. "Reuse Partitioning in Cellular Systems," *IEEE Vehicular Technology Conference*, 1983, pp. 322–327.

Chapter 7
Digital Cellular Radio

7.1 INTRODUCTION

In the less than ten years since its first introduction, analog cellular has met with enormous success through out the world. Today it is common while driving to see the person in the car next to you talking on a cellular phone or while dining in a restaurant to see other people talking over a cellular phone. As shown in Figure 7.1, there are presently more than seven million subscribers in the United States as of 1992, a figure that is rising approximately 33% annually [1]. As of May 1990, there were 13.7 million subscribers worldwide. By the turn of century, the expected number of subscribers for Europe is about 20 million, who will be served by up to 5,000 base stations with the associated network components. Figure 7.2 provides an estimate of the number of subscribers for digital (GSM) service in Europe, along with the total number, which includes many incompatible analog systems. In the United States, which has the highest number of subscribers, cellular mobile radio was introduced in 1982. By the end of 1991 there were more than 6.5 million subscribers and the number has risen steadily. Taking into account the present market penetration of about 1.2% and the forecasted rate of 10% for 1998, about 25–30 million subscribers will be using cellular mobile system in the U.S. by the year 1998. In Japan, the cellular communication market may also offer a major opportunity, where it is expected that at least 10% of all automobiles will have cellular telephones by the turn of the century. As shown in Figure 7.3, about 4 million subscribers are expected by 1994 [2,3].

Last year, one in five telephones installed in the Untied States was cellular. In the United Kingdom, this ratio was one in four. In Sweden, where cellular has highest market penetration, one in two telephones was cellular. Throughout the world, all the forecasts are predicting that growth rates will accelerate dramatically over the next couple of years for three main reasons: enhanced quality of mobile communication; complete area coverage (urban, suburban, and rural); and significantly lower cost of digital systems over analog.

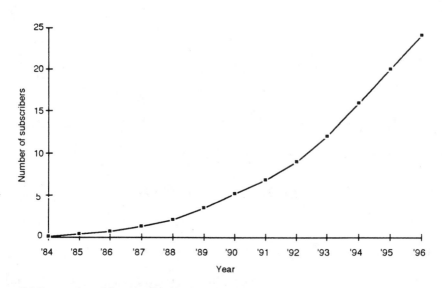

Figure 7.1 Forecast for cellular subscribers in the United States.

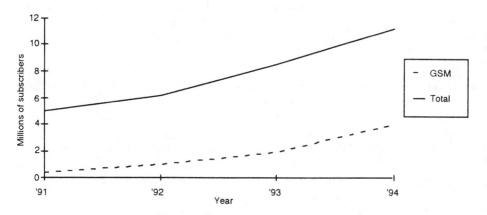

Figure 7.2 Forecast for cellular subscribers in Europe.

The cellular industry is the fastest growing sector of the world's telecommunication industry. Unfortunately, the capacity limitation of the analog system will not permit such a phenomenal growth. In the United States, large cities like Los Angeles, New York, and Chicago each have more than 300,000 subscribers and are rapidly reaching the full capacity of the system. In Europe, there are presently some eight million subscribers, of which 75% are concentrated in the United Kingdom and Scandinavia. The problem is serious in the United Kingdom, where a 66-MHz bandwidth (E-TACS) is inadequate to consistently deliver the desirable 2% grade of service to an estimated 500,000 subscribers. This

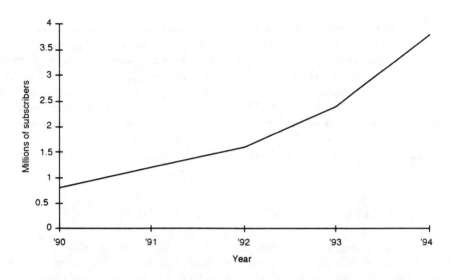

Figure 7.3 Forecast for cellular subscribers in Japan. After [42].

situation has been reached in spite of the system operators' attempts to increase the spectrum efficiency by using techniques such as cell splitting, sectorization, channel borrowing, and the overlaying of cells. It is also expected that the potential number of subscribers for the mobile service in Europe could be 10–20 million or even more. Also, in both Europe and North America the percentage of handheld terminals has reached greater than 20% of total market penetration. In the long run, both Europe and the United States are expected to have 50% or more of the handheld terminals. One essential design parameter of the handheld terminal is the operational duration of the unit before recharging the battery is necessary. The duration is highly dependent on the battery capacity and the average power consumption of the unit. The battery capacity will of course be limited by weight limitations, so the importance allocated to the handheld terminals leads to rather severe limitations on the allowed power consumption and the implementation complexity. Based on the above subscriber requirements, the new system selected should have a large capacity and be designed for easy handheld use.

Chapter 6 showed that the capacity of the analog system remains more or less in the same range, even after reducing the channel bandwidth. In other words, narrower bandwidth does not necessarily improve the spectrum efficiency. On the other hand, digital TDMA has the potential of increasing the capacity and reducing the cost, and will favor the handheld users. In the United States., the North American dual-mode cellular system will be compatible with the existing AMPS system and will share the existing spectrum. In Japan, new spectrum is being opened in addition to the present spectrum for analog cellular.

In view of this, Chapter 7 is devoted to the digital TDMA system. In Section 7.2, we discuss the reasons for choosing digital TDMA for future cellular radio systems.

264

Objectives and the spectral efficiency of American and European digital cellular radios are discussed in Section 7.3. Section 7.3.2 outlines the possible approaches that the operators of analog cellular radios will take to replace their systems with digital. Section 7.4 firmly establishes the requirements of American, European, and Japanese digital cellular radios. The remaining parts of this chapter are an elaboration of requirements of Section 7.4. Sections 7.5 and 7.6 provide the technical and operational requirements of European and American digital systems. Since the Japanese system is very much similar to American digital cellular, only the differences between the two systems are discussed in Section 7.7. In Section 7.8 other proposed systems of North America, such as E-TDMA and CDMA, are discussed.

7.2 ANALOG CELLULAR VERSUS DIGITAL CELLULAR

The driving force behind digital implementation is the increased system capacity, the reduction of mobile unit size, and the average power requirements, which in turn reduces the cost of the subscriber's unit. Of course, digital TDMA has its own problems also. In this section, we elaborate on these advantages and disadvantages by comparing the analog FDMA systems with digital TDMA systems. The details of this discussion are listed in Table 7.1.

A digital mobile radio system was first implemented in military tactical surroundings. In military communication, security in conversation is very important. Therefore, if an analog signal is used, it will be preceded by scrambling techniques. In general, unscrambling is easy so high marks are not given to analog implementation. Anyone with a frequency scanner can easily tune to the channel of interest and listen. In cellular systems the transmission over the radio path poses problems of security in two ways: with respect

Table 7.1
Comparison Between Analog FDMA and Digital TDMA

	Analog (FDMA)	Digital (TDMA)
Channel bandwidth	Narrow	Wider
Band utilization	Limited by IM	Better
Spectrum efficiency	Lower	Higher
Security	Easy to intercept	Encryption and authentication make it secure
Antenna multicoupler	Needed, makes the system complex	Not required for truly wideband TDMA
Power loss at cell site	High	Low
Cell site equipment	Small	Smaller
Main problems	IM at cell site	Transmission speed limited by delay spread of the channel ($R_d \leq 1/\Delta_d$)
	Power and the number of channels are limited	

to listening into conversation and unauthorized access and use of the system. Digital TDMA provides an easy solution to these problems by using encryption for signaling and user data as well as by implementing secure authentication procedures. Digital modulation will also permit ISDN services. A digital cellular user would be able to transmit and receive data files, faxes, and so forth while driving a car [6].

In analog FDMA, the mobile station operates in a single channel occupying 25–30 kHz bandwidth mode. The single-channel-per-carrier mode of operation makes the mobile transceiver easy and simple to work with, but at the base station there are as many transmitters and receivers as there are number of channels. Many of the channels are combined before feeding the signal to the antenna, which requires a combiner (generally a 16-channel combiner can be used). On the receiver side, a multicoupler is required to distribute the signal to individual receiver. Usually the combiner is large in size and consumes high power. As a result, analog system occupies more floor space. The interference tolerance capability is also low. On the other hand, a wideband digital TDMA system at the cell site does not need an antenna multicoupler, thus reducing the power loss and the system complexity. Also, being digital the system can use VLSI, and therefore the size and weight of the cell site equipment is reduced.

As discussed in Chapter 6, the spectrum efficiency of cellular radio is restricted by cochannel interference and the occupied bandwidth per user. With digitally modulated speech, a given voice quality could be achieved at considerably higher cochannel interference levels partly due to the fact that effective forward error-protection techniques can be applied in the digital case to improve the error performance. This in turn allows more frequency reuse of the channels and thus a larger number of channels per square mile can be utilized. In summary, digital technology offers a number of distinct advantages over analog technology, including supporting more traffic per cell, smaller cell radius, enhanced voice quality, improved network security, greater confidentiality, and a large range of data services including ISDN access. Also, digital VLSI implementation will allow smaller, lighter, and less expensive handheld portable units to be developed. In spite of the many advantages of digital TDMA, the system encounters the following problems:

Since the mobile unit may be anywhere in the cell, the distance to the cell site is a random variable. Therefore, signals transmitted by the mobiles to the base station will have different propagation delays. Guard time, $T_g \geq R/C$, must be provided, as shown in Figure 7.4, in order to avoid overlapping by adjacent time slots. Here, R is the maximum distance of the mobile from cell site and C is the velocity of light. Besides guard bits, in each time slot there are synchronizing and overhead bits, which further reduces the actual throughput and hence the frame efficiency. Also, the actual signaling rate is limited by the multipath delay spread, Δ_d. In the case of land mobile environments, delay spread on the order of 3–6 ms is the minimum expected, which limits the data speed to $R_d \leq 1/\Delta_d$. Thus, the data rate will be limited to 150 Kbps. Assuming QPSK modulation, the channel data rate will be limited to 75 Kbps. For 16-Kbps digital voice with 50% overhead bits, the total bit rate per user is 24 Kbps. Therefore, the number of channels per carrier is

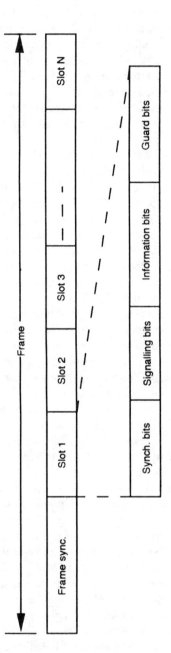

Figure 7.4 Typical wideband TDMA frame.

limited to six. If more than six channels are required, a separate carrier will be needed. For fading mitigation, time diversity and error correcting codes are used. Cell site diversity antennas and rake receivers can be employed for this purpose also. Rake receivers can make use of multipath effects to combine signals [7,8].

7.3 EUROPEAN, AMERICAN, AND JAPANESE DIGITAL CELLULAR OBJECTIVES

Both Europe and the United States are going digital, but their reasons and objectives for doing so are different. In this section we shall elaborate on these reasons. In 1993, a new pan-European digital mobile cellular telephone system should be in operation in sixteen European countries. Eighteen European nations have decided to adopt to the standard. Current GSM recommendations provide country "color code" (country identification code) for a total of 26 European nations. Other countries like Australia and Hong Kong have also expressed interest in GSM implementation. This will conclude the coordinated development and implementation program that was initiated about 10 year ago. In the United States, EIA/CTIA has come up with IS-54: the Dual-Mode Mobile Station–Base Station Compatibility Standard, which requires both the present analog and the new digital system to coexist until digital replaces the analog cellular system completely. The changeover to digital will occur smoothly in two phases. In the first phase, equipment manufacturers will bring out a dual-mode cell-site system, as shown in Figure 7.5, which will operate interchangeably on analog as well as on digital cellular channels. As shown in the figure, the present analog cellular system at the cell site will be replaced or augmented by a dual-mode unit. These units will be able to use analog and digital channels in any combination, even at the same cell site. The dual channel receiver will have two channels of signals in quadrature, namely, I and Q channel. The DSP algorithms will process the balanced inphase, I, and quadrature phase, Q, signals appropriately for analog and digital signals (only the inphase component is required for analog signals, both inphase and quadrature signals are required for digital signals). The processing hardware and the associated overhead makes the transceiver more complex. It is expected that initially most system operators in the United States will use 2.5 MHz of recent FCC allocation so that the capacity of the present analog system is not reduced. In the second phase of the digital conversion, analog subscriber units and channels will be discontinued. This phase is anticipated to occur in the late 1990s when the present analog system will reach obsolescence. We shall discuss this process in detail in Section 7.3.2. Figure 7.6 shows the market forecast up to year 1995 when 80% of the cellular systems will be digital.

The objectives of GSM in Europe and EIA/CTIA in the United States are shown in Table 7.2. The main differences between their objectives involve capacity of the system and orderly coexistence. In North America, it is absolutely necessary to increase the capacity of the system at least tenfold over the lifetime of the cellular system, which is also specified to have a ten year minimum. On the other hand, the GSM system in Europe

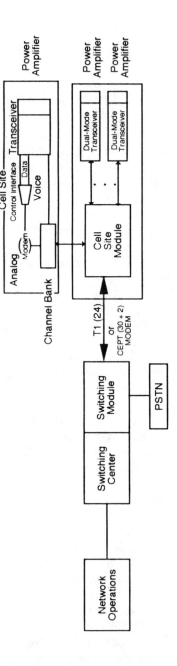

Figure 7.5 Dual-mode cellular system (analog and digital). Source: [5, p. 80].

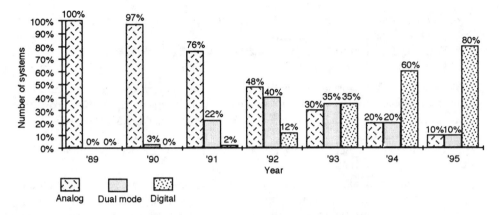

Figure 7.6 Telephone technology per year (U.S. market forecast). Source: [3, p. 73].

Table 7.2
Requirements of GSM and EIA/CTIA Digital Systems

	GSM	EIA/CTIA
System Capacity	About a twofold increase over current analog system	Tenfold increase over current analog system in two phases
Coexistence	Complete replacement of analog systems	Coexistence with analog systems and smooth transition to digital (two stages of seamless transition)
Data	ISDN compatibility	Multiples of 9.6 Kbps (PSTN compatible)
Privacy	Includes key encryption	Degree of privacy included in TR 45.3
Lifetime	Not specified	10 years, minimum
Emission	New standard: CEPT GSM 05	Compatible existing emission standard

will still require 25 kHz per user, and thus will have little more capacity than analog systems currently do. The slight increase in capacity is due to the fact that digital modulation has a lower required carrier-to-interference ratio than analog FM. A little more than a twofold increase in capacity is expected, as discussed in Chapter 6 and below in Section 7.3.2. Here in the United States, analog systems will stay in parallel with digital systems until probably the middle-to-end of the century. In the Pan-European system a common air interface is defined, which allow users to make and receive calls anywhere within the borders of the member countries. Under GSM specifications, each country's system maintains all users as if they were home subscribers and accommodates users from other countries as roamers. The definition of common air interface in Europe is mandatory in view of the incompatibility between systems in different countries. The only compatible systems between countries in Europe are:

- The Scandinavian countries of Denmark, Finland, Norway, and Sweden with their NMT-450 and NMT-900 systems;
- Belgium, Luxembourg, and the Netherlands with their NMT-450 system;
- Systems between the United Kingdom and Ireland.

This makes it difficult for subscribers to keep cellular service when roaming in other countries. For a subscriber living in Helsinki and making frequent trips to New York and London, three systems would be required: NMT-900 for Finland, AMPS for New York, and TACS for the United Kingdom. In order to remove this incompatibility, European PTTs have agreed to a pan-European GSM standard, which we shall discuss in Section 7.5.

Automatic roaming, as describe within IS-41.3, addresses mobile registration and call validation procedures grouped under the term "service qualification." Here, for a roamer a visited mobile switching center (MSC) initiates a service qualification request to the home MSC or to a clearinghouse service before the roamer is allowed or denied the service. The other major objectives for GSM are compatibility with the evolving integrated services digital network (ISDN). EIA/CTIA does not require ISDN compatibility; it only requires that the data rate be compatible with the existing data communications network (PSTN), which requires data up to 9.6 Kbps on a per-channel basis. For user privacy, the GSM specifications are highly secure, incorporating what amounts to a key encryption and authentication system. Some amount of privacy and coding is also being considered by the TIA 45.3 committee. Some secure access and channel encryption schemes are likely to be introduced in the future on a proprietary basis by different manufacturers. Lastly, EIA/CTIA requires that the new digital system to last at least for ten years and be compatible with the present analog emission standard. On the other hand, GSM has come up with a new standard designated as: CEPT GSM 05 [12–14].

Digital system specifications for Japanese cellular radio (JDC) are similar to the American cellular standard. The principal objectives in Japan are an increase in capacity, and voice security. The capacity problem in Japan is similar to that in America. Japan has added new spectrum in the 1.4–1.5 GHz range to alleviate this problem. The JDC system also sends all information in a digital format, making it more secure.

7.3.1 Spectral Efficiency Comparison of American and European Digital Cellular Systems

In this section we compare the capacity of American and European digital cellular systems with their respective analog systems. We shall arrive at the relative spectrum efficiency of these systems by making use of the theory developed in Chapter 6. In particular, we shall apply (6.8), which expresses the capacity in terms of erlangs/MHz/km^2. The following assumptions are made:

- Total number of channels in the United States (A or B) is 416;
- Channel bandwidth is 30 kHz;

- Total available bandwidth (in each direction) is 12.5 MHz;
- Number of control channels is 21;
- Number of voice channels is 395;
- Cell repeat patterns, N, is 12;
- Required C/I for present analog system is 18 dB;
- Required C/I for digital system is 15 dB;
- The total CGSA area is 6,500.

The solution for American system assumptions include:
- Number of voice channels per cell is 32;
- Blocking probability, P_B, is 0.02;
- Offered traffic (using Erlang B formula) is 23.7 erlang/cell;
- Carried traffic is 23.23 erlang/cell.

Thus the spectral efficiency of the analog system from (6.8) in the United States is

$$\frac{23.23 \times 6,500}{12.5 \times 6,500 \times 2.6R^2} = \frac{0.715}{R^2} \text{ Erlangs/km}^2/\text{MHz}$$

where, $(6,500/2.6R^2)$ is the number of cells in the system. Using equation (6.4) we can write:

$$[D/R]^2_{\text{Dig}}/[D/R]^2_{\text{Analog}} = \sqrt{[C/I]_{\text{Dig}}/[C/I]_{\text{Analog}}}$$

or

$$A_{\text{Dig}}/A_{\text{Analog}} = \sqrt{10^{-0.3}} = 0.71$$

Thus the area of the digital cell $\approx 0.71 \times$ area of the analog cell. Therefore, the spectral efficiency of phase 1 of the digital system when the number of channels per 30 kHz is 3 can be computed as:

- Number of channels per cell = 96;
- Offered traffic (erlangs/cell) = 84.1; and
- Carried traffic (erlangs/cell) = 82.42.

Thus, the spectral efficiency of digital phase 1 is

$$\frac{82.42}{12.5 \times 2.6 \times R^2 \times 0.71} = \frac{3.57}{R^2} \text{ erlangs/MHz/km}^2$$

Thus, the relative spectral efficiency of a digital system with respect to the present analog system, phase 1, is $3.57/0.715 \approx 5.13$.

Similarly, the spectral efficiency of phase 2 when the speech coding rate will be reduced to half can be arrived at as follows:

- Number of channels per cell = 192;
- Offered traffic (erlangs/cell) = 178.2; and
- Carried traffic (erlangs/cell) = 174.6.

Thus, the spectral efficiency of digital phase 2 is:

$$\frac{174.6}{12.5 \times 2.6 \times R^2 \times 0.71} = \frac{7.6}{R^2} \text{ erlangs/MHz/km}^2$$

Thus, the relative spectral efficiency of a digital system with respect to the present analog system, phase 2, is $7.6/0.715 = 10.6$.

Plots of the American analog, digital phase 1 and digital phase 2, are shown in Figure 7.7. Similarly, calculations for cellular systems in Europe can be made. The following assumptions are made:

- The required C/I ratio for digital system is claimed to be 6-dB lower than the present analog system.
- Analog channel bandwidth is 25 kHz;
- Digital channel bandwidth/user is 25 kHz;
- Digital channel bandwidth is 200 kHz;
- Cell repeat pattern, N, is 4;
- Number of traffic channels (125×8) is 1,000;

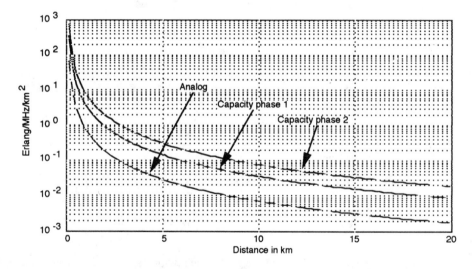

Figure 7.7 Cellular system capacity in the U.S.

- Number of channels per cell is 250;
- Offered traffic (erlangs/cell) is 235.8; and
- Carried traffic (erlangs/cell) is 231.1.

Thus, the spectral efficiency of the present analog system is

$$\frac{231.1}{25 \times 2.6 \times R^2} = \frac{3.55}{R^2} \text{ erlangs/MHz/km}^2$$

Similar to the computation for the U.S. system, the area of the digital cell can be reduced to about 0.5 × area of the present analog cell. Thus, the spectral efficiency of phase 1 of the European digital system is

$$\frac{231.1}{25 \times 2.6 \times R^2 \times 0.5} = \frac{7.1}{R^2} \text{ erlangs/MHz/km}^2$$

Thus, the spectral efficiency of digital cellular with respect to present analog system is nearly equal to 2. Similarly, with half-rate speech coding it can be shown that the spectral efficiency of the digital system becomes $(14.7)/R^2$ erlangs/MHz/km^2, which means that the relative efficiency (phase 2) with respect to present analog system is nearly equal to 4.1. Plots of the European analog, digital phase 1 and digital phase 2 cellular system's capacity are shown in Figure 7.8.

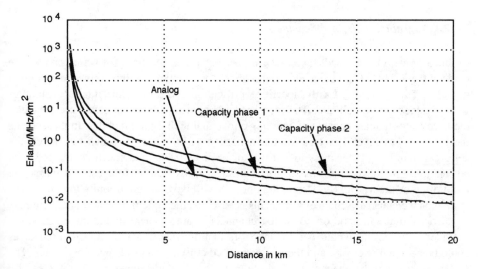

Figure 7.8 Cellular system capacity in Europe.

7.3.2 Integration of Analog Cellular With Digital in the United States

We can divide the present coverage areas into two types: those served by analog cellular and those with no service. Digital will initially be used to increase capacity, and therefore it makes sense to introduce digital soon. Thus, it is essential that the new digital system be able to coexist with analog for some length of time before the system is completely converted into the digital form. This process in the United States will take place in two steps. The first phase of conversion will allow one 30-kHz analog voice channel to be converted to three digital channels, adding threefold increase in capacity. To put this differently, let us consider a single 30-kHz analog channel. Currently, it can handle 25–30 subscribers with the required 2% blocking probability. With digital conversion, this capacity will increase to handle approximately 90–100 subscribers for the same channel with the same blocking probability. In the second phase of conversion it is expected that with half-rate encoding (4 Kbps versus 8 Kbps) six digital voice paths will be provided. This in turn will increase the capacity by sixfold. The process of migration from analog to digital will vary from one supplier to another based on the system's design. Two basic schemes have been proposed in the literature for this transition [9,10–16]: migration and overlay. In the migration scheme, both the analog and the digital system can be fully integrated (e.g. both analog and digital radios will coexist at the cell sites) and in the course of time analog cellular systems will migrate to digital completely. We can therefore consider the cell site to be hybrid. In the overlay scheme, the analog and digital systems are kept independent of each other and these systems are connected either at the PSTN or at MTSO. We discuss below both of these schemes.

7.3.2.1 Migration or Hybrid Scheme

In this scheme, changes will take place in three steps. In the first, software changes are made at the MTSO. Then the analog radio transceiver modules will be replaced by digital modules at the cell site. Lastly, location verification and digital line interface modules will be added in the cell site. The migration (or hybrid) scheme in effect creates dual-mode cell sites that are capable of handling either analog or digital mobiles. In this scheme the operator has several factors to consider, such as the order in which to convert the cells, the number of channels to convert in each cell site, and the time for the total conversion. Obviously, the time factor is influenced by market forecast and the actual monitoring of the traffic. By design, the operator will have the opportunity to convert the system on a channel-by-channel basis. This is based on the system's capability to originate a call on digital and hand off to analog channels in areas where digital coverage is not provided. The design would permit operators to prioritize transition so that core areas could be changed over first and the less congested outlying areas changed over later. The system layout and the channel configurations in this case are shown in Figure 7.9(a, b). Since each cell site has both analog and digital capabilities, their traffic is combined at

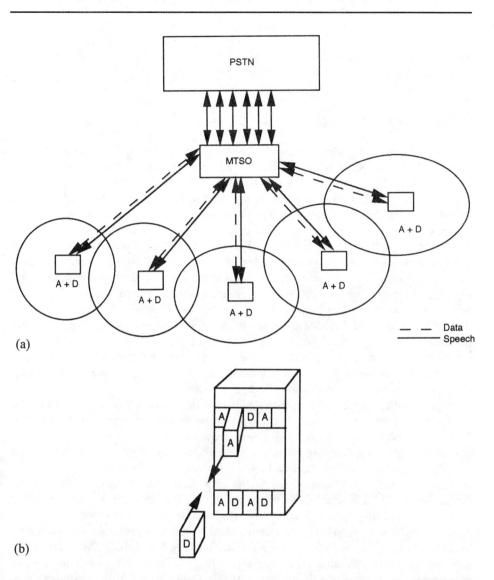

(a)

(b)

Figure 7.9 (a) Integrated analog and digital systems (migration approach). (b) Channel configuration in digital migration at a cell site (each analog channel is replaced by three digital channels in phase 1 and six in phase 2).

the cell site itself before the messages are passed to the MTSO. Since the traffic is combined and carried over, the common trunks are utilized to their highest efficiency. This also permits the base-station control facilities to be utilized to their maximum as there is no duplication of the required hardware and software. Because both the digital and analog systems are collocated, equivalent coverage patterns will be required from both systems. However, MTSO software will have to make a distinction between analog and digital subscribers at the time of channel assignment.

7.3.2.2 Overlay Scheme

The configurations shown in Figure 7.10(a,b) have analog and digital systems interconnected at the PSTN. A call from one of these systems to the other is routed via the central office (class 5 office), which is similar to a call from either one of the systems to the outside world. In this configuration, ticketing and control information are not transferred directly between the two systems because the switching office acts as a sole interface between analog and digital systems. One of the advantages of this scheme is that two systems can be independently designed to interface at the central office. The new digital design can be such that it is not affected by the older analog system. Here, the analog calls will go through the older system, and digital calls will go through the new system. As in the migration scheme, an inverse relationship exists between the buildup of the new system and the scaling back of the old system. For each digital channel added (in steps of three in phase 1 and steps of six in phase 2), a corresponding 30-kHz analog channel will be deactivated in the old system. Here, the operator has to monitor the analog system reduction and digital growth. As the analog service is reduced to a level where there is no longer a justification for operating two systems side by side, the remaining analog channels will be converted to the digital system.

Since both systems interface independently at the central office, trunks will not be utilized to their maximum capacity because both types of traffic are not grouped together. Trunk utilization to analog systems will also decrease with time. There is also a duplication of control facilities in this case because the old analog subscribers operate out of their cell sites, while the new digital subscribers will operate out of the digital cell sites. If the overlay is complete, that is, the entire analog network is overlaid with digital network, then an automatic roaming capability can be provided to subscribers. However, this would require an extra set of control channels and dedicated data links between the analog and digital MTSOs. If these MTSOs are manufactured by different manufacturers, then both MTSOs must have the ability to provide automatic roaming with the other MTSO. Furthermore, with different exchange codes assigned to these MTSOs, PSTN can route the call appropriately. However, one disadvantage of this approach is that the customers who migrate to digital will have to change their cellular phone number. Call processing efficiency will also suffer in spite of the interconnection between these two MTSOs.

A variation of this scheme is shown in Figure 7.10(c) where the analog and the digital cell sites are independent of each other and interconnected at the same MTSO. In this configuration, the MTSO makes the decision as to whether the calling subscriber is digital or analog. Routing to the correct cell site is based on the location of the subscriber as well as the type of the subscriber. Unlike the above case of integration, this scheme provides better switching utilization of the MTSO, and thus provides a higher rate of

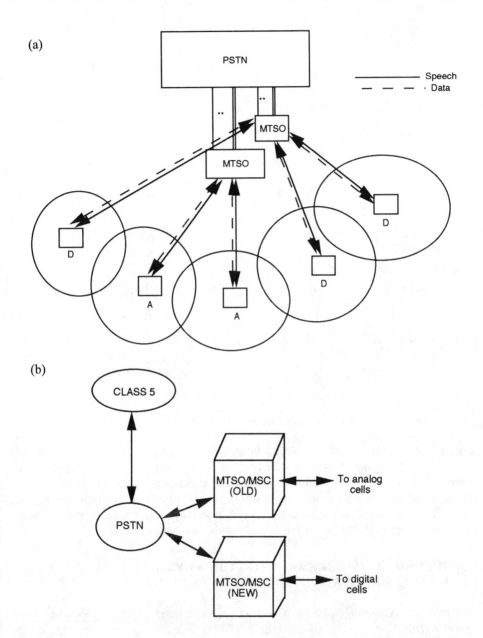

Figure 7.10 (a) Analog and digital systems interconnected at PSTN (overlay approach). (b) Two independent MTSOs serving analog and digital systems separately.

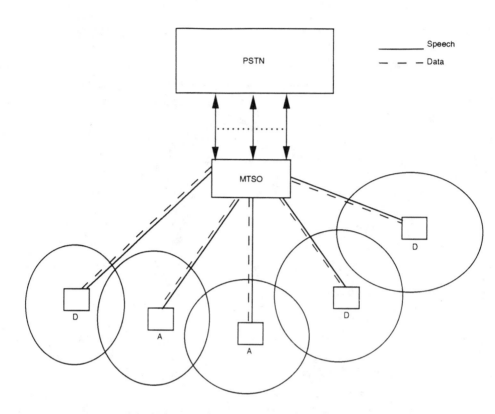

Figure 7.10(c) Analog and digital systems interconnected at MTSO (overlay approach).

return to the system operator. Modifications will, however, be required in the MTSO software to distinguish between digital and analog subscribers. This distinction is also necessary during the handoff process because the digital subscriber should be switched to the digital cell sites, while the analog subscriber should be switched to the analog cell sites. Once again, due to the breaking up of the facilities into analog and digital cells, trunk efficiency to the MTSO is not utilized to the maximum extent.

7.4 TECHNICAL REQUIREMENTS OF THE AMERICAN, GSM, AND JAPANESE SYSTEMS [17, 20–23, 26, 28]

All three digital systems—North American digital cellular (NADC), Japanese digital cellular (NDC), and group special mobile (GSM)—use time-division multiple access (TDMA) and frequency division multiple access (FDMA). As shown in Figure 7.11, these systems packet data at specific times at specific frequencies. Thus, several conversations takes place simultaneously at the same frequency by using different time slots. Systems are also frequency duplexed so that the transmit and receive frequencies are different.

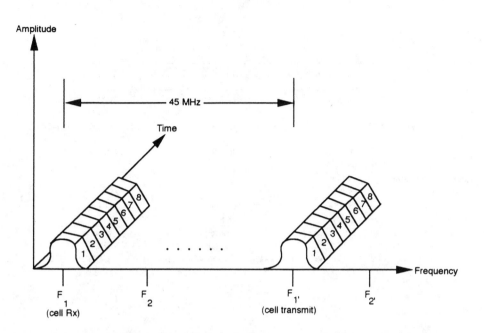

Figure 7.11 Typical TDMA/FDMA frame structure.

Both the American and the European digital systems are based on TDMA technology, but the reasons for adopting digital systems are different. In the United States, the analog system is being replaced due to an acute shortage of capacity, while in Europe the reason is the compatibility between different analog systems. GSM also demands compatibility with ISDN, improved user privacy, and a well-defined open network architecture (ONA). CTIA, on the other hand, does not require ISDN compatibility. It only requires that the data rate be compatible with the existing data communications of the world (up to 9.6 Kbps per user). The differences between European and American systems are summarized in Table 7.3. The technical requirements of the NADC, GSM, and JDC systems are detailed in Table 7.4.

Table 7.3
Difference Between European (GSM) and American (EIA/CTIA) Digital Systems

	GSM	*EIA/CTIA*
System capacity	Twofold increase over current analog, initially	Fivefold increase over current analog system, initially
Coexistence	Replacement of analog systems	Coexistence with analog systems and smooth transition to digital
Data	ISDN compatibility	Multiples of 9.6 Kbps

Table 7.4

Specifications for Different World Systems

	European (GSM)	American (TIA 45.3)	Japanese
Access method	TDMA/FDMA	TDMA/FDMA	TDMA/FDMA
Frequency bands:			
Mobile to cell (MHz)	890–915	824–849	940–956, 1429–1441, 1453–1465
Cell to mobile (MHz)	935–960	869–894	810–826, 1477–1489, 1501–1513
Channel bandwidth (kHz)	200	30	25
Modulation	GMSK	$\pi/4$ DQPSK	$\pi/4$ DQPSK
Bit rate (kbps)	270.833*	48.6	42
Filter	BT = 0.3 (Gaussian)	$\sqrt{}$raised cosine	$\sqrt{}$raised cosine
Voice channel coding	RPE-LTP/ CONVOLUTIONAL 13 kbps	VSELP/ CONVOLUTIONAL 13.2 Kbps (phase 1) 8 kbs (phase 2)	VSELP/ CONVOLUTIONAL 11.2 Kbps
Voice frame (msec)	4.6	20	20
Interleaving (msec)	40	27	27
Associate control channel	Extra frame	Same frame	Same frame
Handoff method	MAHO	MAHO	MAHO
Adaptive equalization	Yes	Yes	Yes
Users per channel	8	3 (phase 1), 6 (phase 2)	3 (phase 1), 6 (phase 2)
Subscriber unit power level (W)	0.8,2,5,8,20	0.8,1.2,3	?
Number of channels	124	832 (phase 1) 1664 (phase 2)	?
Market size	15 million	10–15 million	

*Frequency hopped.

In the United States, it was decided that the American digital cellular standard (ADC), developed by TIA, should be compatible with the present analog cellular system, AMPS. Thus, a 30-kHz channel bandwidth with TDMA was recommended in January 1989 by TIA. This is similar to the analog system in that for each cell served by a radio base station, a number of preassigned frequency/time channels are assigned. The spacing between the carriers in a GSM system is 200 kHz, while in the Japanese system it is kept at 25 kHz and is compatible with their present analog system. Eight time slots carry speech and data in GSM system, while the number of users per channel for both American and Japanese systems is presently three. It is expected that within five years it will be possible to compress the voice and channel coder rate by 50%, and that it will still deliver equivalent quality of voice. Thus, in the long run it is expected that the number of channels per carrier will be six in the NADC and JDC systems. TDMA frames are "physical" frames that are repeated on the radio path.

The bandwidth for GSM systems is 25 MHz, which provides 124 channels with each having a bandwidth of 200 kHz. With 8 users per channel, there are 1,000 actual speech or data channels. The number of channels will double to 2,000 as the half-rate speech coder is introduced. The frequency band used for the uplink is 890–915 MHz (from mobile station to cell site), and 935–960 MHz (from cell site to mobile) for the downlink. With a total bandwidth of 25 MHz in the U.S., and a channel bandwidth of 30 kHz, the total number of carriers will remain at the present level of 832. This will provide a total of 2,500 channels, which could double to 5,000 within five years upon introduction of half rate speech coders. In the proposed Japanese system, in the 800-MHz band the mobile-to-cell transmission frequency is 940–956 MHz, and from cell to mobile the frequency band is 810–826 MHz. This will provide 640 carriers, and with 3 channels per carrier provide a total of 1,920 channels. The number of channels will also double as the half-rate speech coders are introduced. Due to a present growth rate of 50%, it is expected that the present analog cellular system will reach its saturation by 1994–95. In order to cope with this saturation in Japan, a digital cellular system at the 1.5-GHz range has been considered. New bands of frequencies are: 1453–1465 MHz and 1429–1441 MHz (mobile to base); 1501–1513 MHz and 1447–1489 MHz (base to mobile), an Rx/Tx frequency difference of 48 MHz.

The modulation method in GSM is Gaussian minimum shift keying (GMSK), which facilitates the use of narrow bandwidth and coherent detection capability. In GMSK, the rectangular pulses are passed through a Gaussian filter prior to their passing through a modulator, as discussed below. This modulation scheme satisfies the adjacent channel power-level requirement of −60 dBc specified by CCIR. The normalized pre-Gaussian bandwidth is kept at 0.3 (BT produced), which corresponds to a filter bandwidth of 81.25 kHz for an aggregate data rate of 270.8 Kbps. With 200 kHz of carrier spacing, and with this data rate, the spectral efficiency of system is 1.35 bps/Hz (270.8/200). With a bit interval of 3.7 μs, the GSM signal will encounter significant intersymbol interference in the mobile radio path due to multipath (multipath minimal delay spread nearly equal to 3–6 μs in urban areas). As a consequence, an adaptive equalizer is necessary. In the United States, the modulation scheme is linear $\pi/4$ DQPSK, which is a departure from the first-generation constant envelope FM. With a 48.6-Kbps channel data rate over a 30-kHz channel bandwidth, this provides a spectral efficiency of 1.6 bps/Hz, which is about 20% more efficient than a GSM system. Here a raised square-root cosine filter with alpha cutoff of 0.35 is proposed at both the transmitter and receiver, which will provide a spectral null around 16 kHz. Use of raised square-root cosine filters at both ends of the channel provides a matched filter configuration. It should be noted that the lower rolloff reduces the bandwidth, but at the same time also reduces the power amplifier efficiency. Similar to the U.S. system, the modulation scheme in a Japanese digital system is also $\pi/4$ DQPSK with a filter rolloff factor of 0.5, which is once again a compromise between adjacent channel interference and power amplifier efficiency. In order to avoid extreme complexity in amplifier design, the gross bit rate is kept at 42.0 Kbps, which is slightly lower than that of the American system. Further elaboration of this is provided in Section 7.7.

With 270.8 Kbps divided among eight users in GSM, the per-user data rate is 33.85 Kbps. A speech coder is a regular pulse excitation with a long-term predictor for full-rate speech that converts speech to 13 Kbps. Every 20 ms, speech encoder releases 260 bits of raw speech data and the associated channel coder provides 456 bits at the output. This amounts to a data rate of 22.8 Kbps (456/0.02) per user. An interleaver collects the speech information generated over 40-ms time period ($2 \times 456 = 912$ bits) and interleaves over eight frames. The interleaving distance of 8 is found adequate to randomize the burst errors due to multipath. In other words, errors that are clustered in the sequence of 912 channel bits tend to be randomly dispersed in the bit stream presented to the decoder at the receiving end. With an interleaving duration of 40 ms and an interleave distance of 8, the frame length is theoretically limited to 5.0 ms. Thus, the actual frame length is kept at 4.6 ms. The details of this framing are discussed in Section 7.5.1.1. In the United States, with a total data rate of 48.6 Kbps (1,944 bits in 40 ms) divided among three users, the per-user data rate is 16.2 Kbps. As a part of this, 13.2 Kbps represents the coded voice and channel correction information and the balance of 3 Kbps represents system control information. The NADC system assumes a frame of 40 ms with 6 slots, each of 6.667-ms duration. It is expected that within five years, voice and channel coder rates can be compressed by 50%, and will still deliver equivalent speech quality. This would result in a gross data rate requirement of only 8 Kbps per user time slot. As shown in Section 7.3.1, with a 2-dB reduction in the required C/I ratio the system capacity will increase by nearly 1.4 compared to analog cellular. When combined with a sixfold increase of users on the same 30-kHz channel, the long-range channel capacity will increase by more than eight times when compared to first-generation analog cellular systems. This capacity increase by about tenfold will be realized in two seamless transitions, namely, the transition from analog to three-channel digital and then to six-channel digital. The details of framing and slot format are described in Section 7.6.1.3. The Japanese digital cellular radio frame is similar to American. A frame is 20 ms and is made up of three time slots, each having a 6.667-ms duration. Two of these frames placed together represent an American frame duration of 40 ms. There are a total of 280 bits per time slot providing a total bit rate of 42 Kbps. The individual bit rate of 14 Kbps is divided between coded speech (11.2 Kbps) and control information (Kbps).

There are five different categories of mobile telephone units specified for the European GSM system: 0.8W, 2W, 5W, 8W, and 20W. The power level can be adjusted to vary between 3.7 mW and 20W. To optimize cochannel interference, each base station (BS) individually directs each mobile station (MS) to use the minimum power setting necessary for reliable transmission. This setting is determined by the BS and provided to the MS. The mobile power units for American system are substantially reduced and are 0.8W, 1.2W, and 3W. In all these systems, mobile-assisted handoff is used. The level can be adjusted to vary between 2.2 mW and 6W. In all three systems, a mobile-assisted handoff scheme is used. The remaining part of this chapter is an elaboration on these three digital systems of the world, based on the information in Table 7.4.

7.5 EUROPEAN GSM SYSTEM [20–24]

Based on the tests conducted during the experimental phase of the system evaluation, GSM decided to select a digital TDMA/FDMA scheme. Speech is digitally encoded at a low data rate and encrypted. To provide robustness under fading, forward error correction and interleaving are provided. An efficient modulation scheme with a low C/I ratio is adapted. For further interference reduction, frequency hopping, transmitting only during active speech, and power control have been included in the system. The main features of the system are:

- Each TDMA channel has eight full-rate channels on a single carrier.
- Full-rate digital speech is at 13.0 Kbps, with provisions for a half-rate capacity for the future.
- Authentication and encryption for privacy and protection against unwanted detection.
- Overall transmission delay < 80 ms.
- Optional use of frequency hopping to provide diversity and to randomize the interfering effects.
- Constant envelope GMSK modulation with good performance under cochannel interference.
- Transmission scheme capable of taking advantage of voice activity detection (VAD) by means of discontinuous transmission (DTx) to reduce interference level.

Details of each of these are provided in Section 7.5.1.

7.5.1 Technical Details of the GSM System

A block diagram of a digital system is shown in Figure 7.12(a). Analog speech is digitized and passed through a residual pulse-excited linear predictive coder, which takes a 20-ms sample of speech and converts it into 260 bits at the output. Thus, the encoded data rate is 13 Kbps. The coder passes the residual information (most important bits) through a half-rate channel encoder and provides 456 bits at the output in 20 ms. Thus, the convolutionally coded data has a rate of 22.8 Kbps. The 456 bits of data is interleaved with an interleaving depth of 8 to combat the effects of burst noise. At this stage, data is arranged in a frame along with other overhead bits. Interleaved data is passed through a special form of minimum shift keying (MSK) modulation known as GMSK. Here, the data is first filtered by Gaussian filter before applying it to a modulator. The modulated data passes through a duplexer switch where filtering is provided to isolate between the transmitted and the received signal. On the receive side the signal is demodulated, and deinterleaved before the error correction is applied to the recovered bits. A demodulator

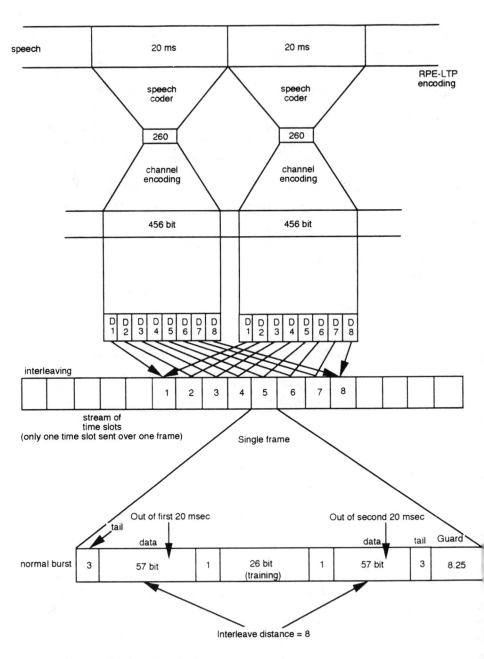

Figure 7.12(a) The digital mobile radio system.

box down-converts the signal using the signal from a frequency synthesizer. Only one synthesizer per transceiver is required as the equipment does not transmit and receive simultaneously. In order to optimize the performance of the receiver, the demodulator contains an equalizer where distortions due to multipath are removed. The equalizer is a high-speed digital signal processor.

7.4.1.1 Speech and Channel Coding [23–25,28]

To understand how the basic information is compiled into a frame, consider the speech coding mechanism shown in Figures 7.12(b,c,d). Basic speech is sensed by a coder for 20-ms segments and produces 260 bits at the output. Thus, the output data rate of a speech

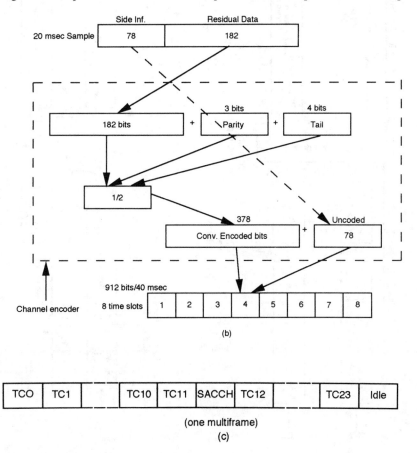

Figure 7.12(b,c) (b) Speech coding in GSM. (c) Channel coding in GSM.

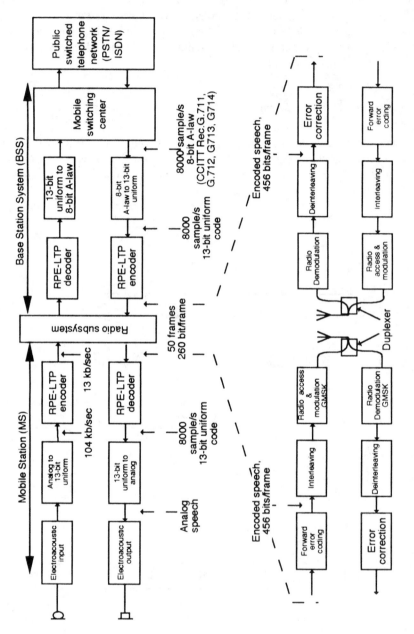

Figure 7.12(d) Time sharing of the GSM traffic and control channels. Source: [23].

coder is 13 Kbps. Every 20 ms the channel coder releases 456 bits. The 260 bits input to the channel coder are divided into 182 bits of residual (important) data and 78 bits of side information. Side information remains uncoded and is simply bypassed. Three bits of parity and a four-bit tail are added to 182 bits of residual before the bits are passed through a half-rate convolutional coder. A 40-ms segment of speech is interleaved with a distance of 8 to combat errors in the multipath surroundings. This distribution of 2×456 bits over eight frames is shown in Figure 7.12(c). Thus, a total of 114 bits of speech data per user is sent per frame and the total information is sent over eight frames. Since a total of 456 bits are generated in 20 ms, the user data rate is 456/0.02, or, 22.8 Kbps, which includes 13 Kbps of raw data and 9.8 Kbps of parity, tail, and channel coding. One can also compute this rate by noting that the user data is sent over 24 frames out of a total of 26 frames per multiframe, lasting 120 ms. This amounts to a total number of user-transmitted bits of 2,736 bits (114×24) over a period of 120 ms or a user data rate of 22.8 Kbps, which is the same as that computed above. In order to support various monitoring and control functions, each speech traffic channel is associated with a so-called slow associated control channel (SACCH), which is transmitted over one frame of a multiframe, as shown in Figure 7.12(d). Thus, the data rate of a SACCH is 0.95 Kbps (114/0.12 = 950 bps). The SACCH message contains 184 information bits in 480 ms, or 383 bps. Error control coders process this information to produce 456 bits, which are interleaved and distributed over four time slots. With one time slot per multiframe, it takes four multiframes to send the message. This corresponds to a data rate of 0.95 Kbps, as arrived at above (456/480). The multiframe also has an idle frame for future use. Thus, the bit rate for guard time, ramp-up time, synchronizing sequence, training sequence, and accounting for one idle frame per multiframe is 10.1 Kbps {[42.25(guard + ramp + synch) $\times$ 25 + 156.25(idle frame)]/0.12}. Thus, the total user data rate can be divided into following components:

- Speech code, 13.0 Kbps;
- Parity, tail, and channel;
- Encoding, 9.8 Kbps;
- SACCH, 0.95 Kbps;
- Guard time, ramp-up, synchronization, 10.1 Kbps;
- Total data rate per user, 33.85 Kbps.

Since the total number of users per RF channel is eight, each 200-kHz channel carries data at the rate of 270.8 Kbps.

7.5.1.2 Delay Requirement

Transmission delay in any network has to be controlled in order to avoid speaker annoyance, which is dependent on the magnitude of the returning signal and the amount of delay involved. On short connections, delay is small enough that the echo appears as a sidetone and the talker feels this as a natural coupling to the ear. As the round-trip delay increases, attenuation to the echoes are desired to reduce annoyance to the talker. For this reason, a delay requirement has been imposed in the GSM network as follows [23]:

- Speech coding algorithmic delay < 20 ms;
- Interleaving delay < 37 ms;
- Processing delay < 8 ms;
- Other delay in the radio-subsystem < 15 ms;
- Overall delay ≈ 80 ms.

Ideally, one can assume that higher the delay in the speech coding, better the quality of the speech, or alternatively, less the speech data rate for the same quality of service.

As discussed above, the 2 × 456 bits of coded and error-protected speech frame are interleaved by a factor of 8. With a frame duration of 4.6 ms, the total interleaving delay amounts to 37 ms. This delay is the tradeoff between speech quality reduction due to burst noise caused by multipath and annoyance due to delay. On the other hand, an even deeper interleaving delay would cause speech quality improvement, but this was decided against because of annoyance to the receiver.

Processing time in a CMOS VLSI mobile transceiver is the function of clock rate. A higher clock rate does reduce the processing time but increases the power requirement. On the other hand, increased processing time is required for lower power consumption. The delay figure of 8 ms is merely to satisfy the power consumption requirement of hand held mobile unit.

Assuming a miscellaneous delay of 15 ms in the radio path, the overall delay of 80 ms represents the outcome of the various tradeoffs described above. The value of 80 ms is that it is supposed to be short enough to avoid any conversational delay problems for majority of calls. For connection to PSTN, an 80-ms delay will come as an add-on to the existing delay. Consequently, there is a need for echo control in all calls originated or terminated by a GSM network. For this reason it is proposed to include an echo canceler at the PSTN interface of the GSM system to remove any echoes being returned from the PSTN to the mobile user. The echo control is not supposed to cause any problems to the mobile, since the radio system is essentially a four-wire telephone system. Calls to the satellite system should be avoided as the CCITT requirement for a total delay of 400 ms may be exceeded.

7.5.1.3 Data Interleaving

In order to combat the effects of error due to interference and noise, error correction techniques are used. The redundancy introduced due to error correcting codes increases the data rate. For example, the raw data rate, due to speech coding, is only 260 bits over a period of 20 ms, as shown in Figure 7.13. However, after block and channel encoding, the number of bits is increased to 456/20 ms. Error correcting codes are better at correcting

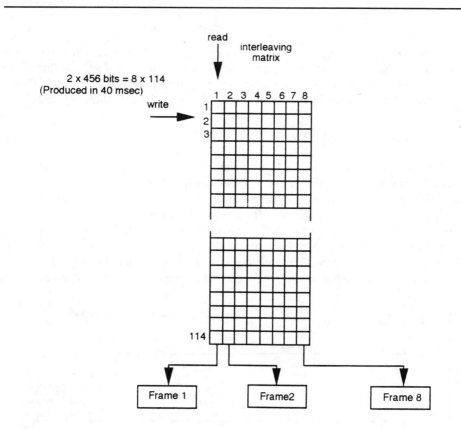

Figure 7.13 GSM interleaving approach [28].

randomly distributed errors, but do not work well when the errors occur in bursts. For bursty errors, interleaving of the data [28] is recommended. Interleaving is the process of recording and distributing data bits in a different order from which they are generated. Recording data, bits distribute the burst errors in the receiver, thereby helping in speech recovery in the presence of multipath errors. A simple example of bit interleaving, with the interleaving depths of three, is shown below. Obviously, as this depth is increased, original speech has more protection against burst errors.

- Original coded speech bit order: 1, 2, 3, 4, 5, 6, 7, 8, 9, 10, . . . , 18, ...
- Transmitted order: 1, 4, 7, 10, 13, 16, 19, 22, 25, 28, ...
- Burst of errors: _1, _4, _7, 10, 13, 16, 19, 22, 25, 28, ...
- Reordered bits: _1, 2, 3, _4, 5, 6, _7, 8, 9, 10, . . . , 14, ...

Due to burst error, the adjacent bits 1, 4, and 7 are affected. However, in the reordered bits they are separated. Since the error correcting codes improve performance for distributed errors, reordered bits will have a higher probability of being correct at the receiver.

The speech encoding process is shown in Figure 7.12(b). Speech produced over two adjacent 20-ms intervals are first interleaved into eight time slots, and then these slots are transmitted over eight frames. The rearrangement approach is shown in Figure 7.13. In this scheme, each frame carries speech from the adjacent 20-ms segments.

7.5.1.4 Frame Structure [23–25, 28]

The GSM system has a total available bandwidth of 25 MHz divided among 124 channels of 200 kHz each, as detailed in Table 7.4 and shown in Figure 7.14. Each of these 124 channels operates at a different carrier frequency. An individual channel is time-division multiple accessed by users at different locations within a cell site. The frame duration is 4.615 ms divided among 8 time slots. Each of these time slots is a physical channel occupied by an individual user. Each time slot or physical channel carries control and traffic data in the burst form. The time duration of an individual channel is 0.577 ms. Higher order frames such as multiframe consist of 26 frames and have a duration of 120

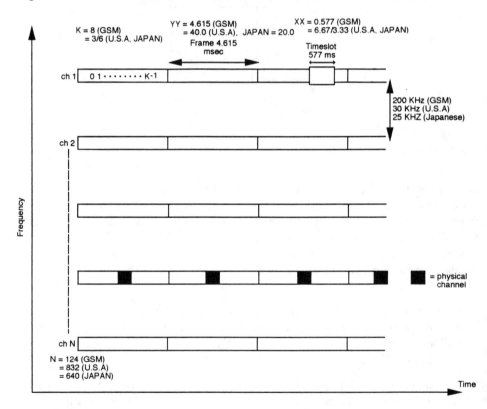

Figure 7.14 Time and frequency division multiple access for cellular radio.

ms (26 × 4.615 ms). One superframe consists of 51 traffic multiframes or 26 control multiframes. The highest order frame is hyperframe and consists of 2,048 superframes or 2,715,648 frames (2048 × 51 × 26). The time duration of a hyperframe is 3:28:52.76. To organize the information transmitted on each carrier, GSM defines several time intervals, ranging from 0.9 μs (a quarter of one bit in the guard interval of the basic frame) to a hyperframe interval of over 3 hours. Details of frame, multiframe, superframe, and hyperframe are provided in Figure 7.15.

7.5.1.5 Speech and Traffic Channels

The radio subsystem provides a certain number of logical channels, which can be separated into two parts: speech channels and control channels. On these channels, a physical layer (layer 1) provides the actual transmission of information and supports other higher layers. Functions for other higher layers will be discussed in Chapter 8. Here we only stress on the physical layer and the basic functions performed by these channels. Traffic channels are exclusively used to transfer two types of user information: speech and data. Two data rates, full rate at 22.8 Kbps (encoded) and half rate at 11.4 Kbps, are used, as shown in Figure 7.16. The main function of control channels is to transfer signaling information. They are subdivided into three categories: broadcast control channels; common control channels, which are shared among many users; and dedicated control channels assigned to a single mobile user.

The broadcast control channel (BCCH) is a unidirectional channel from the base to the mobile and is used to broadcast system information regarding the mobile's present cell as well as neighboring (up to 16) cells. BCCH may include a frequency correction channel (FCCH) to allow MS to accurately tune to BS and a synchronization channel (SCH) to provide frame synchronization data to the mobile.

A common control channel (CCCH) is used for both uplink and downlink communications. For uplink communications, this channel is used for random access by the mobile, while for downlink communications the channel is used for paging and assigning a dedicated control channel to the mobile. No acknowledgment is provided from either sides in this operation.

Dedicated control channels are either autonomous (standalone) control channels or associated to a dedicated channel. The standalone dedicated control channel (SDCCH) is an autonomous control channel. The slow associated control channel (SACCH) is assigned to a traffic channel or to an SDCCH. This channel in particular is used for power and frame adjustment, control information, and for measurement of data. The fast associated control channel (FACCH) is assigned to a traffic channel and is achieved by frame stealing. FACCH information bursts are identified by a ''flag'' known as a stealing flag. For an active traffic channel, this is used for handover messages.

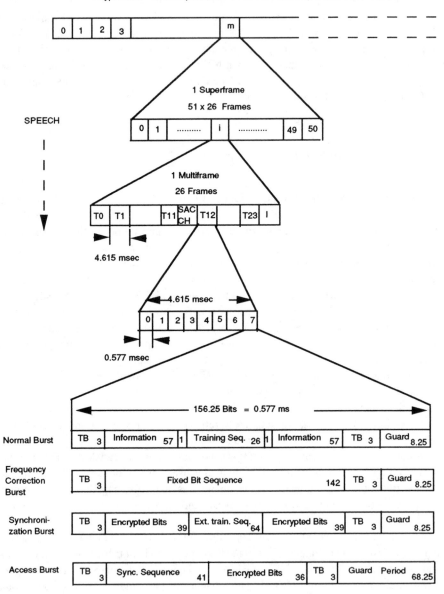

1 Hyperframe = 2048 Superframe = 2 715 648 Frames (3H 28m 53 s 760 ms)

| 0 | 1 | 2 | 3 | | m | |

SPEECH

1 Superframe

51 x 26 Frames

| 0 | 1 | .... | i | .... | 49 | 50 |

1 Multiframe

26 Frames

| T0 | T1 | | T11 | SAC CH | T12 | | T23 | I |

4.615 msec

4.615 msec

| 0 | 1 | 2 | 3 | 4 | 5 | 6 | 7 |

0.577 msec

156.25 Bits = 0.577 ms

Normal Burst

| TB 3 | Information 57 | 1 | Training Seq. 26 | 1 | Information 57 | TB 3 | Guard 8.25 |

Frequency Correction Burst

| TB 3 | Fixed Bit Sequence 142 | TB 3 | Guard 8.25 |

Synchroni- zation Burst

| TB 3 | Encrypted Bits 39 | Ext. train. Seq. 64 | Encrypted Bits 39 | TB 3 | Guard 8.25 |

Access Burst

| TB 3 | Sync. Sequence 41 | Encrypted Bits 36 | TB 3 | Guard Period 68.25 |

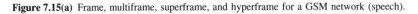

Figure 7.15(a) Frame, multiframe, superframe, and hyperframe for a GSM network (speech).

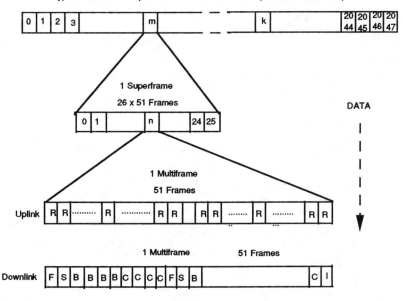

Figure 7.15(b) Frame, multiframe, superframe, and hyperframe for a GSM network (data).

7.5.1.6 Modulation and Frequency Hopping [29–30]

Spectrally efficient digital modulation requires prefiltering the digital data before modulation. By rounding off the sharp edges of the data bit, the signal bandwidth after modulation is substantially reduced. The GMSK system is a specific type of MSK system where the rectangular data bits are prefiltered by a low-pass Gaussian filter. GMSK has the following features that make it suitable for mobile radio applications:

- Constant envelope, which allows the utilization of efficient transmitters using power amplifiers in a saturation mode (class C mode of operation).
- Compact output power spectrum, which means narrower main lobe and lower sidelobe peaks keep the adjacent channel interference at a low level.
- Good bit error performance.

With a normalized BT product of 0.3 (B = Gaussian filter of 3-dB bandwidth and T is the data symbol time), GMSK nearly satisfies the adjacent channel interference requirement of -60 dBc. As shown in Figure 7.17 a normalized filter bandwidth of 0.24 is sufficient to meet the spectral containment requirements of the adjacent channel. Here, the x axis is the normalized frequency difference, which is equal to $(2\Delta f)/(1/T)$. With Δf = 200 kHz and $T = 274 \times 10^3$ bps, the normalized frequency difference is $(2 \times 200 \times 10^3)/(274 \times 10^3) \approx 1.5$.

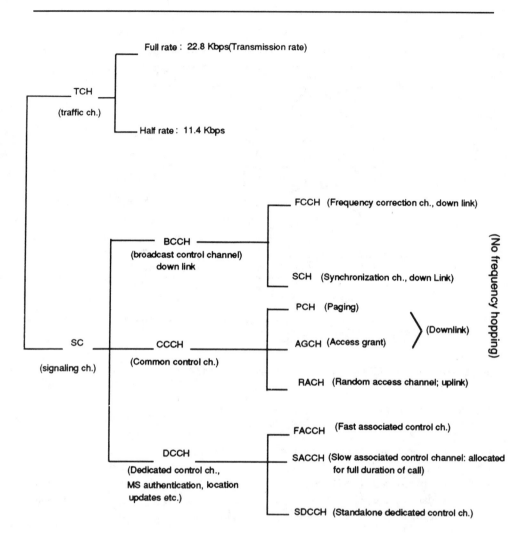

Figure 7.16 Logical channels in GSM.

In view of this, GMSK modulation has been recommended for the GSM system. A detailed discussion of this modulation is beyond the scope of the present discussion. We will confine ourselves to differential GMSK, which is a preferred implementation under fading channels where exact coherent detection is difficult to achieve.

The GMSK transmitter consists of a differential encoder, a Gaussian premodulation LPF, and an FM modulator with a modulation index of 0.5, as shown in Figure 7.18. Both one bit and two-bit differential detectors are shown in Figure 7.19(a,b). The BER curves of one and two-bit differential detection of GMSK with different BT products are

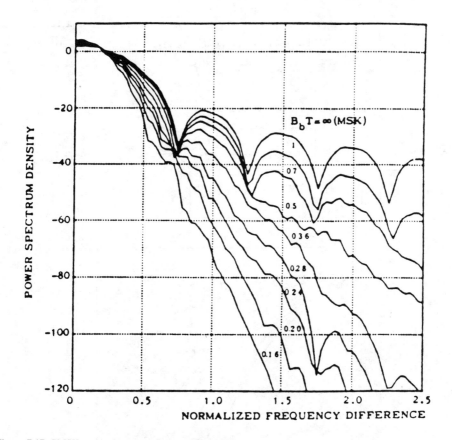

Figure 7.17 GMSK power spectrum. Source: [30, p. 15].

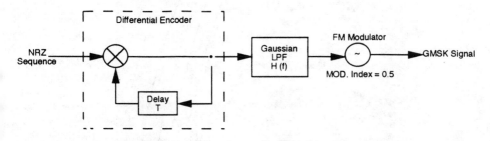

Figure 7.18 GMSK modulator.

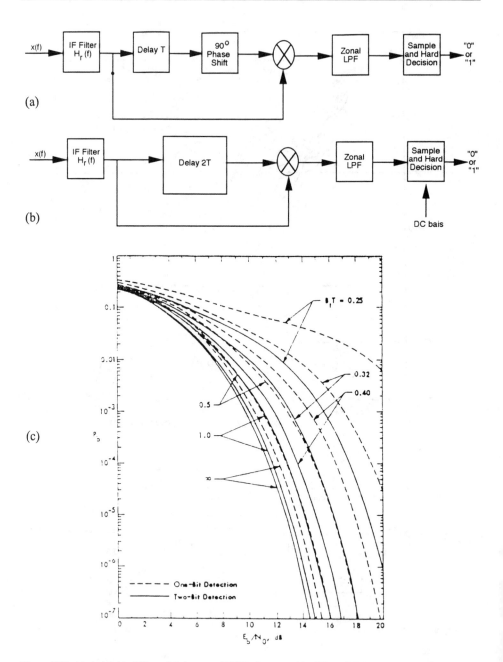

Figure 7.19 (a) A one-bit differential detector GMSK; (b) a two-bit differential detector; (c) performance of one and two-bit differential detection.

shown in Figure 7.19(c). Since the collected energy over $2T$ intervals is larger, the performance of a two-bit differential detector is always better than that of a one-bit.

7.5.1.7 Frequency Hopping

A GSM system base station will use frequency hopping (each time slot is transmitted on a different frequency) optionally. Thus, it will be mandatory for mobiles to have frequency hopping capability. There are two main reasons of using frequency diversity. First, to provide interference diversity, and second, to provide error correction capability. Frequency diversity will provide interference reduction on both non-BCCH control channels and on traffic channels (TCCHs). Hopping is performed on a TDMA frame-to-frame basis, that is, every 4.6 ms. Hopping patterns are selected such that the hops are orthogonal (coherent) within the same cell and uncorrelated within different cells. Orthogonality within the same cell assures no collisions within a cell. However, collisions may occur between different cells. Hopping is done in such a way that, according to the calculated sequence on both ends of the link, the mobile sends and receives each time slot on a different frequency. Thus, this is an example of low-frequency hopping where the carrier frequency remains constant for a duration of one frame, as shown in Figure 7.20. This allows all data bits in a frame to be at the same frequency. In this case, processing gain equals 10 log(hopped bandwidth/information bandwidth). Frequency-hopping assignments provide both the frequency-hop sequence and the start time with respect to the hyperframe sequence number.

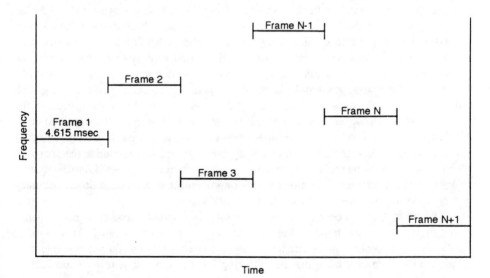

Figure 7.20 Frequency-hopped signal in GSM.

At least at the beginning of the system operation, frequency hopping will be used occasionally in response to a significant interference. It is expected that as time passes, frequency hopping will be used on a regular basis. Frequency diversity will also provide error correction, especially useful at low speed. This will improve the performance of the system for users encountering fading of a longer duration. BCCH and CCCH are not hopped.

7.5.1.8 Discontinuous Transmission

In order to maximize the spectrum efficiency of the GSM system, voice-activated transmission was adopted. The basic technique, termed discontinuous transmission (DT_x), is to switch the transmitter on only for those periods when there is active speech to transmit. In this way the average interference in the air will be reduced, thus allowing a smaller frequency reuse cluster size, and thus a higher capacity. It was found that given an average voice activity of 50%, the spectrum efficiency can easily be doubled. Therefore, transmitting when no information is passed was inefficient both spectrally and in terms of power consumption. Such transmission not only increases the noise level of the GSM spectral region, but also wastes the battery power of handheld units, which is why GSM voice uses DT_x. Implementing an adaptive threshold voice activity detector (VAD) algorithm, the GSM equipment determines when a speaker is not talking and interrupts transmission. At the receiving end, the GSM equipment detects the DT_x and fills in the empty frames with "comfort noise." The basic problem of DT_x, is the potential degradation of the voice performance by speech clipping and by noise contrast effects. The design of a speech detector has to weigh the risk of clipping the talk spurts against the risk that noise is incorrectly classified as speech. Clipping of the speech is a serious impairment that will reduce the overall quality of the system. On the other hand, false classification of noise as speech must be minimized since an increased activity factor would increase the interference on the air. When the VAD is used to switch the transmitter on and off, the effect will be a modulation of the background noise at the receiving end. When speech is detected, the background noise is transmitted together with the active speech to the receiving end. As the speech burst ends, the connection is broken and the received noise drops to a very low level. This step modulation of noise is perceived as very annoying and may, in high contrast cases, reduce intelligibility. In order to minimize the annoying effects of the receive-channel noise modulation, a comfort noise is inserted into the receive channel when the connection is broken. This improvement is dependent on the accuracy by which the inserted noise matches the transmitted noise.

The noise transmitted from a mobile station is normally acoustical noise that is picked up by the microphone of the handset or loudspeaking terminal. This noise is significant from location to location and also a function of time during a single call. Solutions to these problems, using the following functions, are shown in Figure 7.21.

- A VAD algorithm, which utilizes a number of the parameters of RPE-LTP speech codes in its discrimination process of speech with noise.

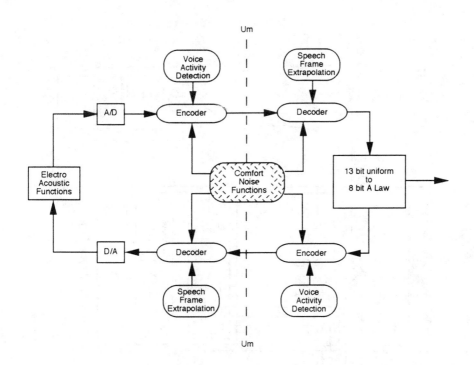

Figure 7.21 Speech processing functions.

- A mechanism that estimates background noise characteristics at the transmit side.
- An inband transmission scheme for updating the comfort noise parameters utilizing the normal speech codes frame.
- A mechanism for generation of constant comfort noise between updates.

7.5.2 System Architecture

As shown in Figure 7.22, the system architecture [20–24] of a GSM mobile radio network mainly consists of two subsystems: a base station subsystem and a network subsystem. Additionally, an operations and management center, which links all the components of the network, performs the functions needed to operate and maintain the radio part of the GSM network.

A base station subsystem consists of base transceiver systems (BTS) and base station controllers (BSC). Each cell is controlled by several BTSs, and several BTSs may be under the management of one BSC. The BSC and its BTSs constitute a base station subsystem (BSS), which is connected to a mobile service switching center (MSSC). A BSS on one side interfaces to the mobile over the air interface and on the other side it provides the interface to the MSSC over PCM channels. The speech between the BSC

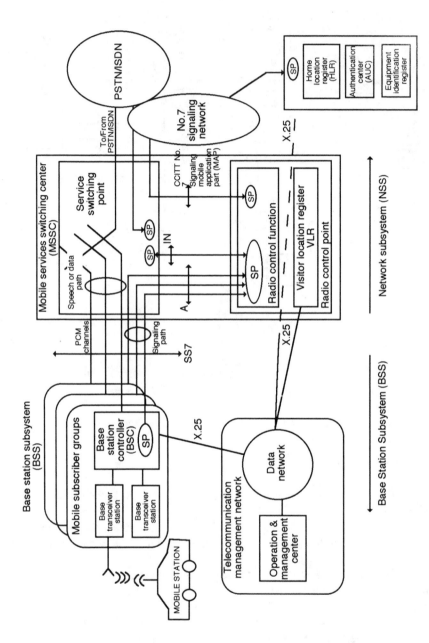

Figure 7.22 GSM mobile radio network. Source: [20, p. 391].

and the MSSC can be carried either on 16 Kbps or 64-Kbps links. In the first case, the coding of speech is the same as on the radio interface, with the addition of some data bits for synchronization and transcoders at the MSSC site. This allows up to 120 channels of parallel speech on a single 2-Mbps link, thereby reducing substantially the cost to the operator. In the second case, speech is usually transcoded from 64-Kbps A-law to GSM 13-bit linear coding. The MSSC is the core of the network and controls one or many BSSs, mainly performing normal switching functions. They provide the switching functions for calls when only mobile subscribers are involved as well as when subscribers in the fixed (PSTN and ISDN) and packet-switched networks are involved. The MSSC is the point of interconnection between the GSM and ISDN/PSTN networks. An important feature of the GSM system is the automatic international roaming, which means that the mobile station is able to make and receive calls anywhere in the GSM service area without any special actions taken by the mobile subscriber, that is, anywhere in those countries implementing the GSM system. This implies that the mobile station must update its location to GSM network and the network must be able to handle this information effectively. The location information is handled by two different data bases, a home location register (HLR) and a visitor location register (VLR).

The HLR is the database that contains all data concerning the subscription of the mobile subscriber, that is, their access capabilities, subscribed services, and supplementary services (or features). The HLR also contains information about the VLR that is currently handling the mobile station. When the mobile changes location, the HLR is updated accordingly. In addition, the HLR provides the MSC with information about the MSC area where the mobile is actually located (dynamic data) to allow incoming calls to be routed immediately to the called subscriber. The VLR stores all the information about mobile subscribers that enter its coverage area, allowing the MSC to set up the incoming and outgoing calls. It can be considered a dynamic database of the subscriber that exchanges considerable amount of data with the HLR.

The mobile station is always handled by a VLR, which is associated with the geographical area where the mobile is currently roaming. When there is an incoming call for the mobile, the HLR is interrogated about the present address of the VLR. The call will then be routed to the correct MSC, which extends the call to mobile through the BSS over a radio path.

The authentication center (AUC) stores information that is necessary to protect communications through the air interface against any intrusions to which the mobile is vulnerable. The legitimacy of the subscriber is established through authentication and ciphering, which protects the user information against unwanted disclosure. Authentication information and ciphering keys are stored in a database within the AUC, which is protected against unauthorized access. We shall discuss this comprehensively in the following paragraphs.

There are two types of identification performed by the GSM network. The first one, connected with the subscriber, is known as the international mobile subscriber identity (IMSI) and has nothing to do with the equipment. This is strictly subscriber-related information and is stored on the ''smart card'' to enable the subscriber to make calls from taxi cabs and other public places and be billed directly. Thus, the subscriber may not necessarily subscribe for a· mobile unit. The mobile station itself includes its own identity information, known as the international mobile station equipment identity (IMEI), which is used to prevent the use of mobile stations that have been stolen or are not of the type approved by the system.

From the above discussion, it should be clear that a large amount of information is exchanged between the mobile and the network that is processed at the MSSC and does not necessarily lead to the establishment of a speech path. This also implies that the mobile station must update the system about its location when in the idle mode. Thus, the signaling path is completely separate from the speech path. The signaling exchange between the MSSC and the network is more separate than that between the MSSC and the mobile. As shown in Figure 7.22, the radio control function directly exchanges and processes the signaling information with the base station controller over dedicated semipermanent links, which pass through the service switching point. Similarly, the radio control function exchanges and process the signaling information with the network using the mobile application part (MAP) of the CCITT no. 7 signaling standard. The details of signaling requirements, architecture, and the functions of CCITT no. 7 are covered in the next chapter.

7.5.2.1 Location Updating

As explained above, the system must know the location of all active mobile stations in real time as they roam about. As seen in Figure 7.23, each cell is served by one BTS. Each location area is divided into many cells, which may be served by one or more BSCs. A VLR may serve one or more location areas. An inactive mobile is ignored by the system. As soon as the mobile switches its power, it retrieves its stored location-area identity and compares it with the one being broadcast within its present cell. If they match, the mobile does not have to do anything as the subscriber is already correctly located. However, if it does not match, the mobile identifies itself by transmitting its IMSI together with the identities of the previous and present location areas. The BSC transmits this information to the associated VLR. The IMSI assigned to each subscriber consists of MCC (mobile country code) + MNC (mobile network code) + PLMN (unique mobile code for subscriber identification (MSIC).

Each time a mobile station moves into a new location area, the corresponding VLR is informed. If both the present and previous areas are served by the same VLR, the mobile station is given a new TMSI and its location is updated in the VLR memory. On the other hand, if the mobile enters a new VLR area, its HLR, the old VLR, and the new VLR are informed. The old VLR erases the data for the mobile and the new VLR records the relevant parameters needed to process calls.

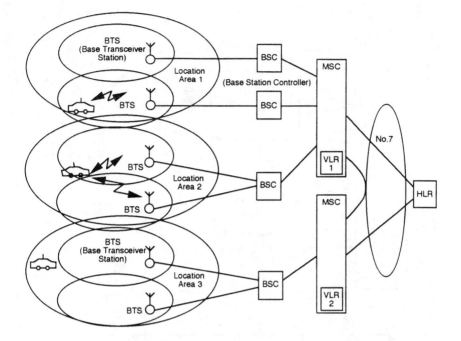

Figure 7.23 Mobile station location identification. Source: [20, p. 394].

7.5.2.2 Different Types of Calls: Incoming Call

Figure 7.24 shows the sequence of steps for incoming calls [20–24] to the mobile station. To call a mobile subscriber, the calling party simply dials the subscriber's ISDN or PSTN number. The initial address of the HLR is derived from the ISDN or PSTN number, designated as step 1 in the figure. The call is routed to the gateway MSC, which interrogates the HLR to obtain the address of the MSC within the area where the subscriber is currently located. Steps 2 and 3 cover these actions. The gateway MSC sends the initial address message to the MSSC, which in turn retrieves the location of the subscriber from its VLR. Steps 4 through 6 cover this action. The mobile station is then paged through all base stations (BSS) that are served by that MSC, since the exact location of the mobile is uncertain. Steps 6 and 7 cover the paging commands to the base station and mobile. This uncertainty reduces the frequency of location updates for the mobile. The paging to the mobile is done by the TMSI over the common control channel to determine in which cell the mobile is presently located. The mobile responds to paging to the base station, as shown in step 8. The base station in turn provides this information to the MSSC (step 9). The VLR also authenticates the called mobile station and prepares for traffic channel encoding. The called party is alerted while the originating party is being informed that the mobile station is being called. When the mobile station answers, a traffic channel is

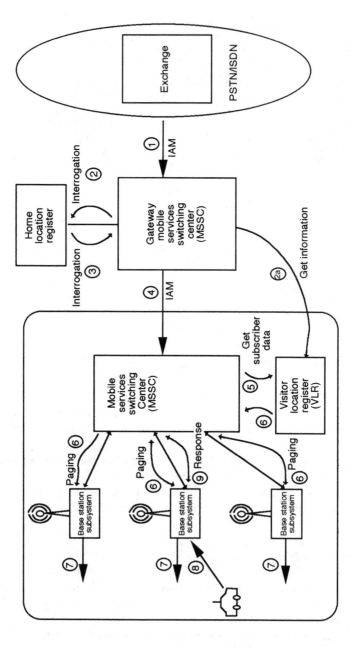

Figure 7.24 Steps for incoming calls to mobile. Source: [21, p. 402].

selected and connection is established to the previously established incoming circuit. At this point, the call is declared successfully completed.

7.5.2.3 Handover

It has been pointed out in Section 7.4 that the system performs mobile assisted handoff. This means that the mobile measures the signal strength of its own cell and sixteen other adjoining cells. When a mobile travels from one cell to another during a conversation, its measurement will indicate that the signal is getting weak in the present base station to which it is linked. The handoff process is shown in Figure 7.25. Assume that the initial mobile call is established through MSC (a) to BSC (1), through link 1a to cell 1. If the measurements by the mobile and the BSC indicate that the adjoining cell 2 is a better choice, the BSC selects link 1b and a new voice channel and informs the mobile to tune to this new channel. The new speech path becomes MSC (a) to BSC (1), through Link 1b to cell 2. The initial link 1a is released. Further movement of the mobile may bring the mobile to cell 3, which is controlled by a different base station but still within the same MSC. The process is the same, but the new path becomes MSC (a) to BSC (2), through link 2a to cell 3. Finally, the mobile may come to a cell that is controlled by a different BSC and MSC, for example MSC (3). In this case MSC (a) will extend the fixed telephone link to the new MSC (3), but retain supervision of the call. The MSC (a) delegates activation and deactivation to MSC (3) during future handover to a new MSC. The voice channel path in this case becomes MSC (a) to MSC (3) to BSC (3), through link 3a or 3b to the mobile. When the mobile station enters a cell that depends on another MSC to MSC link (like link 4), a new transit link is prepared by MSC, in this case link 4 (from MSC (a)), and replaces the old link 3 between itself and old MSC (3).

7.5.2.4 Authentication and Ciphering

Authentication identifies the right mobile and ciphering prevents listening-in by third parties. The ciphering key encodes both the speech and data on the traffic channel. We shall illustrate the mobile authentication and ciphering mechanism by taking up an example of a mobile originating a call, as shown in Figure 7.26(a). The following sequence of steps takes place at various points in the network.

1. The mobile requests to make an outgoing call. The VLR is informed through the MSSC of mobile's action. The VLR takes over control of the authentication procedure.

2. The MSSC requests the mobile's IMSI and passes it to the VLR. The VLR identifies the mobile station HLR. At this point, the VLR conveys the IMSI together with its own identity and the identity of the current MSSC to the HLR so that the incoming call can be directed to the mobile's present location. This action is shown as step 1 in Figure 7.26(b). Once the HLR has received the subscriber's IMSI, it makes a

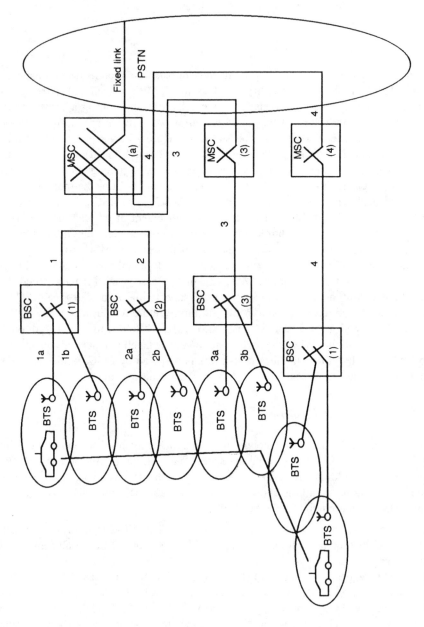

Figure 7.25 Handover of a roaming mobile.

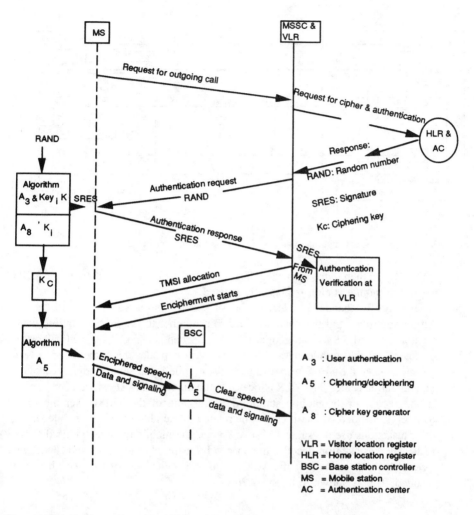

Figure 7.26(a) Sequence of steps for mobile-originated call.

request to the authentication center for the subscriber's ciphering key K_i. This key is present at the mobile station as well as at the authentication center.

3. The mobile subscriber's ciphering key K_i and RAND are used in the authentication center in algorithms A_3, and A_8 to generate signature or signed response (SERS) and the ciphering key K_c for traffic channel encoding. The random number, the signed response, and the K_c make up a triplet for a mobile station, which can be used for further ciphering. The triplet is used for only one communication and then destroyed. The authentication center calculates several triplets and sends them to

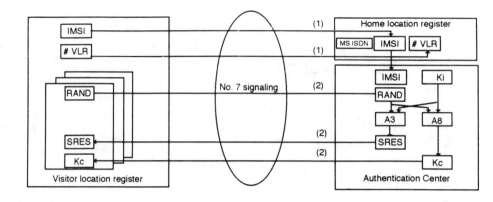

Figure 7.26(b) Ciphering and authentication keys for mobile stations. Source: [20, p. 393].

the HLR, and then to the associated VLRs, which in turn forward them to the MSSC where the mobile station is presently located.

4. When a traffic channel is selected, the VLR conveys the value of K_c to the base station controller and the value of a random number to the mobile station, which calculates the SERS by using algorithm A3. Speech is encoded by using algorithm A5 and K_c at the mobile station. The VLR will consider the authentication to be complete when the SERS response from the mobile is the same as that received by authentication center. The VLR allocates to the mobile a TMSI, which is transmitted to and record in the mobile station. Transmission of the TMSI over the traffic channel is encoded so that it cannot be related to the identity of the mobile station. All transactions use the same TMSI so long as the mobile remains within the same VLR. The base station decodes the speech by using algorithm A5 and sends clear speech and data signals to the MSC. Thus, the radio part of the speech is only encoded. This is shown in Figure 7.26(c).

7.6 AMERICAN DIGITAL CELLULAR (EIA/TIA IS-54)

Due to the acute shortage of analog cellular capacity in large U.S. cities, it was decided by the EIA/TIA committee (TR45.3) to adapt the TDMA technology. In 1989, the committee issued an interim digital cellular standard. Field tests (lockdown tests) were conducted, in 1991–92, to verify that the base stations and mobile units manufactured by different companies met the IS-54 (system), IS-55 (dual-mode mobile), and IS-56 (base station) standards. A limited number of systems were put into operation last year (1992). It is expected that the system providers will initially allocate a few MHz of bandwidth for digital radio out of the total of 25 MHz available for cellular. By the late 1990s, it is expected that most systems will be digital.

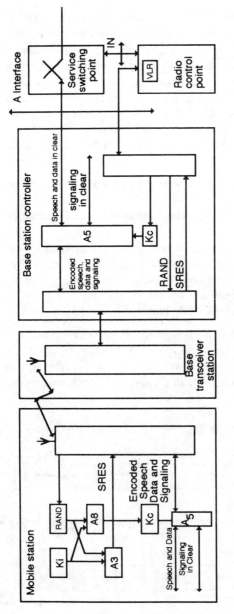

Figure 7.26(c) Authentication and encoding of the traffic channel. Source: [20, p. 393].

7.6.1 Technical Details

The block diagram for digital cellular phone is shown in Figure 7.27. Analog speech is digitized and passed through a vector sum-excited linear predictive (VSELP) coder. The technique uses two codebooks to vector quantize the excitation (residual) signal. The coder reduces the data rate substantially at the expense of a higher demand on the processing speed. The coder models speech into one long-term and two short-term "vector" blocks based on the codebooks. These vector blocks are coded according to standardized "code-books." Since some elements of speech change slowly over time, fewer bits are applied to the long-term block. More bits are used to represent short-term voice as they change rapidly. Coded data passes through a channel coder and a modulator block, where the following functions are performed:

1. Error correction and detection bits are added to most the important data bits.
2. Control channel data is added.
3. Training sequence data is added for applying equalization filtering to account for multipath fading.
4. Guard bits are added to account for timing inaccuracy in the TDMA frame.
5. For full-rate speech coding (phase 1 of development), a channel coder interleaves the data over two time slots for better interference rejection.
6. Channel-coded output data stream becomes input to a $\pi/4$ DQPSK modulator and up-converted to the 900-MHz band.

Finally, the RF-modulated data is transmitted as fixed-size packets.

On the receiving side, the transmitted carrier is received, filtered, and down-converted to an IF frequency where it is filtered again. This IF signal is then passed through a DQPSK demodulator and the data is recovered. a channel decoder strips off the coding

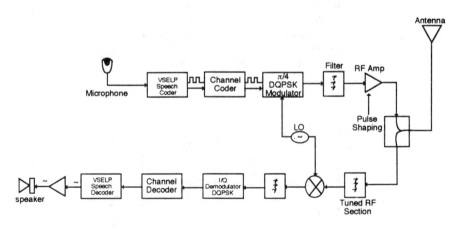

Figure 7.27 Simplified block diagram of mobile transceiver.

information before the burst data passes through VSELP speech decoder for the original speech recovery. Finally, the speech is amplified and fed to the speaker.

Speech is divided into 20-ms segments where 159 coded bits are generated, providing an output data rate of 7.95 Kbps. These bits, as shown below and in Figure 7.28, are divided as follows:

- Short-term filter coefficients, $\alpha_i's$, 38 bits/frame;
- Frame energy, $R(O)$, 5 bits/frame;
- Lag, L, 7 bits/subframe, 28 bits/frame;
- Codewords, I, H, 7 + 7 bits/subframe, and 56 bits/frame;
- Gains β, $\gamma1$, $\gamma2$, 8 bits/subframe and 32 bits/frame;
- Total number of bps, 159/20 ms = 7,950.

The transfer function of the synthesis filter is given by:

$$A(z) = \frac{1}{1} - \sum_{i=1}^{N_p} \alpha_i z^{-i}$$

where $\alpha_i's$ are the short-term filter coefficients, which are derived from the input speech and are related to the reflection coefficients, and the order, N_p, of the predictor is 10.

Frame energy value is computed and encoded once each frame period of 20 ms. Frame energy value, $R(O)$, reflects the average input signal power over 20 ms. This frame

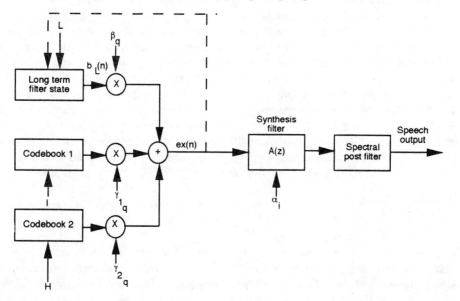

Figure 7.28 VSELP speech coder.

energy is encoded into five bits. The coder obtains long-term predictor coefficients at four subintervals of 5-ms duration each. The lag, L, for each of the four subframes can take on the value of 20–146, which corresponds to 127 possible codes. Thus, seven bits are required to encode each lag during a 5-ms subframe duration, contributing a total of 28 bits over a 20-ms frame interval. The VSELP uses two excitation codebooks where each code vector inputs, I and H, are seven bits each and are generated every 5 ms. Thus, a total of 56 bits of code words are generated per 20-ms frame cycle. The speech coding algorithm is of the class of speech coders known as code-excited linear predictive coding (CELP), stochastic coding, or vector-excited speech coding. These techniques use codebooks to vector quantize the residual or excitation signal. The actual algorithm is known as VSELP. Figure 7.28 shows the block diagram of the speech coder, which generates three gain terms, β_q, γ_{1q}, and γ_{2q} every subframe of 5-ms duration. Here β_q is the quantized gain associated with long-term filter state. γ_{1q} and γ_{2q} are the scaling factors associated with two codebooks.

7.6.1.1 Channel Encoding [17]

In order to mitigate the effects of channel errors, three types of channel protection are employed: half-rate convolutional coding to the most vulnerable bits of the speech coder data stream, cyclic redundancy checks used for the twelve most perceptually significant bits of coded speech, and interleaving the transmitted data of each frame over two time slots to mitigate the Rayleigh fading.

The channel coder block diagram and the error scheme for the NADC system are shown in Figure 7.29(a,b). The first step in the error correction process is the separation of 159 coded bits into class 1 and class 2 bits. There are 77 class 1 bits and 82 class 2 bits in the 159-bit speech coder frame. The class 1 bits represent that portion of the speech data stream to which the convolutional coding is applied. A 7-bit CRC is used for error detection purposes and is computed over the 12 most perceptually significant bits of the class 1 bits for each frame. Class 2 bits are transmitted without any error protection. A total of 89 class 1 bits pass through the half-rate convolutional encoder, which cranks out 178 encoded data bits at the output. Combining the encoded 178 class I bits with 82 class 2 bits provides a total output of 260 bits over 20 ms, or a data rate of 13.0 Kbps.

7.6.1.2 Data Interleaving

To improve the error performance, 260 bits of convolutionally encoded data are spread across two time slots, as shown in Figure 7.30. In the figure, the speech sample y is separated into two halves. One-half of the information is transmitted over slot 1, and the other half over time slot 4. The next 20 ms of the speech sample, z, is divided into slot 4 of the present frame and slot 1 of the next frame. By separating the coded speech into

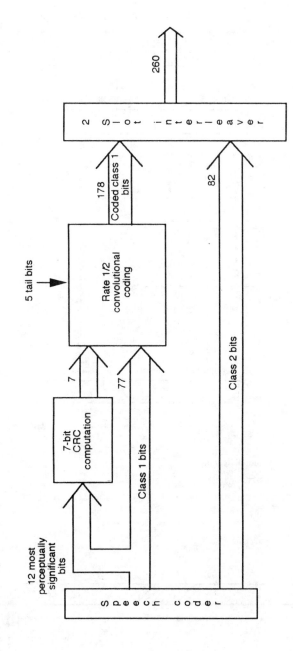

Figure 7.29(a) Channel coder.

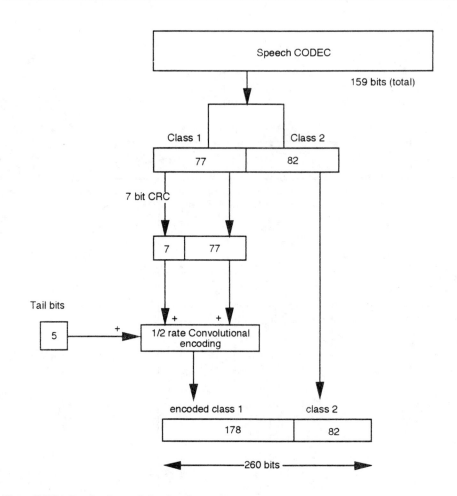

Figure 7.29(b) Details of convolutional coding.

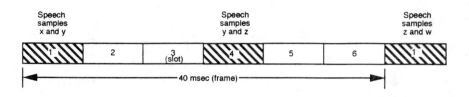

Figure 7.30 Data interleaver.

two time slots, if one slot is lost due to Rayleigh fading, part of the speech can still be recovered by the other slot, thereby, providing better interference rejection.

7.6.1.3 Frame Structure

The frame structure [17] of each TDMA channel is 40 ms in duration. Each frame consists of six equal duration time slots (1–6), having 162 symbols (or 324 bits). Each full-rate traffic channel utilizes two equally spaced time slots of the frame (1 and 4, 2 and 5, or 3 and 6). Each half-rate traffic channel at a future date will utilize one time slot.

At the mobile station, the offset between the reverse and the forward-frame timing, with no time advance applied, is one time slot plus 88 bits. Thus, slot 1 of frame N in the forward direction occurs 412 bit periods after time slot 1 of frame N in the reverse direction. Frame and individual time slots from mobile to cell and from cell to mobile are shown in Figure 7.31.

Formats from the mobile and cell sites are different. The cell site does not use guard and ramp-up time because the cell site carrier is always on. Unlike the cell site, the mobile is always pulsed for the individual time slot only. The synchronization field consists of 28 bits of data for slot synchronization, equalizer training, and time slot identification. Phase changes for six unique synchronization sequences are shown below:

- Sync1 $-\pi/4$ $\pi/4$ $-3\pi/4$ $-3\pi/4$ $3\pi/4$ $-3\pi/4$ $-\pi/4$ $-3\pi/4$ $\pi/4$ $-3\pi/4$ $\pi/4$ $\pi/4$ $\pi/4$ $\pi/4$
- Sync2 $3\pi/4$ $\pi/4$ $3\pi/4$ $3\pi/4$ $-3\pi/4$ $-3\pi/4$ $\pi/4$ $-3\pi/4$ $-3\pi/4$ $-\pi/4$ $-\pi/4$ $\pi/4$ $\pi/4$ $\pi/4$
- Sync3 $-3\pi/4$ $\pi/4$ $-3\pi/4$ $-\pi/4$ $-\pi/4$ $3\pi/4$ $-\pi/4$ $-3\pi/4$ $3\pi/4$ $3\pi/4$ $3\pi/4$ $-\pi/4$ $\pi/4$ $\pi/4$
- Sync4 $-3\pi/4$ $\pi/4$ $\pi/4$ $-3\pi/4$ $3\pi/4$ $3\pi/4$ $\pi/4$ $-\pi/4$ $-3\pi/4$ $3\pi/4$ $-\pi/4$ $-3\pi/4$ $-\pi/4$ $\pi/4$
- Sync5 $\pi/4$ $-\pi/4$ $\pi/4$ $-\pi/4$ $3\pi/4$ $3\pi/4$ $\pi/4$ $-3\pi/4$ $-3\pi/4$ $-\pi/4$ $-\pi/4$ $3\pi/4$ $3\pi/4$ $-\pi/4$
- Sync6 $-3\pi/4$ $-3\pi/4$ $-3\pi/4$ $3\pi/4$ $\pi/4$ $\pi/4$ $-\pi/4$ $-3\pi/4$ $-\pi/4$ $\pi/4$ $-3\pi/4$ $3\pi/4$ $\pi/4$ $-3\pi/4$

These synchronization words have good autocorrelation properties. The cross-correlation properties are also good so that the six time slots are well identified. The training sequence portion of the synchronization data is used for applying equalization to the received RF signal so that the channel impairments due to multipath can be compensated. Since the synchronization word is known to the receiver a priori, the receiver finds the channel inverse filter (inverse filter coefficients) by processing the received signal and comparing it against the stored sequence. Thus, the inverse filter coefficients model the multipath effects of the channel for the training sequence. This inverse filter coefficient is then applied to the whole received packet for the slot. Each TDMA slot also contains twelve bits of signaling message (supervisory and control) in every time slot. This field is used in both directions of the transmission and is designated as SACCH. A coded digital verification color code (CDVCC) provides the same function as the SAT tone in analog cellular. It is used to identify the base station that the mobile is presently using. This is a 12-bit field, permitting 255 distinct values of CDVCC. The same CDVCC may be used for all base and mobile transmissions in the same cell (or sector, where this is appropriate). DVCC is an 8-bit word that is coded using a shortened (12, 8) Hamming

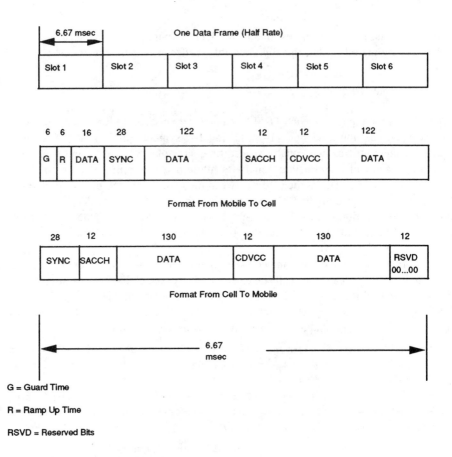

Figure 7.31 Slot and frame structure for American digital cellular radio.

Code to form a 12-bit CDVCC. The procedure of forming a CDVCC from an 8-bit DVCC is as follows. Eight bits of DVCC are represented as the coefficients of the polynomial $d(x) = d_7 + d_6x + d_5x^2 + d_4x^3 + d_3x^4 + d_2x^5 + d_1x^6 + d_0x^7$, where d_7 is the most significant bit and is received earliest in time by the receiver; $d(x)$ is multiplied by x^4 and then divided by $(1 + x + x^4)$ to find the remainder $b(x)$.

$$\frac{d(x)x^4}{1 + x + x^4} = a(x) + \frac{b(x)}{1 + x + x^4}$$

CDVCC is then represented as: $(b_3, b_2, b_1, b_0, d_7, d_6, d_5, d_4, d_3, d_2, d_1, d_0)$. Here bit d_0 is transmitted first. Field RSVD in the slot format from cell to mobile is simply reserved for future use.

As explained above, a TDMA frame consists of six slots, each with a duration of 6.67 ms, and contains 324 bits or 162 symbols. Thus, the total data rate is 48.6 Kbps shared equally among three users (16.2 Kbps per user). Per-user encoded speech has a rate of 13.0 Kbps [260/(6.67 × 3 10^{-3})]. For signaling between the base and the mobile station, 12 bits are allocated per slot, which amounts to a data rate nearly equal to 0.6 Kbps [12/(0.00667 × 3)]. Additionally, a 12-bit CDVCC is sent in each time slot to and from the mobile and base stations. As explained above, this field is used to indicate that the correct (rather than cochannel) data is being decoded. The per-user CDVCC data rate is also 0.6 Kbps. The data rate per user due to guard time, ramp time, and synchronization amounts to 2 Kbps [40/(6.67 × 3)]. Thus, a breakdown of the total user data rate is:

- Speech code, 13.0 Kbps;
- SACCH, 0.6 Kbps;
- CDVCC, 0.6 Kbps;
- Guard time, ramp up, synchronization, 2.0 Kbps;
- Total data rate per user, 16.2 Kbps.

It should be noted that no energy is transmitted during the guard time, which is six bits long and that the ramping period of six bits or three symbols in the reverse channel is due to the finite time the mobile takes to bring its transmitter to full power condition.

7.6.1.4 Modulation

The modulation is $\pi/4$ shifted, differential encoded quadrature phase-shift keying ($\pi/4$ DQPSK). DQPSK is similar to quadrature phase-shift keying (QPSK). The main difference is that in $\pi/4$ DQPSK, the transmitted symbol phase change is referred to the previous phase state. Thus, no absolute phase is required at the receiver, which makes the receiver design easy. A phase transition comparison between DQPSK and $\pi/4$ DQPSK is shown in Table 7.5. The ideal constellation for $\pi/4$ DQPSK is shown in Figure 7.32. The use of gray coding limits the error to a single bit in a symbol consisting of two bits. Though there are eight constellation points at any given instant (for a pair of bits), transition is limited to four constellation points, as shown in Figure 7.33(a). Assuming that at T_0, the constellation point is *, then at the next cycle (transition) any of the four phase states

Table 7.5
Comparison Between DQPSK and $\pi/4$ DQPSK Modulation Formats

Symbol	DQPSK Phase Transition	$\pi/4$ DQPSK Phase Transition
00	0°	45°
01	90°	135°
10	−90°	−45°
11	−180°	−135°

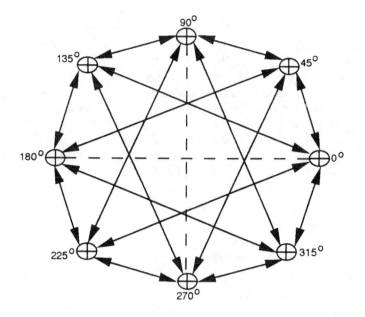

Figure 7.32 $\pi/4$ DQPSK constellation.

marked by $\odot$ are possible, based upon a combination of the pair of bits. Assuming 01 is the combination of bits at T_0, the next constellation point marked $\odot$ is the start of transition T_1.

Serial baseband data, b_m, is split into data bits X_k and Y_k before being differentially encoded. Differential encoding avoids the phase ambiguity at the receiver. Starting from bit 1 of stream b_m, all odd-numbered bits form stream X_k, and all even-numbered bits form stream Y_k. This is shown in Figure 7.33(b).

The digital sequences (X_k) and (Y_k) are encoded onto (I_k) and (Q_k) according to:

$$I_k = I_{k-1} \cos[\Delta\Phi(X_k,Y_k)] - Q_{k-1} \sin[\Delta\Phi(X_k,Y_k)]$$
$$Q_k = I_{k-1} \sin[\Delta\Phi(X_k,Y_k)] - Q_{k-1} \cos[\Delta\Phi(X_k,Y_k)]$$

where I_{k-1} and Q_{k-1} are the amplitudes at the previous pulse time. The phase change $\Delta\Phi$ is determined according to Table 7.5.

7.6.2 Different Types of Calls [17,32]

7.6.2.1 Mobile Initialization

At the time of subscribing for service, the mobile either selects system A or system B and this information is permanently stored in its memory. If the mobile user subscribes

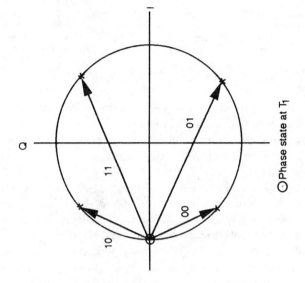

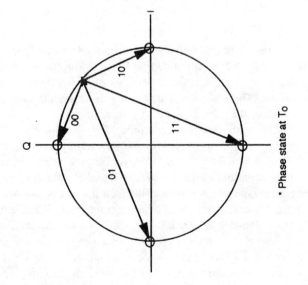

Figure 7.33(a) Unfiltered phase transition for $\pi/4$ DQPSK.

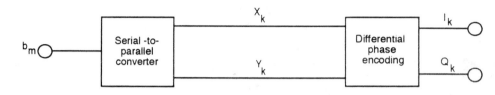

Figure 7.33(b) Differential encoding of $\pi/4$ DQPSK.

for system A, the serving system status field is set. Otherwise, he or she is a subscriber to system B and the serving system status field is reset. As soon as the mobile switches on, the user scans the associated primary set of dedicated control channels for either system A or system B (f_c for first dedicated control channel of system A-834.099 MHz/ 879.99 MHz, f_c for first dedicated control channel of system B-835.02 MHz/880.02 MHz; with a total of 21 control channels in system A and system B), and examines their signal strengths to find the strongest one. At this point the mobile updates the digital overhead information and determines paging channel numbers, scans them, and locks on to the strongest paging channel. The mobile reads the system ID, and if the received ID (SID) is the same as the internally stored ID in its memory, then the mobile is in its home system; otherwise the mobile is declared a roamer. From the paging channel, the mobile also determines the access channel numbers. At this stage the mobile simply starts monitoring the paging channel; where the information is repeated every 46.3 ms. This sequence is shown in Figure 7.34(a).

7.6.2.2 Paging

As described above, after the mobile is initialized it goes into an idle mode and starts monitoring the strongest paging channel continuously. When the incoming call is received by the MTX, it sends commands to the cell site indicating that an incoming call is directed towards a particular mobile. The cell sites transmit a page message over the paging channels. The mobile detects the incoming call by matching the transmitted mobile identification number (MIN_r) with its internally stored MIN_s. At this point the mobile finds the strongest forward access channel (cell to mobile), retrieves the access attempt parameters, and checks for a busy-idle status bit of the reverse access channel (from mobile to cell). The mobile transmits a page response message over a selected access channel and waits for the cell site to respond over the forward access channel. After the cell site receives the page response, it makes a landline connection with the help of the MTX, selects an idle voice channel, and sends the voice channel assignment message over the forward access channel. The mobile receives the voice channel assignment and sends the acknowledgment over a slow associated control channel (SACC). The SACC is a 12-bit control channel and is a part of the voice channel. At this stage, the mobile turns off its transmitter; adjusts its power level; sets its dual-mode transceiver to the digital mode, the CODEC rate, the time slot number, and the time alignment offset; sends a

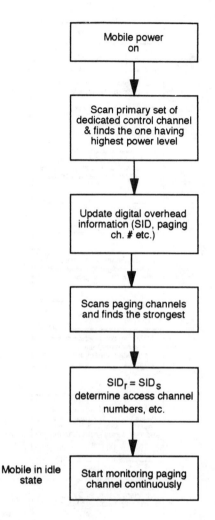

Figure 7.34(a) Mobile initialization sequence. After [42].

coded digital verification (CDVCC); and enters into the waiting state for the cell site to respond. After the cell site receives the CDVCC, it sends an alert message over the voice channel. The mobile, after receiving an alert message, activates ringing. At this stage, the user can press the SEND key and enter into the conversation mode. A flowchart representing this sequence of operations is shown in Figure 7.34(b).

7.6.2.3 Call Origination

The user can dial offline and, when satisfied with the accuracy of dialed digits, can press the SEND button on the phone. At this time the mobile scans the reverse access channels

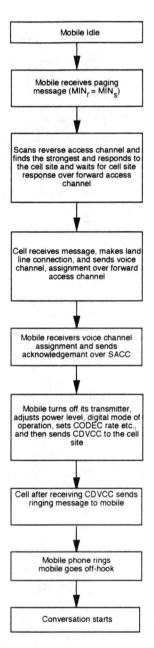

MIN$_r$ = Mobile Identification Number received
MIN$_s$ = Mobile Identification Number stored

Figure 7.34(b) Paging. After [42].

and finds the strongest one. Retrieving the access attempt parameters from the forward access channel allows one to make sure the reverse access channel can be used. The mobile transmits the desired telephone number along with its own identity over the reverse channel and waits for the cell site to respond. The cell site receives the origination message, makes a landline connection through the MTX, assigns a digital channel, and sends the channel assignment message over the access channel. The mobile sends an acknowledgment, turns off its transmitter, adjusts its power level, tunes to an assigned RF channel, selects digital mode for its transceiver, sets the speech CODEC rate, and so forth. Later, when synchronization is complete, the mobile enters into the conversation mode. This sequence of events is shown in Figure 7.34(c). The reader should find functional similarities between this and call origination in the analog system.

7.6.2.4 Handoff

Handoff is the process used when the mobile moves out of an area served by a cell to another area served by an adjacent cell. In the digital mode of operation, the mobile estimates the channel quality by measuring the forward-link received signal strength (RSSI) and the BER, and providing this information to the serving cell site. This information is sent either over the SACCH or FACCH channels. The reader should note that the power measurement is totally a responsibility of the cell sites in the analog system. The process is known as mobile-assisted handoff (MAHO). When the cell site senses that the signal strength is below a certain threshold value, the MTX is alerted. The MTX commands adjacent cell sites to make measurements on the channel presently being used by the mobile. In turn, the cell site asks any mobile presently being served to make an estimate of the signal strength on a specified channel. It should be noted that the mobile, in digital mode, can make measurements on any channel during the offtime of the frame when not receiving data. Thus, the MTX knows which is the best cell site for mobile. At this point, the MTX commands the old cell site to send the handoff message to the mobile (new channel assignment, power level, DVCC, and so on), and alerts the new cell site. Upon receiving the order from the MTX, the old cell site transmits a handoff message to the mobile in the forward traffic channel by replacing data control information over a FACCH. After receiving the handoff message, the mobile turns on the signaling tone for 50 ms; turns off the signaling tone and the transmitter; adjusts the power level; tunes to the new channel; sets the transmitter and receiver to the digital mode, the time slot to that indicated by message field, the speech CODEC rate, the time alignment offset, and so forth. Once the mobile is synchronized, the transmitter is turned on. The flowchart for a handoff operation is shown in Figure 7.34(d).

7.7 THE JAPANESE SYSTEM COMPARED TO NADC

Since the JDC system is similar in many ways to the NADC system, we shall only describe the Japanese system briefly to bring out the similarities and differences between the two

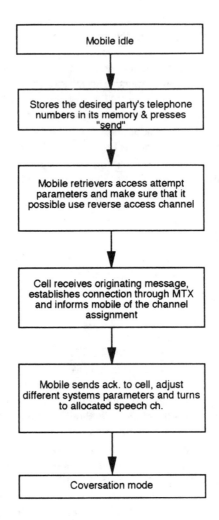

Figure 7.34(c) Call origination. After [42].

systems. Table 7.6 provides a comparison between the two systems. One major difference between the two is that the JDC system is digital only, while the NADC system is both analog and digital. In other words, the North American system works in a dual mode. A JDC channel is 25 kHz wide, while a NADC channel is 30 kHz wide. The JDC system uses frequencies in 800–900 MHz and 1,400–1,500 MHz bands, while the NADC system only uses the 800–900 MHz band. The JDC frame structure is shown in Figure 7.35. The frame is 20-ms long and accommodates three time slots compared to a NADC frame length of 40 ms with six time slots. When two JDC frames are put end to end, the structure becomes that of NADC. Since the 20-ms frame duration is divided among three time

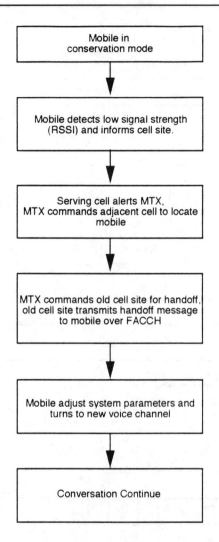

Figure 7.34(d) Mobile-assisted handoff. After [42].

slots, the slot duration for JDC and NADC is the same. Since the frame duration is 20 ms, there are 50 frames per second. Each frame contains 840 bits of data, which corresponds to a data rate of 42 Kbps. Both JDC and NADC use $\pi/4$ DQPSK as the data modulation scheme and square-root raised cosine filters in both the transmitter and receiver. However, the JDC system uses a Nyquist filter with a rolloff factor of $a = 0.5$, while NADC uses $a = 0.35$.

Many details of the Japanese system are not presently available. The expected time-slot structure for the traffic and control channels is shown in Figure 7.36. This is based

Table 7.6
NADC and JDC System Comparison

	NADC	*JDC*
Frequency range (MHz)	824–849 up (mobile to cell)	810–826 down
	869–894 down (cell to mobile)	940–956 up
		1,429–1,441 down
		1,447–1,489 up
		1,453–1,465 down
		1,501–1,513 up
Slot/frame	6	3
Modulation data rate (Kb)	48.6	42.0
Channel spacing (kHz)	30	25
Filter factor	0.35 square root Nyquist	0.50 square root Nyquist
Modes	Analog/digital	Digital only

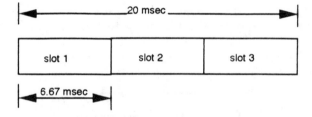

Figure 7.35 JDC frame structure.

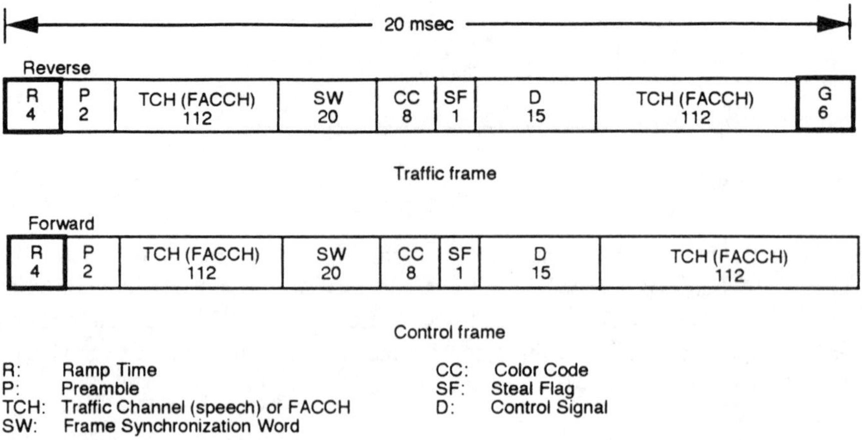

Figure 7.36 Frame structure of JDC.

on the TDMA mobile radio system, presently under development by NTT, and is supposed to satisfy the Japanese digital cellular radio standard. Here, the guard time between TDMA bursts is six bits (three symbols) when no energy is transmitted. Additionally, four bits are assigned for both the up and down ramp burst amplitudes. Twenty bits for synchronization and training words are assigned in the middle of the burst. Eight bits for the color code are used to identify the base station. A steal flag of one bit distinguishes the FACCH from the traffic channel. Fifteen bits are used either for housekeeping or for real-time control functions. Per-user traffic data rate is 11.2 Kbps (224/0.02). The data rate per user is 14 Kbps (280/0.02), providing a total data rate of 42 Kbps, which is lower than the 48.6 Kbps for the ADC system. The difference of 2.8 Kbps is allocated between preamble, ramp, guard bits, color code, steal flag, and so forth.

7.8 OTHER PROPOSED NORTH AMERICAN DIGITAL CELLULAR SYSTEMS

Improvements to enhance digital cellular capacity is an extension of the TIA/IS-54 TDMA system by Hughes Network Systems and of a code division multiple access (CDMA) system by QUALCOMM [33–35]. Both these proposed systems provide dual-mode operation with the existing analog cellular system. The CDMA technology developed by QUALCOMM calls for a wideband channel having a bandwidth of 1.25 MHz to be used in every cell ($N = 1$, reuse), which is equivalent to 42 30-kHz channels. Each channel is shared by many users with different codes. QUALCOMM claims that it can offer 10–20 times the capacity of the currently available analog system. On the other hand, E-TDMA uses the presently recommended IS-54 system having a channel bandwidth of 30 kHz, but uses digital speech interpolation (DSI) in conjunction with a half-rate voice coder to achieve 10 times the analog cellular radio capacity. The technical objectives of the QUALCOMM CDMA, Hughes E-TDMA, and IS-54 TDMA systems are shown in Table 7.7. In this section we shall elaborate on the technical characteristics of the E-TDMA and QUALCOMM CDMA systems. The North American digital cellular system, based on IS-54, has already been discussed in section 7.6.

7.8.1 E-TDMA

As shown in Table 7.7, the main technical objective of E-TDMA is to enhance the proposed capacity of digital TDMA (IS-54) by using a half-rate vocoder and digital speech interpolation. The half-rate speech coder is based on a code-excited LPC algorithm. It is claimed that subjective speech quality (mean opinion score (MOS)) is the same as that achieved by a 8-Kbps speech coder. The algorithm uses a more advanced DSP chip that allows for real-time analysis of the speech. Time assignment speech interpolation (TASI) is a voice-operated switching and channel assignment system to interpolate additional talkers onto communication facilities. In the past, TASI systems have been used with submarine cable voice-channel facilities.

Table 7.7
Comparison of Different Digital Systems in the United States

	TDMA (Ericsson, AT&T, NTI)	CDMA (QUALCOMM)	E-TDMA (Hughes/IMM)
Characteristics	Divides 30-kHz channel into six time slots, yielding three equivalent voice channels based on IS-54 standard (phase 1)	Calls for 1.25-MHz wide channel or 42 30-kHz equivalent channels.; uses soft handoff to increase voice quality; uses rake receiver to take advantage of multipath fading; low C/I ratio increases capacity	Uses a 30-kHz channel, same as IS-54, but uses six slots for six discrete channels rather than three for TDMA; each channel uses a half-rate vocoder; uses digital speech interpolation to increase capacity
Digital or analog	Digital	Digital	Digital
Capacity (vs. analog)	3–5 times	10–20 times	8–10 times
Vocoder rate (Kbps)	8	8*	4.8 Kbps
Initial spectrum conversion required (%)	5	10	10
Data-handling capability	Yes	Yes	Yes
Privacy	Yes	Yes	Yes

*Variable.

In a normal telephone conversation, each subscriber speaks less than half the time. The remainder of the time is used for listening, gaps between words and syllables, and pauses. Considering a large subscriber group, the average talk cycle is about 3.5 seconds, in which the actual time when the speaker is active is about 40%. See Figure 7.37.

The implementation of TASI involves expenditures for switching and control equipment, but when compared to the cost of an additional channel, the system should provide a substantial reduction in cost per channel. The complexity involved is in recognizing

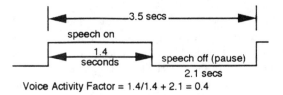

Figure 7.37 Speech activity.

speech energy in a few milliseconds and assigning appropriate physical channels and slots to the mobile user.

The basic concept of DSI is shown in Figure 7.38. A typical five-second speech segment, in which conversations on six voice channels are occupied only 40% of the time, is compressed into three voice channels that are occupied for about 80% of the time. From the figure, one should note that the consecutive speech segments of a mobile are not necessarily on the same radio channel. This is also a form of frequency hopping, which helps in improving performance in multipath surroundings.

Figure 7.39 provides two examples of DSI pools composed of three and eighteen radio channels. Here, the time slots labeled C are the control channels for the assignment and release of speech channels, which are labeled as V. As claimed by Hughes Network Systems, three radio channels can carry twenty conversations, and nineteen channels can carry 208 conversations. These cases provide capacity gains of about seven and eleven, respectively, over analog FM. Since each channel contains both the time slots C and V, channel-by-channel conversion from analog FM to E-TDMA, or from TDMA to E-TDMA, can take place.

In the forward direction, the base station control monitors the voice activity detector. The slot assignment is then sent to the mobile over the control slot. Mobile then turns to the assigned speech slot and retunes to the control slot during the idle period of the TDMA frame. In the reverse direction, the mobile sends the slot assignment request over a control channel to the base station controller. BSC sends assignment to the mobile over the control channel, and then the mobile tunes to the assigned slot.

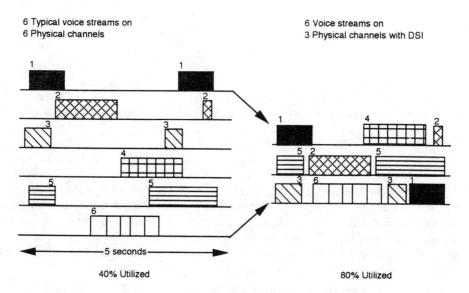

Figure 7.38 DSI operations [33–34]. Source: [44].

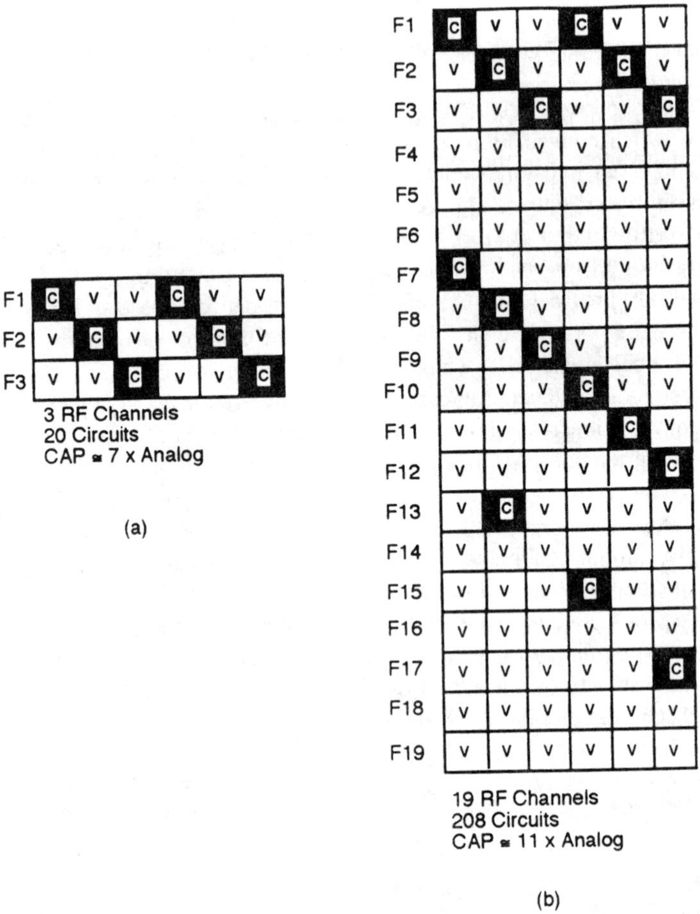

3 RF Channels
20 Circuits
CAP ≈ 7 x Analog

(a)

19 RF Channels
208 Circuits
CAP ≈ 11 x Analog

(b)

Figure 7.39 DSI pool with (a) 3 E-TDMA channels and (b) 19 E-TDMA channels. Source: [44].

7.8.2 Proposed CDMA

The second proposed system in the United States to increase the subscriber's capacity is digital CDMA by QUALCOM [36–39]. The basic concept behind increased capacity is the use of a wideband channel (proposed channel BW = 1.25 MHz) where many subscribers can talk together without interfering with each other. The isolation between users is due to their code words, whereby the individual user can intelligibly hear the speech correctly encoded with his or her assigned code, while speech with other codes behaves as background noise. Factors that affect the channel capacity are:

- Use of wideband spread-spectrum waveform;
- Low C/I ratio;

- Use of variable rate vocoder;
- Voice activity factor; and
- Power control in the forward and reverse channel.

Some claimed advantages of this system are:

- No need for an equalizer;
- No guard time required as in TDMA;
- Easy transition from analog to digital;
- No need for channel frequency management like in FDMA and TDMA;
- Soft handoff assures smooth transition of a mobile from cell to cell; and
- Graceful degradation.

This section elaborates these points.

The proposed system is a QPSK-modulated direct-sequence spread system having ten channels, each having a bandwidth of 1.25 MHz. It is claimed that 118 subscribers can simultaneously talk over one channel. Assuming all channels are available to all cells of a seven-cell system, the total number of simultaneous users that can be served is 118 × 10 = 1180. Assuming equal channel assignment in each cell for analog cellular, the total number of subscribers per cell is 56 in analog FDMA. Thus, theoretically the capacity of a CDMA system is over twenty times the capacity of an analog system. This of course assumes perfect isolation between users, that is, though there are 118 simultaneous users per channel, they do not interfere with each other. Before we elaborate on the above points, let us briefly review the basic fundamentals of DSSS.

The basic spread spectrum is shown in Figure 7.40. Each subscriber uses the same carrier frequency, ω_c, and occupies the same bandwidth. Two steps are involved in the generation of a transmitted signal in a multiple access environment. First, the carrier is phase modulated by data bits with a rate $R_i = 1/T$, where T is the bit duration. As a second step, the phase-modulated data is spread by a spreading function $p_1(t)$ with a chip rate of $R_c = 1/T_c$ before transmission. The resulting signal, $p_1(t)s_1(t)$, is passed through the channel where it is combined with other spread signals having different codes.

Assuming M transmitters, the received signal is given by:

$$r(t) = \sum_{i=1}^{M} p_i(t)s_i(t) + n(t) + I(t)$$

where $n(t)$ is the additive noise and $I(t)$ is the interference. The correct receiver will have $p_1(t)$ as its despreading code, and other receivers will have different despreading codes. Only the original waveforms, which were spread by $p_1(t)$ at the transmitter, will effectively be despread; others will appear like noise to the demodulator.

One measure of spread-spectrum performance is the system processing gain, G_p, and is given by the ratio of channel bit rate to information bit rate, R_c/R_i. Thus, if $R_c = 1.25$ Mbps and $R_i = 8$ Kbps, the processing gain G_p is nearly equal to 22 dB. Clearly, the input and the output signal-to-noise ratios are related as:

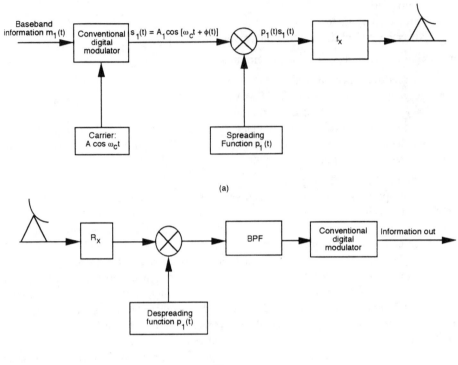

Figure 7.40 Spread-spectrum (a) transmitter and (b) receiver.

$$(S/N_0) = G_p(S/N_0)_i$$

We shall now go back and elaborate on the channel capacity and other advantages of CDMA, as stated above.

7.8.2.1 Use of Wideband Spread Spectrum Waveform

As reported in [37], the number of diversity branches, M, for a wideband system having a bandwidth of B Hz, can be computed by:

$$M = \frac{\Delta + 1/B}{1/B} = B\Delta + 1$$

where Δ is the time delay due to multipath fading, and $1/B$ is the pulse duration. Assume $\Delta = 0.5$ for suburban and $\Delta = 3$ μs for an urban area. Considering the proposed bandwidth of 1.25 MHz for CDMA and $B = 30$ kHz for the present analog system, the number of diversity branches in both these cases is shown in Table 7.8. As can be seen, the wider the bandwidth, the less the fading. For $B = 1.25$ MHz, the fading of its received signal is reduced and receiver behaves as if it had a diversity branch, $M = 1.625$ (between a single branch and two branches) for suburban areas, and $M = 4.75$ (between four and five branches) for urban areas. Thus, a wideband signal would provide more diversity gain in urban areas. For $B = 30$ kHz, no effective diversity gain is noticeable.

7.8.2.2 Low Value of C/I ratio

Being self-interference limited, the CDMA capacity is determined by the average signal-to-interference ratio resulting from all subscribers within a system after accounting for attenuation and antenna isolation. As shown in (6.26a), when power control is exercised, the average carrier-to-interference ratio is: $C/I = 1/(N - 1)$. Thus, as C/I decreases, the number of subscribers, N, can be increased, which enhances the system's capacity.

7.8.2.3 Variable Rate Vocoder

By using a variable rate vocoder of 4.8–16 Kbps, different grades of service can be obtained. On the other hand, the vocoder rate can be made to vary as a function of the speech activity factor. For a duration of higher speech activity, one can use a 16-Kbps vocoder and reduce down to a 4.8-Kbps vocoder for a durations of very low speech activity. This will provide a higher speech quality and still maintain a low average data rate. This will permit more users to share the spectrum, and thus, increase the capacity.

7.8.2.4 Voice Activity Factor

Human speech has an activity factor of about 40%, that is, a speaker typically talks only about 40% of the time and the remaining part of the time is either listening or there is a pause. When users assigned to a channel are not talking, it reduces the interference to

Table 7.8
Diversity Branches in FDMA and CDMA

	Number of Diversity Branches, M	
Environment	*B = 30 kHz* *(FDMA)*	*B = 1.25 MHz* *(CDMA)*
Suburban	1.015	1.625
Urban	1.09	4.75

those who are talking. Thus, voice activity alone reduces the interference by 60%, and increases the channel capacity by 2.5 times. CDMA takes natural advantage of this phenomenon without any design expense. The technique can also be used in FDMA and TDMA systems, but these systems require the design of a demand-assigned protocol controller.

7.8.2.5 Forward Link Power Control

The forward link power control is used to reduce the necessary interference outside its own cell boundary. Exact orthogonality of codes within the cell is difficult to maintain, thus the interference is always introduced while despreading the desired user's RF signal. This is not true in TDMA and FDMA, where the orthogonality can be preserved by providing guard time and guard band. In order to minimize this interference, power control is required so that the power levels of the mobiles located close to the cell arrive at the same level as those mobiles located at the boundary of the cell. However, this will still leave the power levels of the adjoining cells uncontrolled, which may arrive at a level high enough to cause performance degradation to the signal of interest. With this brief introduction to direct sequence spread-spectrum systems, let us return to the proposed CDMA system.

7.8.2.6 Equalizer

Both FDMA and TDMA require an equalizer to combat the affects of intersymbol interference caused by bit spreading due to multipath. However, a CDMA system only requires a despreader instead of an equalizer when the symbol duration is greater than the delay spread. For an 8-Kpbs channel rate, the symbol duration is nearly equal to 125 μs, which far exceeds the average delay spread of an urban area (typically nearly equal to 3–6 μs). Thus, a correlater will only be required that is simpler to design than to use an equalizer.

7.8.2.7 Guard Time

A TDMA system always requires guard time to combat the clock inaccuracies among users. Obviously, guard time reduces the frame efficiency of the system. On the other hand, there is no existence of guard time in CDMA systems.

7.8.2.8 Transition from Analog to Digital

Since the recommended bandwidth of an individual CDMA channel is 1.25 MHz, only 10% of the total bandwidth (0.1 × 12.5 MHz) is required to change over the existing FM analog channels to digital CDMA. It has been observed that only 5% of the total users make up more than 30% of the total traffic. System providers can give incentive to those users willing to change over from analog FM to digital CDMA. Thus, the capacity of the existing FM system can be increased by about 25%. At 1.25 MHz, the total number of analog channels are: $(1.25 \times 10^6)/(30 \times 10^3) = 41.6 \approx 41$ channels. Using a cluster size of 7, the number of channels per cell is nearly equal to 6. Assuming 1.25 MHz can carry 118 users per cell, the capacity of the CDMA system is: $C_{CDMA} = 118/6 \ C_{FM} \approx 20 \ C_{FM}$. Assuming a cell reuse pattern of 4 in TDMA, the total number of channels at 1.25 MHz is: $(1.25 \times 10^6)/(10 \times 10^3) = 125$. Thus, the number of channels/cell = 125/4 $\approx$ 31, so, $C_{CDMA} = 118/31 \ C_{TDMA} \approx 4C_{TDMA}$.

7.8.2.9 Channel Frequency Management

Since a large number of subscribers use the same channel, frequency management is to a large extent eliminated. For example, if all cells have 10 channels, each having a bandwidth of 1.25 MHz, and each channel can serve up to 118 users, simple frequency management is required. Obviously, no frequency management is needed between cells and within an individual cell. Once a channel is fully loaded (118 users), a different channel has to be selected, which is simple to achieve. Since all cells have all channels, no dynamic channel assignment is required.

7.8.2.10 Handoff

CDMA employs a soft handoff, that is, a connection is made to the new cell while still maintaining the original connection with the old cell. This ensures a smooth transition between cells and requires no action on the part of the mobile. The new cell site switches to the code the mobile is using. A slight disadvantage of the technique is an increase in the noise level; as the mobile can have multiple connections to different cell sites. Only when a mobile is firmly established in the cell connection can the old cell site be dropped. This technique reduces the chance of a dropped call in CDMA, as the basic handoff takes place by making a connection to the new cell site before dropping the connection to the old cell site. In comparison, the present FM system uses a break-before-make system, which increases the handoff noise and the chance of a dropped call. Also, the technique of make-before-break will allow diversity combing.

7.8.2.11 Degradation

For analog FDMA, only one user can be served by a single channel. Depending upon the vocoder rate, either three or six users can be served per 30-kHz channel bandwidth for TDMA. Thus, in both these cases there is a limit on the number of users that can be served by the system. However, for CDMA, if more than a certain number of users are normally permitted access to the channel, only the noise level will go up slightly, which in turn will degrade the BER performance in proportion to the percent increase in overload. In other words, there is no hard limit on the number of users that can be served by a single channel or by the system. This is in sharp contrast to the present FDMA and the forthcoming digital TDMA system. On the same line of reasoning, the CDMA system can dynamically shift traffic among unequally loaded cells to balance the whole system. As the traffic in an adjacent cell is reduced, interference is reduced and thus, temporarily higher loads can be served by the cell of interest if a higher load is demanded.

7.9 CONCLUSIONS

In order to serve increased user traffic, it is necessary to convert first-generation cellular systems based on analog FDMA technology to digital TDMA systems. The increased capacity of digital systems is based on the proper selection of modulation with high spectral efficiency. The choice of continuous phase modulation, such as GMSK adopted by the European GSM system, reduces the C/I requirement by as much as 8 dB compared to the present 25-kHz analog FM systems. The motivation for choosing GMSK modulation is also due to limited power availability at the mobile, which forces one to choose a constant envelope modulation. In the United States, the modulation scheme is linear $\pi/4$- DQPSK and is possible due to an improvement in the fabrication techniques of high-power linear amplifiers. Compared to the European GSM system, the NADCS is about 20% more spectrally efficient due to a more complex speech algorithm, which provides encoded speech at 7.95 Kbps compared to 13 Kbps in GSM. Due to compatibility requirements in the U.S., both analog and digital cellular systems have to coexist. Several techniques for the gradual transition from analog to digital will be implemented by system operators. As a result of digital implementation, improved security will be provided to the subscribers. Unauthorized interception will be difficult due to frequency hopping and the ciphering of data. Fraud by users will also be minimized due to the authentication requirements of the system. Improvements to the basic EIA/TIA IS-54 digital system have been proposed by Hughes and Qualcom in the U.S. Hughes proposed E-TDMA, which enhances the basic capacity due to DSI and half-rate speech coding. Qualcom CDMA, if adopted, may increase the capacity of the basic system by as much as 20 times over the analog system.

PROBLEMS

7.1 Why is the data rate over a channel limited by multipath delay spread? If delay spread is limited to 5 ms, what is the maximum data rate one can transmit in analog FDMA using FSK modulation? What is the maximum number of channels in digital TDMA, assuming an individual user data rate of 4.8 Kbps with half-rate convolutional encoding? You can assume QPSK modulation with no channel bandwidth constraint.

7.2 Table 7.1 provides an incomplete list of advantages and disadvantages of analog FDMA and digital TDMA. Complete the table by providing at least three more entries in the table.

7.3 Table 7.2 provides the requirements of GSM and EIA/CTIA American digital cellular systems. Think of at least two more requirements that you can add to this table to make it more complete.

7.4 Find the spectral efficiency as a function of cell radius of NADC systems for all combinations of cell repeat pattern, N, and number of sectors per cell, and plot them (total of six plots)

- Cell repeat pattern, $N = 4$, 7, and 12;
- Number of sectors/cell = 3, 6;
- Required C/I = 14 dB;
- Total coverage area = 5,000 km^2.

9.5 Several schemes have been suggested in Section 7.3.2 that will allow the system operator to gradually integrate digital cellular with the present analog system. Can you suggest one more scheme that is not covered in this section? State clearly its advantages and disadvantages.

7.6 Find the frame efficiencies for GSM and NADCS systems shown in Figure 7.41.

7.7 Answer the following questions with respect to Figure 7.12(a): (a) the need for tail and guard bits; (b) the need for training bits. As can be seen in the figure, the training bits are placed at about the middle of the frame. Can you think of some advantages

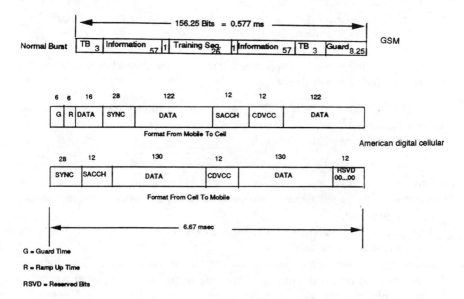

Figure 7.41 Frames for GSM and ADC.

to this approach? What impact will it have on the system's performance if you place the training bits at the beginning or at the end of the frame?

7.8 Let us assume that the speech coder produces 200 bits every 25 ms. Of these, 150 bits are residual data and 50 bits are side information. Assuming that we use convolutional coding with 3/4 rate, what is the resultant channel data rate? You can assume that a total of six additional bits are required as parity and flushing bits, and that there are a total of six channels per carrier. How much transmission bandwidth would you need if the data is $\pi/4$ DQPSK modulated. Assume the required bandwidth to be from null to null.

7.9 For example 8 let us assume that we interleave data. The interleaving will take three adjacent 25-ms intervals, and interleaved over six time slots and then transmitted over six frames. Assuming that the frame duration is 5 ms, find the data transmission rate and the delay encountered.

7.10 Explain the following terms explicitly with respect to digital TDMA, GSM, and NADC systems:
- Full and half-rate speech channel;
- Broadcast control channel;
- Common control channel;
- Dedicated control channel;
- Comfort noise;
- Speech activity detector;
- Authentication;
- HLR and VLR;
- Network management center;
- X.25 network.

7.11 From the following data, conclude whether the adjacent channel interference requirement of −80 dBc is met or not.

- Normalized 3-dB bandwidth of premodulation filter = 60 kHz;
- Channel data rate = 40 Kbps/user;
- Channel bandwidth = 250 kHz.

Assume the modulation to be GMSK form.

7.12 Find the frequency hopping rate for a system having a 5-ms frame duration shared by ten users. Assume that the mobile sends and receives on two different frequencies during a time slot.

7.13 Draw a sequence of steps similar to Figure 7.26(a) for the mobile originating a call and mobile-assisted handoff.

7.14 Assuming half-rate speech coder producing 80 bits every 20 ms. Half of these are class 1 bits and other half are class 2 bits. Assuming half-rate channel coding, what is the output data rate of the channel encoder?

7.15 We have named the Figure 7.30 frame an interleaver. Justify this.

7.16 Assume that the initial phase of the modulator is 0 deg, decode the six synchronizing sequences given in Table 7.6.

7.17 Give reasons to justify why $\pi/4$ DQPSK should not be regarded as 8-PSK. What are the reasons for using differential phase encoding in the NADC system?

7.18 Convert the flowcharts for mobile initialization sequence, paging, call origination, and handoff in terms of the sequence of interactive messages between base station, MSC and mobile. Represent them in the form of a control sequence of operations similar to the analog mobile calling sequences described in Section 2.3.2.

7.19 Explain the basic concept of DSI. Can the channel utilization be made 100%? Explain.

7.20 Find the capacity of the E-TDMA systems shown in Figure 7.42, and express it in terms of analog system capacity in the United States.

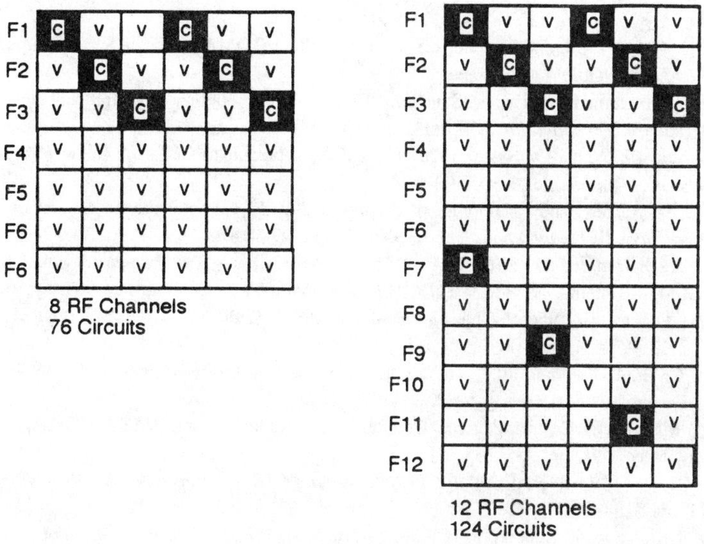

F1	C	V	V	C	V	V
F2	V	C	V	V	C	V
F3	V	V	C	V	V	C
F4	V	V	V	V	V	V
F5	V	V	V	V	V	V
F6	V	V	V	V	V	V
F6	V	V	V	V	V	V

8 RF Channels
76 Circuits

F1	C	V	V	C	V	V
F2	V	C	V	V	C	V
F3	V	V	C	V	V	C
F4	V	V	V	V	V	V
F5	V	V	V	V	V	V
F6	V	V	V	V	V	V
F7	C	V	V	V	V	V
F8	V	V	V	V	V	V
F9	V	V	C	V	V	V
F10	V	V	V	V	V	V
F11	V	V	V	V	C	V
F12	V	V	V	V	V	V

12 RF Channels
124 Circuits

Figure 7.42 E-TDMA systems. Source: [42].

REFERENCES

[1] Lissakers, E. "Switching to Digital," *Cellular Business,* Dec. 1989.

[2] Stewart, A. "Growing Pains in the Networking," *TE&M,* Sept. 1990, pp. 66–68.

[3] Madrid, J. "The Difference Between Mobile and Mobility," *TE&M,* Sept. 15 1990, pp. 72–76.

[4] Bohm, M and H. H. Schulz. "Commercial Introduction of the ECR 900," *Electrical Communication,* Vol. 63, No.4, pp. 383–388.

[5] Slekys, A. G. "Exploring the Digital Wave," *Cellular Business,* Feb. 1990, pp. 72–94.

[6] Jiang, T. "A Comparison Between the Three Mobile Communication Systems," *IEEE Vehicular Technology Conference,* 1987, pp. 359–362.

[7] Cook, C. E., and H. S. Marsh. "An Introduction to Spread Spectrum," *IEEE Communication Magazine* Vol. 21, No.2, 1983, pp. 8–16.

[8] Proakis, J. G. *Digital Communications,* McGraw-Hill Book Company, 1983.

[9] Chan, H. and C. Vinodraw. "The Transition to Digital Cellular."

[10] Tarallo, J. A. and G. I. Zysman. "A Digital Narrow-Band Cellular System," *IEEE Vehicular Technology Conference,* 1987, pp. 279–280.

[11] Franklin, W. J. "Going Digital," *Cellular Business,* Aug. 1989.

[12] Losee, M. "The Coming Age of Digital," *Cellular Business,* July 1988, pp. 48–51.

[13] Huff, D. "The Transition: Moving into Digital," *Cellular Business,* April 1991, pp. 38–46.

[14] Uddenfeldt, J. "Krister Raith, Bo Hedberg;" *IEEE Vehicular Technology Conference,* 1988, pp. 516–519.

[15] Hoff, J. "Mobile Telephony in the Next Decade," *IEEE Vehicular Technology Conference,* 1987, pp. 157–159.

[16] IMM Ultraphone 100. "Wireless Digital Loop Carrier: System Description."

[17] EIA/TIA IS-54. "Dual-Mode Mobile Station-Base Station Compatibility Standard," Dec. 1989.

[18] Uddenfeldt, J., and B. Persson. "A Narrowband TDMA System for New Generation Cellular Radio," *IEEE Vehicular Technology Conference,* 1987.

[19] Stzernvall, J.-E., and J. Uddenfeldt. "Performance of a Cellular TDMA System in Service Time Dispersion," *IEEE Vehicular Technology Conference,* 1987.

[20] Ballard, M., E. Issenmann, and M. M. Sanchez. "Cellular Mobile Radio as an Intelligent Network Application," *Electrical Communication,* Vol. 63, No. 4, pp. 389–399.

[21] Weib, W., and M. Wizgall. "System 900, The ISDN Approach to Cellular Mobile Radio," *Electrical Communication,* Vol. 63, No. 4, pp. 400–408.

[22] Rahier, M., D. Rabaey, and J. Dulongpont. "Advanced VLSI Components for Digital Cellular Mobile Radio," *Electrical Communication,* Vol. 63, No. 4, pp. 409–414.

[23] Natrig, J. E., S. Hansen, and J. de Brito. "Speech Processing in Pan-European Digital Mobile Radio System (GSM)—System Overview," *IEEE Vehicular Technology Conference,* 1989, pp. 1060–1064.

[24] Hellwig, K., P. Vary atl. "Speech Codec for European Mobile Radio System," *IEEE Vehicular Technology Conference,* 1989, pp. 1065–1069.

[25] Southcott, C. B., D Freeman atl. "Voice Control of the Pan-European Mobile Radio System," *IEEE Vehicular Technology Conference,* 1989, pp. 1070–1073.

[26] Feher, K. "Modems for Emerging Digital Cellular Mobile Radio System," *IEEE Transaction on Vehicular Technology,* May 1991, pp. 355–365.

[27] Raith, K. "Capacity of Digital Cellular TDMA System," *IEEE Transaction on Vehicular Technology,* May 1991, pp. 323–331.

[28] Goodman, D. J. "Second Generation Wireless Information," *IEEE Transaction on Vehicular Technology,* May 1991, pp. 366–374.

[29] Simon, M. K., and C. C. Wang. "Differential Detection of Gaussian MSK in a Mobile Radio Environment," *IEEE Transaction on Vehicular Technology,* Vol. VT-33, No. 4, Nov. 1984, pp. 307–320.

[30] Hirade, K., and K. Murota. "A Study of Modulation for Digital Mobile Telephony," *29th IEEE Vehicular Technology Conference,* March 1979.

[31] Raith, K., B. Hedberg, atl. "Performance of a Digital Cellular Experimental Test Bed," *IEEE Vehicular Technology Conference,* 1989, pp. 175–177.

[32] RF Communication Forum, HP presentation, June 1992.

[33] Kay, S. "E-TDMA," *Cellular Business,* June 1992, pp. 60–64.

[34] Hughes Network Systems, Inc. "High capacity Digital cellular for Wireless Telephony," March 1992.

[35] Hughes Network Systems Inc., "Cellular Facts," 1992.

[36] Crawford, T. R. "CDMA," *Cellular Business,* June 1992, pp. 50–56.

[37] Lee, W.C.Y. "Overview of Cellular CDMA," *IEEE Transactions on Vehicular Technology,* Vol. 40, No. 2, May 1991, pp. 291–301.

[38] Gilhousen, K. S., atl. "On the Capacity of a Cellular CDMA System," *IEEE Transactions on Vehicular Technology,* Vol. 40, No. 2, May 1991, pp. 303–311.

[39] Pickholtz, R. L., atl. "Spread Spectrum for Mobile Communication," *IEEE Transactions on Vehicular Technology,* Vol. 40, No. 2, May 1991, pp. 303–311.

[40] Ziemer, R. E., and R. L. Peterson. *Digital communication and Spread Spectrum Systems,* Macmillan Publishing Co. 1985.

[41] Bhargava, V. K., atl. "Digital Communications by Satellite," Wiley-Interscience Publication, 1981.

[42] Hewlett-Packard, Spokane Division. "Introduction to NADC and JDC Digital Cellular Systems," presentation.

[43] Ahola, K. "Europe's GSM: Passage to Digital," *TE&M*, September, 1990, p. 54.

[44] Slekys, A. G., "High Capacity Digital Cellular for Wireless Telephony," *Hughes Network Systems*, March 1992, p. 6.

Chapter 8
Cellular Networking

8.1 INTRODUCTION

The potential of cellular technology is just being tapped. With increased portable usage, the ability to easily make and receive calls while moving between different cellular systems is becoming increasingly critical. Several interim solutions have been identified and are being offered to provide validation, call delivery and handoff for cellular users. Today, during a conversation the RF power level of the mobile is routinely monitored within an active cell site. If the mobile power level drops below a certain level, indicating that the mobile is leaving the cell, an automatic sequence of operations takes place that includes finding the best cell site for the coverage, mobile reallocation of a new channel in the most desired new cell site, and handover to the new cell. These procedures are carried out within a system where all adjacent cells are within the control of one mobile switching center (MSC). However, when a mobile subscriber enters the area of another MSC (adjacent system), the handover capability does not exist in general. Thus, the mobile is forced to hang up and redial on the new system. Providing a handoff capability between systems requires more than just the protocol to exchange the data between systems. The protocol must also account for the service that a mobile requiring handoff, expects. The protocol should also be able to accommodate the future addition of new services.

The above description is only half the story of cellular networking. The other half is the requirement for call delivery by one system to another that are widely separated geographically. As the industry matures in the United States and the rest of the world, a cellular "supersystem" will be required that will cover large geographical areas, preferably the whole world, by networking together a large number of smaller scattered systems. Such networked systems will provide to subscribers a greatly expanded service area in which calls can be placed and received without the hassles of roaming that accompany today's cellular service. This will also make certain that customers will be provided the same features and service, irrespective of home area or foreign location. The benefits that these networked systems provide to system operators include the marketing advantage of a large service area, a reduction in fraud losses, an increase in revenue by improving the percentage of calls successfully completed to mobile customers, and an improved quality of service at the boundary of two independent-but-networked systems.

Thus, cellular networking is a mobile telecommunications architecture that provides disjoint or overlapping coverage areas with a coherent call processing environment. In other words, cellular networking is the ability to extend basic cellular service from one CGSA to another CGSA, as shown in Figure 8.1. As can be seen, when the boundaries of the two systems overlap, the active call can be handed off between the cell sites associated with the different systems [1–4]. For widely separated systems (systems A and B from system C), call delivery is the problem that will require mobile location identification, paging, and call delivery.

The networking of systems can provide for independent systems joining together to form a complete corridor of service, such as Seattle to San Diego or Boston to Miami. While this is not a practical example, the illustrated capability identifies the main features that the cellular networking has to provide.

In view of the above discussion, this chapter is developed into the following sections. First we provide some basic definitions used in networking, customer requirements, and carrier needs, followed by call delivery to adjacent and nonadjacent CGSAs. A conceptual network configuration for minimizing fraudulent calls by roamers is discussed in Section 8.4. For efficient call delivery, a separate signaling network is required. Networks adapted in both Europe and the United States are SS7 and X.25, and their protocols are based on the open system interconnection (OSI) model. Thus, we discuss the protocol requirements in Section 8.5. Lastly, we extend the discussions on these networks in the United States from a standardization point of view.

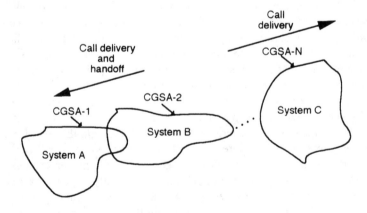

Figure 8.1 The call delivery system and handoff between systems.

8.2 DEFINITIONS USED IN CELLULAR NETWORKING

Cellular networking is the ability to extend the cellular service to mobiles moving from one CGSA to another. When the boundaries of two CGSAs overlap, an active call can be handed off between cell sites associated with two different operating systems. Intersystem handoff will solve two different problems: service deterioration due to a low C/N ratio at the boundary of two systems and excessive cochannel interference.

The first problem causes excessive noise to be sensed by the mobile due to a low carrier-to-noise ratio at the cell boundary of two systems. This will eventually cause the mobile to drop the call and force it to reoriginate in the new system. This eventual drop of the call is also due to the fact that the different systems do not have the facility to determine the best cell for call handling. Excessive cochannel interference may be felt by those mobiles that might have gone inside the boundary of the other system before the call is finally dropped. Before we discuss cellular networking, let us first discuss the various terms used in networking [5–6]:

- Home system: the system to which a mobile customer subscribes. The cellular system transmits mobile system identification (SID) and if it matches with the one programmed within the subscriber's terminal, then the system is a home system. Associated with the home system is the home location register (HLR), which contains all the pertinent information about the mobile.
- Visited system: a different system to which the mobile is traveling. This may be the adjoining system or a system that is widely separated from the home system. Customers who have been given cellular networking privileges can obtain comparable services in these areas. Associated with the visited system is the visited location register (VLR), which will have pertinent information transferred from the HLR. The transfer of information should preferably be automatic and without the knowledge of the mobile. At what stage the information from HLR should be transferred to the VLR is dependent on the system design, and will be discussed subsequently in this chapter.
- Call originations: cellular networking will allow the mobile user to originate calls in any visited system in an identical fashion to his or her home system.
- Termination of calls: cellular networking will allow a subscriber to receive calls in the visited system. This feature can be an optional subscription service or can be provided to all subscribers.
- Controlling switch: for calls originated by the subscriber, the controlling switch is the one connected to the nearest cell site from where the mobile receives the highest signal strength. For calls terminating to a subscriber, the controlling switch is the home system of the mobile. The other name of the controlling switch is the anchor switch.
- Candidate switch: a system networked to the serving system that contains cell sites that may be candidates for the handoff of calls served by the boundary cells of the adjoining system.

Before we discuss the technical approach to cellular networking, let us briefly discuss the requirements of the customer and the system operator.

8.2.1 Customer and Carrier Requirements

The basic customer need is the ability to make and receive calls anywhere and anytime, easily, sometimes referred as "personal communications" or "follow-me communications." The basic idea is that the customer should not be required to remember a long sequence of numbers. He or she should be able to make calls easily and with the same network access codes irrespective of location, and also be able to receive calls throughout the country. Those trying to reach a subscriber should not be required to remember the long sequences of key presses, access codes, and so forth. One of the drawbacks of the current cellular systems is that the roaming procedure varies from system to system. In some systems the roaming feature is automatically activated and in others one has to call customer service ahead of time so that it can be activated. The features do not go with the subscriber when traveling in other cities. Others, who want to call a subscriber, must know the ten-digit access code for the city the subscriber is in, and then the subscriber's ten-digit telephone number. Then, after dialing the twenty-digit telephone number, and if the called phone is powered and not in use, it will ring. After going through all these long procedures, if the phone is busy, the calling party pays for the long distance call to the roamer access number.

The objective from the system operator's point of view is to serve their customers in the best possible way and maximize their profit. These objectives can be met if the following conditions are satisfied: fraud reduction; increased air time usage; and competitive edge. By reducing the fraud, the system operators can maximize their profit. Here, the fraud simply mean the unauthorized use of the system. Increased air time usage by customers can be achieved by directing calls to the roaming customers. By providing networking for the system, the operator provides a virtual increase in coverage, and thus can achieve a competitive edge over the other cellular system operator in the region.

8.3 TECHNICAL APPROACH TO CELLULAR NETWORKING

One of the biggest problem in cellular networking is due to nonstandard interfaces from the cell site to the MSC within one system, and the MSC-to-MSC interface between two systems. These interfaces in general depend upon the supplier, and therefore are likely not to match with different systems. On the other hand, the interface between mobiles and the cell sites has been standardized, which allows any vendor's mobile units to communicate with any other vendor's cell sites. Also, the interface between the mobile MSC and the PSTN is well defined. Keeping in mind these incompatibility problems, cellular networking considerations include the nearness of the networked system, and whether the systems to be networked are manufactured by the same vendor or by different manufacturers. The nearness of the system to be networked is determined by whether the systems are adjacent to each other or widely separated. If the two systems are adjacent to each other, an incoming call may be delivered directly to the adjoining system provided the mobile is visiting that system. Additionally, a mobile crossing the boundary of one system has to be handed off to the adjacent system. If the mobile is visiting the CGSA, which is widely separated from the home office, incoming calls may have to be delivered to the mobile directly. No requirements for the handoff exist in this case. Tests for both adjacent and nonadjacent systems were conducted by AT&T from April 1986 through December 1986. For both these cases of adjacent and nonadjacent systems, we shall discuss the technical requirements and their solutions below.

8.3.1 Call Delivery and Handoff to an Adjacent System

The interconnection between two adjoining CGSAs is shown in Figure 8.2. Voice and data link interconnections between two MSCs have been in existence since 1985. Here, the data links carry the necessary intersystem location requests and their replies, as well as page requests and their responses. Let us take up an example of an incoming call to a mobile whose home area is CGSA 1, and who is temporarily located in CGSA 2. For an incoming call to the subscriber from the network interface point (NIP) in CGSA 1, MSC 1 will request paging the subscriber in all cells of CGSA 1. In addition, a paging

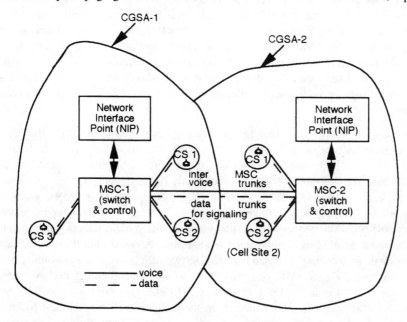

Figure 8.2 Layout of two adjacent systems. After [1, p. 33].

in CGSA 2 can occur at the same time as in the home system or can occur subsequently if the subscriber is not located in CGSA 1.

Simultaneous paging may burden the signaling channel excessively, but the mobile will be located fast, especially if it is visiting the other CGSA. On the other hand, sequential paging may not be acceptable to the calling party due to the additional delay. If the subscriber in the visited system responds to the paging, the call will be set up by the visited system over an inter-MSC voice trunk back to the NIP of the home system. A valid network subscriber can originate the call in either systems. Mobile calls to the PSTN will be completed though the NIP of the system in which the subscriber is presently located (home or the visited CGSA's controlling switch).

A subscriber located in a geographical area where two systems overlap will be handed off between the cells of each system . This handoff will take place through the process of requesting location measurements from a neighboring cell in the different system, and by the selection of the best candidate cell and system for the mobile. The handoff process that takes place between two systems is similar to the handoff that takes place within a single system except there is an additional phase which establishes the voice circuits between the two MSCs. The following sequence of operations takes place before a handoff between two adjoining systems is complete.

- Location phase. The system with an active subscriber will request level measurement from the adjacent system. This may be used in parallel with its own internal handoff measurement or can happen sequentially after the internal handoff fails.
- Circuit negotiating phase. Upon determining that the adjacent system is the best candidate for the subscriber, the current system must coordinate to establish a voice circuit between two systems (MSC-to-MSC voice circuit).
- Handoff phase. When the voice circuit is established between the two systems, the current serving system will command the mobile to switch over to the new voice channel.

For an ordinary mobile call where two different systems are not involved, there are two links that together determine the quality of transmission. These links are over the air from the mobile to the cell site and the link between the serving cell site to the MTSO. The quality of the RF link from the mobile to the cell is controlled by frequent level measurements. The telephone circuits, used to interconnect the cell site equipment to the MSC, are typically maintained by the operating company to have a consistent loss throughout the system. The handoff within a system will go unnoticed by the mobile if both parts of the link meet the signal level objectives. When a second MSC is involved in the handoff process, an additional voice channel is added between the two MSCs so that the continuity between the mobile and the new MSC is carried over to the original MSC. For example, if a mobile subscriber located in CS 1 of system 1 (CGSA 1) is talking through an MSC 1 network interface point to a land subscriber, the talk path will be: Mobile-CS 1 through MSC 1 to NIP to land subscriber, as shown in Figure 8.2. Now, if the mobile moves to CGSA-2, the new talk path will include a voice trunk from MSC 2 to MSC 1. If the links between MSCs are well maintained, negligible degradation to the speech will occur. However, if the handoff protocol is not robust, several anomalies can result that may affect the quality of connection. This degradation, while annoying to the voice caller, can be catastrophic for data transmission. The following two examples illustrate cases where degradation may be experienced as a result of intersystem handoff not done correctly.

Example 1: The "Shoelace" Effect

This problem involves a mobile user whose signal strength while traveling along the boundary of two systems will alternately favor one system and then another, as shown in Figure 8.3. This can lead to a number of repeated handoffs between the two systems. If the system is not sophisticated, new MSC-to-MSC circuits will establish with each handoff with no concern that circuits between the MSCs may already exist. This excessive waste of facilities results in reduced traffic capacity. To overcome this problem, intelligence must be provided to allow the receiving MSC to determine if an MSC-to-MSC circuit already exists. If the circuit already exists between the two MSCs, only a new voice channel has to be assigned by the receiving MSC to clear the existing voice circuit at the old MSC following completion of hand off.

Example 2: The "Trombone" Effect

Short calls are more prevalent in cellular systems, primarily due to the limited number of available RF channels and high air-time charges. As the cost of cellular calls comes

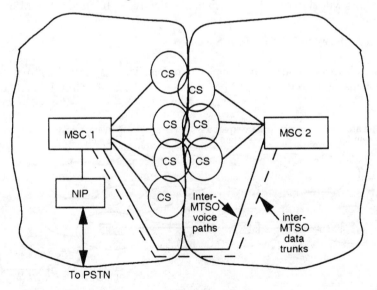

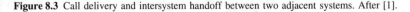

Figure 8.3 Call delivery and intersystem handoff between two adjacent systems. After [1].

down, the average duration of calls is likely to increase. This will be particularly true for digital systems with data transmission facility. For a long duration call the user is likely to cross one or more service area boundaries, which would result into repeated handoffs [7]. Repeated handoffs between systems will lead to repeated extensions of the landline side of the connection from one MSC to another MSC, as shown in Figure 8.4. Here, it is assumed that the mobile's original location is in system 1 (serving area 1). Subsequently, the mobile travels to a cell site controlled by system 2, and lastly to a cell controlled by system 3. The talk paths in these three cases are:

- Voice path when mobile is in system 1: $A_1A_2A_3A_4$
- Voice path when mobile is in system 2: $A_1'A_2'A_3'A_4'A_5'$
- Voice path when mobile is in system 3: $A_1''A_2''A_3''A_4''A_5''A_6''$

The resulting trunking configuration is known as the "trombone" effect. This configuration has several undesirable effects:

- Circuit quality will degrade as additional MSCs are added.
- Both the incoming and outgoing trunks in the intermediate MSC (mobile served by a cell site located in system 3) that is not used at the radio side of the connection are occupied, reducing the traffic carrying capacity of the system.

The problem can be overcome if the connection between the target and the anchor MSC is established in the most efficient or direct path, as shown in Figure 8.5. This will require that the target MSC have knowledge of the anchor MSC's directory number before using the PSTN or the private trunk for the new connection. Establishing the inter-MSC circuit in this fashion will eliminate the tandem connection through intermediate MSCs.

8.3.2 Call Delivery to Nonadjacent Systems

Incoming calls to a mobile can be completed to nonadjacent systems by providing a path from the home MSC to the visited MSC through PSTN, as shown in Figure 8.6. In this

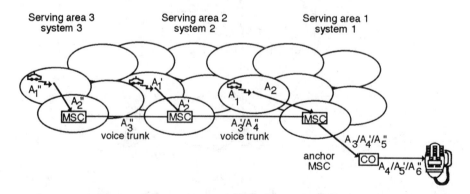

Figure 8.4 The "trombone" effect.

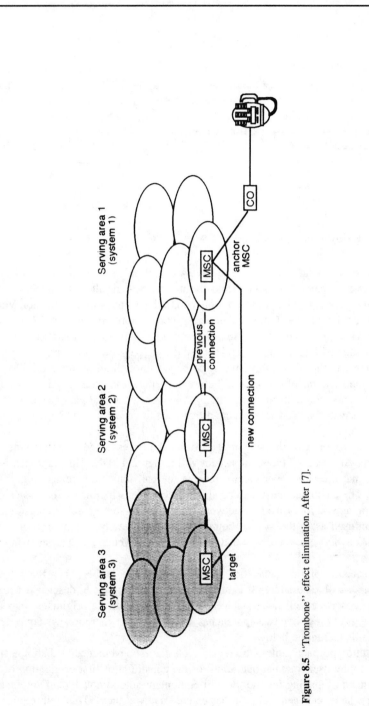

Figure 8.5 "Trombone" effect elimination. After [7].

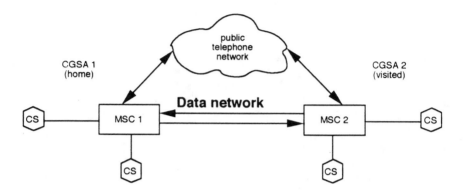

Figure 8.6 Call delivery between two nonadjacent MSCs.

scheme, a subscriber who has traveled to CGSA 2 will be paged in the same manner as described above. A paging response from the mobile will result in the visiting system assigning a temporary local directory number (TLDN) to the mobile and notifying the home system of the TLDN. Once the home system is notified, all future calls can be routed to the visiting system. The home system routes the call via the PSTN using the TLDN, which the visiting system recognizes and uses to page the subscriber. The problem with this scheme is that the mobile has to be paged through all systems initially, since there is no identification of the mobile before it responds to paging. Also, the paging through all systems simultaneously will be prohibitive. Sequential paging will add delay and will be considered a drawback. There are at least three ways in which this drawback can be eliminated.

In the first approach, the customer can dial the access code upon arriving at the visited system. At that time the local system will assign a TLDN. This approach has the disadvantage that the subscriber cannot receive the call until after registering with the local system. The customer must also be able to recognize when he or she has entered a new system or moved to a different networked system. In some cases the customer can simply be confused as to the system being operated. Generally, a mobile user will not have enough knowledge to recognize that he or she has arrived at a new system and needs to register.

In the second approach, the local system assigns the TLDN when the subscriber first originates a call. The situation is similar to the first approach, described above. In this case the subscriber will not be able to receive calls until he originates. In view of the shortcomings of the above two approaches, an automatic registration system is desired. We describe this technique below.

In the third approach, automatic registration, the inconvenience of dialing a special code by the mobile user (manual operation) or the constraint of first origination of a call can be eliminated by having the mobile unit equipment autonomously and automatically register itself whenever it senses that it has entered a new system. This will require some nonvolatile memory in which the subscriber stores the previous system's identification code, which was used during his or her last incoming or outgoing call. As soon as the subscriber moves to an area served by other system, a system code is seen other than what is stored in the unit's memory. The cell sites broadcast the system code regularly, which can initiate the process of self registration with the local system and thus a new TLDN assignment. The automatic registration technique is considered to be superior to the other two techniques discussed above. One of the problems the system owners face is fraudulent calling by roamers. Substantial loss of revenue can occur to the system operators if these users are not tightly controlled. We shall take up the study of fraud minimization below.

8.4 FRAUD MINIMIZATION

As cellular service becomes more and more popular, an increasing number of fraudulent users are causing a substantial loss of revenue by the service providers [9–13]. Cellular fraud is primarily done by roaming users. Disconnected customers and customers who have either suspended service or have been forced to suspend due to nonpayment may travel to another service area and attempt to receive service. This is easy to accomplish as the cellular phone can be programmed to change the subscriber ID, and thus a fraudulent user can make it appear that the phone call is coming from someone else's phone number.

Because of this, a roamer management system is required that will detect fraudulent users and block them in the real time. The roamer validation process assures that the roamer will pay for the calls. The system diagram shown in Figure 8.7 is presently used by some systems in the United States. The validation process determines whether a roamer's number is valid in his or her home system and can be initiated when the mobile makes the first call or when the unit is initially turned on. In either case the serving telephone switching office (MSC) will first check an internal database file to determine if the roamer should be denied the service. Unless the roamer is identified in the resident negative file at the MSC, mobile use will be permitted to continue and the serving MSC will query the clearinghouse for any negative information about the mobile. The clearinghouse will check its own database for this mobile unit. If the clearinghouse finds this number in its negative file list, it will inform the serving MSC so that the carrier can deny any further service to fraudulent user.

If the clearinghouse does not contain any information about a particular mobile, it will query the mobile's home MSC to determine the actual status of the roamer. If the home MSC identifies the mobile to be an invalid customer, the clearinghouse will add this number to its own negative file list and will notify the serving MSC. If the mobile number verification process starts after the mobile initiates its first call, the system will not be able to deny the service to a fraudulent user until after the first call has been made. Depending upon the speed of validation and the amount of backlog at the clearinghouse, the fraudulent user may be able to make several calls before finally being blocked from

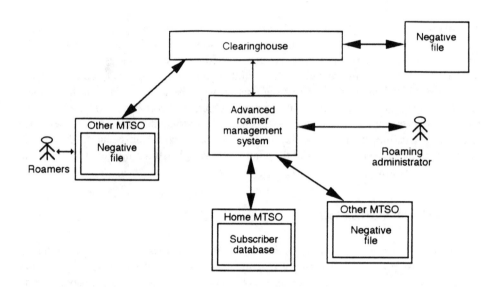

Figure 8.7 Roamer fraud-management network. After [9].

the system. This problem can be overcome if the system is based on a mobile's validation upon registration. Although the system will catch the fraudulent roamers more quickly, it imposes a large burden on the serving MSC and often causes long queues for validation requests at the home MSC. In the end, no matter what scheme one comes up with, it will fail against cheaters who can reprogram their phone with a new ID.

Follow-me roaming (FMR) is a specific implementation of the above call delivery concept. The system requires complete control by the roamer for activating and deactivating the call delivery system. Upon arriving in a new city, the roamer dials a code *18 (211 in Texas) for activating the FMR. The FMR system processes the request, establishes the linkages, and then notifies the subscriber through a voice response that the FMR has been activated. For deactivation, the subscriber simply dials *19 (311 in Texas), which tears down the linkages. By a default mechanism, the linkage will be removed if the subscriber forgets to dial *19 within a certain period (24 hours, at midnight say). Figure 8.8 provides the FMR system configuration. Upon roamer activation of the FMR, a local processor extracts the billing data stream from the home system processor and makes a decision as to whether the roamer should be allowed to make calls.

From the above discussion, it is clear that for call delivery to a roamer it is essential to locate the mobile first, which in turn requires paging. Paging a subscriber is carried over the data trunks. A properly designed data trunk for signaling has to follow a set of rules governing the exchange of data between MSCs. These rules are known as protocols. Thus, we shall discuss the desirable characteristics of protocols based on which this part of the system can be designed.

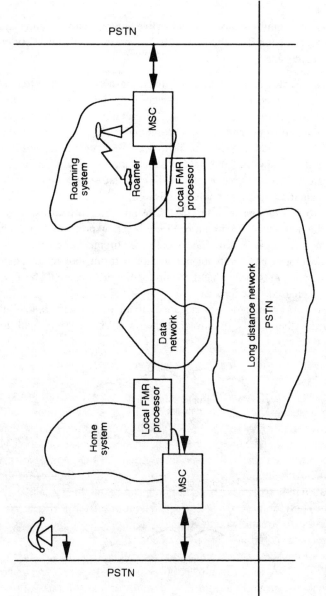

Figure 8.8 Follow-me roaming system. After [19].

8.5 PROTOCOL REQUIREMENTS AND THEIR STRUCTURE

We discuss below the requirements of an effective intersystem signaling system, both what it needs presently, what it will require in the future to grow and satisfy both present and newly evolving services.

- The ability to evolve as the data communication network. The protocol structure should be so designed as to allow a graceful evolution from existing to future communication networks. The two networks of major interest are X.25, a dedicated and public data network, and signaling system 7 (SS7), as shown in Figure 8.9.
- Flexibility. The protocol must be flexible to accommodate the changes in the existing field type and the length of the field. Future addition of new fields must also be accommodated. The addition of private fields (value-added service) for possible revenue-enhancing purposes should be allowed.
- The ability to incorporate new services. The protocol structure should be able to accommodate new services without resulting in a new protocol. For example, if the protocol has to include billing data transfer requirements, it should not result in having to develop a new protocol nor it should result into major changes in the existing protocol. On the other hand, the protocol structure should be able to accommodate the increased functionality.
- Compatibility. As future new services are evolved, the present compatibility should not be lost. In other words, the protocol should not be rewritten with the addition of each new service.

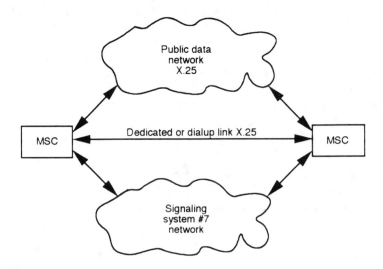

Figure 8.9 Network interconnection.

- Standardization. The resulting protocol should have an international compatibility. The SS7 signaling network, if adapted in the United States, will allow compatibility with Europe and with the rest of the world.

8.5.1 Protocol Structure

The above requirements of the protocol can be satisfied by the seven layers of the open system interconnection (OSI) reference model, which defines a particular set of functions for each layer. The OSI model allows a more common understanding of the layers and the ability to use software developed by different organizations. The logical structure is shown in Figure 8.10 and the functions of each layers are briefly described below.

- The physical layer describes the electrical, mechanical, functional, and procedural characteristics of a physical link to exchange bits of information. This is similar to RS-232 for serial computer communications.
- The data link layer is concerned with the reliable transfer of octets (bytes) over a physical connection. Data is sent with the necessary synchronization, and error control bits added to the message. SS7 networks use the message transfer part (MTP) level 2. The SS7 MTP Level 2 protocol is the key to providing the robustness of the SS7 signaling networks. The X.25 networks use the link access protocol balance (LAPB). We shall describe these in detail in Section 8.6.
- The network layer provides the switching and routing functions required to establish, maintain, and terminate a connection for the transparent delivery of data. It hides the underlying link protocol from the network users. The SS7 network layer is provided by two services: MTP level 3, which provides the basic message routing services; and the signaling connection control part (SCCP), which is used for advanced message routing services.
- The transport layer defines the protocol for reliable exchange of data between two entities and maintaining end-to-end data integrity. The transport layer allows a large block of data to be broken down into smaller blocks (packets) for efficient transmission, and for reassembly at the destination in a proper sequence. In this respect, an end-to-end error recovery is done at the transport layer.
- The session layer defines the protocol for establishing relationships between two applications and synchronization between them for the exchange of information.
- The presentation layer defines the protocol for converting data for different representations. For example, this may be used for converting a data file (from ASCII to binary, say) from one format to another.
- The application layer defines the protocol for exchanging information between two applications. The SS7 network uses a set of procedures at this level known as terminal capabilities application part (TCAP). TCAP, in turn, is based largely on an OSI concept called remote operations, which allows involving the operations and passing parameters required to perform such operations, and to return their results.

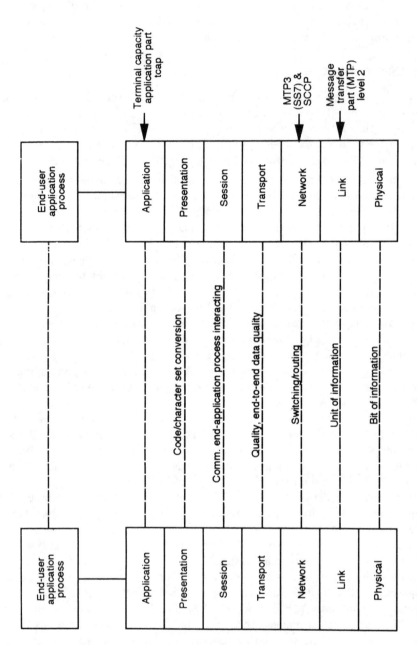

Figure 8.10 OSI model for cellular networking.

The two main layers that are concerned with the end users are the application and the network layers. The application layer defines the services available for communication, which is ultimately transformed into the services that can be offered to mobile subscribers. The application layer answers the question "What can I do?" The network layer defines the connectivity available and answers to the application layer question in terms of "Where can I do this?" For details of the OSI model, the reader is referred to some excellent books on the subject [23–24].

Before we discuss further the protocol standardization in the United States, let us look at the protocol procedure as recommended in X.410. The procedure invokes an operation in a remote system and then reports the outcome of this operation.

8.5.2 Protocol Operations Model

The problem of protocol complexity has been addressed by the message handling systems (MHS) group in the CCITT. As a result of their efforts, X.409 and X.410 were defined. Recommendation X.409 defines a standard notation, or language, that is used for describing complex data structures from which the transmitted bit sequence can be obtained. Recommendation X.410 specifies a general protocol procedure to invoke an operation in a remote system and report the outcome of that operation. The application protocol for the cellular intersystem protocol is interactive, that is, a request for operation is made by a remote system and the outcome is reported to the system originating the operation. X.410 makes use of four types of protocol data units (PDUs), as shown in Figure 8.11 and described below, to perform intersystem operations [9].

- INVOKE is used to invoke an operation remotely. The operation conveys three pieces of information: the operation value, which identifies the operation to be invoked; the invoke identifier x; and the parameters of the operation (operation identifier).

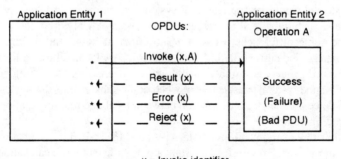

Figure 8.11 Protocol data unit for remote operation.

- RETURN RESULT is used to return the result of an operation to the invoker. It conveys two pieces of information: the invoke identifier x, which identifies the invocation to which this result applies; and the result.
- RETURN ERROR is used for an application-specified error to the invoker. It conveys three pieces of information: the error value, which identifies the error being returned; the invoke identifier x, and the parameter of the error.
- REJECT is used to signal application-independent errors or problems, such as protocol errors, format errors, congestion problems, and so forth.

8.6 PROTOCOL STANDARDIATION IN THE UNITED STATES [12–17]

In the United States, a cellular radio standard supplied by Electronic Industries Association (EIA), EIA TR45.2, has produced an interim standard and cellular radio telecommunication communications intersystem operations: IS-41. The IS-41 standard, a five-part series numbered IS-41.1 through IS-41.5, was approved by the TR45.2 plenary in Washington, D.C., in October 1987. IS-41 is a standard for communication between cellular switches. It is the outcome of a concentrated effort by EIA and Telephone Industries Association (TIA). The intent of the protocol is to provide a vehicle for implementing seamless roaming. Seamless roaming is universally desired and is supposed to be achievable through IS-41 standardization. It is intended to solve most of the roamer problems.

Seamless roaming will allow a subscriber to roam freely throughout the country, using the cellular phone in the same way irrespective of location throughout the country or, perhaps, the world. The subscriber would have to dial no special codes, and make no especial arrangements with the visited system. Calls will simply be delivered wherever the subscriber is presently located. All the features for which he or she has subscribed would work identically on all systems accessed by a roamer. If the roamer has to dial some special code for the activation of features, the same code will be dialed everywhere. If a feature involve tones or announcements to indicate acceptance of a feature command, the same response would be applied to all systems. In summary, a feature would require precisely the same number and the sequence of steps anywhere within the U.S. However, there will be some differences between a subscriber at home and visiting some other system as roamer. The current dialing plan standard, TIA IS-52, allows a local subscriber to dial a seven-digit access number in its own local area, while ten digits have to be dialed in the visited system, irrespective of whether the subscriber is dialing a local number or a long-distance number. This distinction between local and roaming dialing has already been accepted universally.

IS-41 has three versions: IS-41-0, IS-41-A, and IS-41-B. The original issue is called IS-41-0 or simply the revision zero. This standard provides for basic call handoff, allowing active calls to be passed from one system to another adjacent system. IS-41-A adds the basic roamer validation, roamer registration, and call delivery services. The two protocols are not interoperable, though the messages share some common names and functions.

The change allows alignment with the American National Standards Institute specifications of SS7. This will also permit network concepts some what similar to the European GSM system. IS-41-B is the revision of IS-41-A and adds special features to the call delivery process. We shall discuss each of these sequentially.

8.6.1 Handoff to an Adjacent System

IS-41-0, provides for basic call handoff between two dissimilar switches. Intersystem handoff is the procedure required to switch an in-progress call transparently from one MSC to an adjoining MSC. Figure 8.12(a) shows a mobile X using system A for communication to a land-based subscriber Y. The details of the talk path in system A consist of a radio link between a mobile X to cell site, CS, and a fixed connection between cell site CS and the MSC of system A. Figure 8.12(b) shows the sequence of steps for a handoff operation between two adjacent systems. Here, system A requests over the data circuit (shown as step 1) a signal-quality value from one or more adjoining MSCs for the mobile subscriber whose signal is getting weak in system A. The adjoining MSC, B, and the other MSCs (not shown in the diagram) tune into the channel used by the subscriber and report its signal quality value back to MSC A. MSC A determines the best signal-quality value from the data sent back from the other MSCs. The MSC with the highest quality of signal reported is chosen as the candidate system for handoff. Assuming system B to be the candidate in this case, MSC A seizes an unused trunk between MSC A and MSC B and an unused voice channel number is assigned by MSC B for the mobile to switch over to. This sequence is shown in step 2 of Figure 8.12(b). MSC A then instructs the subscriber's mobile unit to tune to the specified channel. The original voice channel to the mobile is released and the new channel through the cell of system B, through MSC B and MSC A to PSTN, is established. Though this example is oversimplified, the actual transition process is complex and may take as much as 150–300 ms. A limitation of this system, as discussed in Section 8.3.1, appears when a subscriber zigzags between two systems. Figure 8.12(c) shows the sequence of messages for measurement of the signal level when the mobile is handed off to cell C, controlled by the third system. The resulting overall connection is the radio link from mobile X to cell C, through three MSCs, C, B, and A, and finally to the PSTN and on to the land subscriber Y. It should be noted that path minimization does not occur in this case. Though the ideal path is a direct trunk between C and A, a longer path consisting of two trunks, A to B and B to C, have been chosen. Path minimization, to prevent redundant inter-MSC trunk setup when a mobile travels between several service areas, is currently being provided only when a handoff back occurs. Figure 8.12(d) shows the setup steps when the mobile is handed back to system B. A similar diagram can be drawn when the mobile is handed back to system A. (See problem 13.) In all of the above cases, upon going on-hook by either of the two parties, the MSC that controls that party will notify the adjoining MSC that the inter-MSC trunk is ideal and should be released.

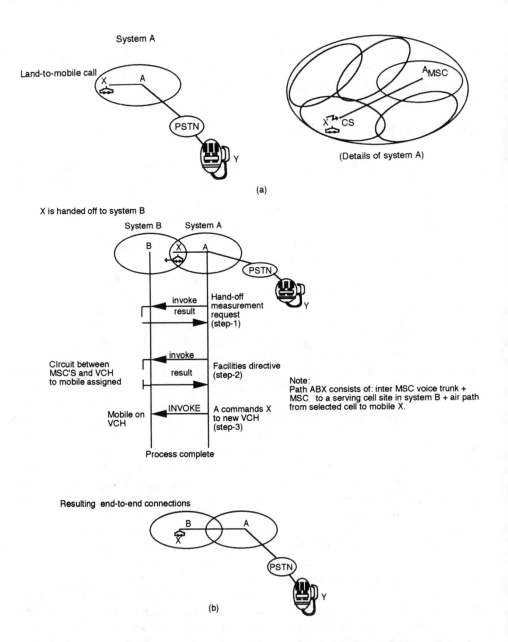

Figure 8.12 (a) Original talk path between mobile X and the landline subscriber Y. (b) Sequence of steps for mobile X handoff to system B and the resulting end-to-end connection (after [14, p. 22]).

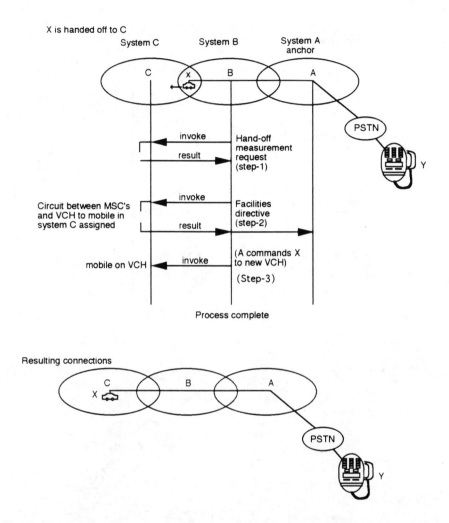

Figure 8.12(c) Sequence of steps for handoff of mobile X to system C and the resulting end-to-end connection. After [14, p. 22].

8.6.2 Roamer Application Service

IS-41-A is the revision of IS-41-0 and retains the basic handoff capabilities of IS-41-0 and adds features for roamer validation, call delivery, and remote feature access [18–21]. The basic network now includes two functions associated with two storages, one with the home system of the subscriber and the other with the system the subscriber is visiting. The HLR is the depository for the system data associated with the home area subscriber. The HLR is a network function that is used to store information about subscribers who belong to that network. The storage will have information about whether the subscriber

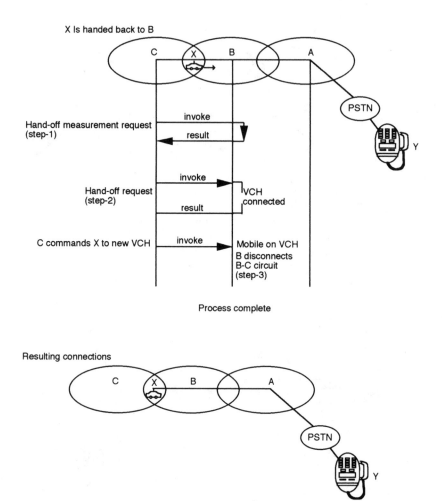

Figure 8.12(d) Mobile handed back to cell B. After [14, p. 24].

has paid his or her bills (validity of the subscriber), what types of features the subscriber is allowed to access, and where the subscriber is temporarily located if it is moved out of his home office. The VLR, on the other hand, carries the information about subscribers who are currently being served. The VLR is associated with the mobile switching center. Layout and interconnection between HLR, VLR, and MSC are shown in Figure 8.13. The VLR and the MSC can be collocated. The following general steps occur in the delivery of the call from the roamer's home system to the visited system.

- The roamer registers with the visited system automatically;

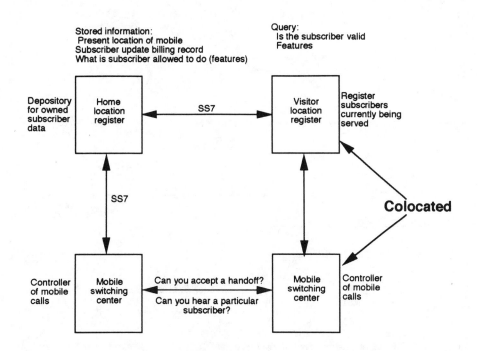

Figure 8.13 Layout and interconnection of HLR, VLR, and MSC. After [18, p. 25].

- The visited system informs the home system of the roamer's whereabouts over the data circuit;
- The home system validates the roamer with all optional subscriber profiles, if applicable;
- The incoming call to the home system is forwarded to the roamer over PSTN.

The initial registration of the mobile is automatic upon entering a new system area. Roamer registration is the process used to inform the HLR of which system a roamer is visiting and obtain permission from the HLR for the roamer to make calls, as shown in Figure 8.14. After registration with the local system, outgoing calls from the roaming subscriber can be made, as a subscriber profile and qualification resides in VLR of the serving MSC.

Call delivery uses the roamer's current location to deliver incoming calls to the roaming customer. When the incoming call arrives at the subscriber's home MSC, the MSC asks the HLR where the subscriber is at the present time. The HLR knows which VLR is presently serving the subscriber, but does not know the temporary local directory number (TLDN) of the mobile. The HLR asks the visiting VLR about the temporary number for the mobile and the VLR in turn asks its MSC to make the actual temporary directory number assignment. The assignment is made by the MSC, and this number is stored in its switch so that an incoming call to the visiting mobile can be identified at the MSC. This assignment is also forwarded to VLR, which in turn forwards the temporary

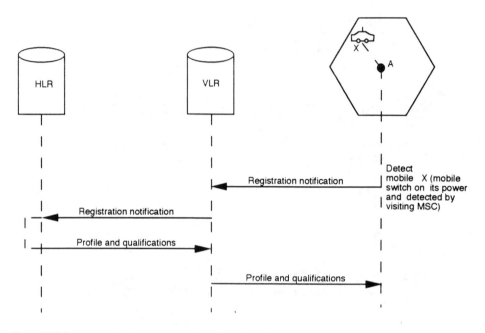

Figure 8.14 Roamer registration process. After [18, p. 28].

directory number to the HLR. The HLR returns this number to the requesting MSC, which forwards the incoming call to the destination MSC over the public telephone network. At the destination MSC, the call is detected as a call to the temporary telephone number. The subscriber is alerted of the incoming call. At this time the temporary number is released so it can be used for another roaming subscriber. The setup process for incoming calls to the roamer is shown in Figure 8.15.

8.6.2.1 Present and Future Planned Improvements in Call Delivery

The TIA TR-45.2 committee is presently working on version B of IS-41, which should add features to the present IS-41-A. Some features will be added in the immediate future and some features will be added later [7]. Features to be implemented soon are:

- Enhanced call handoff and call delivery to handle call conferencing, call waiting, and call transfer features. Standardized feature access code and feature operation sequences so that a roamer can use the feature in the same way.
- Consolidation of subscriber billing records.
- Path minimization to call handoff to eliminate "shoelacing" effects by conserving the interconnecting trunks. The shoelacing phenomenon and its effects have been discussed in Section 8.3.1.
- Enhanced intersystem networking protocol.

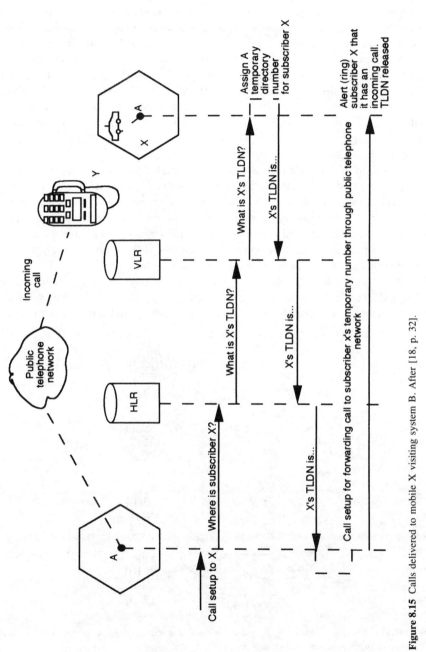

Figure 8.15 Calls delivered to mobile X visiting system B. After [18, p. 32].

IS-41-B requires that the enhanced call feature must operate in the same way in the visiting system as in the subscriber's home system. As an example, Figure 8.16 illustrates the case where a subscriber currently active in a call involving call waiting is about to be handed off to the adjacent system. For satisfactory operation, the feature must continue to operate in the same way even after the handoff has occurred. For three-party conferencing, different manufacturers must agree on the methods of billing the data and controlling the call. One controversial point for controlling the call is which system should be allowed to maintain the bridge after the call is handed off to another system. Here in the U.S., IS-53 describes the different cellular features. IS-53 will standardize cellular feature codes so that call forwarding, for example, will be activated in the visited system in the same way as in the home system.

As a mobile subscriber travels in a single system, the MSC correlates the individual airtime segments from each cell to arrive at the total billable airtime. However, when the mobile travels between systems several discrete records are generated. With an intersystem handoff, the call records residing in different systems must be assembled and rated before being sent to the subscriber. This can be done when billing records are transferred between MSCs, or between MSCs and data processing facilities. If the processing is done within individual systems it may add to several weeks of delay to billing also.

Path minimization is the process of reducing the number of voice circuits between MSCs for calls between systems. Path minimization, as discussed in Section 8.6, will be required for a mobile to hand off from system A to system B and subsequently to system

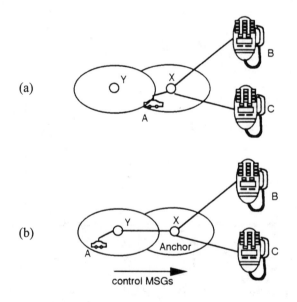

Figure 8.16 (a) Call-waiting between A, B, and C (A controlling the call). (b) A handed off to Y (feature control message sent back to anchor).

C. The voice path will make a long loop involving three cascaded trunks between MSCs. Assuming that there are voice and data circuits between systems A and C, then it would be more efficient to directly connect the trunk from A to C and completely bypass system B. The following features are supposed to be included in the U.S. digital cellular systems at a later date:

- Fraud control enhancements. The HLR will be equipped with intelligence to detect some roamer fraud, such as; two or more subscribers having the same identification, subscribers moving between two service areas faster than physically possible, subscribers roaming beyond preauthorized geographic area limits, and so forth.
- Credit control enhancement. The cost of a given call will be estimated in real time with sufficient accuracy, and a subscriber exceeding his or her preauthorized limits will be cut off. The only difference between the estimate in real time and the final estimate of the cost is the volume discounts, which can only be applied once the complete monthly usage of the mobile is known.
- Real-time posting of billing data. Call billing information will be delivered to the billed party's HLR immediately after call disconnect. From the HLR, the billing information can be posted to clearinghouse systems for intersystem billing and to internal billing systems for immediate estimates of mobile usage. Calls charged to the customer credit card will also be posted immediately. It is the responsibility of the HLR to consolidate billing for multiple billing segments.
- Location service. When a call arrives to a subscriber number, the call can be forwarded to another location, such as to the car, a hotel, a pager, or to the subscriber's home, based on the subscriber's last known location.
- Incoming call screening. Calls from a certain number can be blocked, accepted, forwarded to another number, or a message can be taken.
- Outgoing call screening. This feature will either block or allow outgoing calls to specific numbers.
- Advanced voice mail. Different messages for incoming calls can be played back, based on specified numbers. This will allow lower cost delivery of messages to roamers.

8.6.3 Signaling Network Requirements

The following requirements are anticipated for the present and future growth of the signaling networks in the United States: nationwide connectivity, high availability, real-time response, flexibility for future growth, ability to support multiple controlling agencies, and affordability.

To achieve seamless roaming, nationwide connectivity of the signaling network is desired so that the subscriber data can be accessed by all carriers. This does not mean that all carriers have to access the centralized database. The database can be distributed, that is, each carrier can have its own database. The only requirement imposed is that the database be connected to a signaling network node. The individual database of the subscriber, as discussed above, can be termed the home database and be resident in the HLR.

The network is of little use if all its nodes are not available at all the times. This means that there should be enough alternate paths between nodes so that a catastrophic failure in one or more of the paths does not disable a node completely. Complete 100% availability of a node may be prohibitively expensive, but by judicious design it is possible to approach near 100% availability by providing enough paths between nodes. Node availability can also be enhanced by duplicating the same database (HLR) in more than one node. Thus, if a node is destroyed by natural or manmade disaster, the backup node can be accessed by other carriers.

One of the most important requirements of the signaling network is quick response, which means a minimum delay in the signaling network. This will enable instant verification of mobile data from its home office, which is mandatory for the minimization of fraud. Obviously, the signaling part adds up the delay in the setup of a call to a roamer. Further, an addition of two seconds or less is considered to be acceptable in this application. To achieve this responsiveness, all network components and the signal processors attached to the network must be fast.

Due to the continuing addition of subscribers, it is expected that the requirements for signaling will grow with time. Thus, more nodes will be added in the future and reconfiguration of the existing equipment will be done from time to time. Therefore, the signaling network should be so planned that it is possible to add more nodes, modify signaling routes, and add more signaling links from time to time.

There are two major domains of control, national and local. At the national level, a signaling network operator is responsible for the operation of the overall network. It is the national operator who will control the operation, connections, and traffic in the overall network. However, at the local level, the individual carrier should be allowed to control and maintain their own subnetwork for signaling. This will allow the carriers to maintain tighter control over their local network operations, thus making it easier to maintain.

The design of the network should be affordable by the carriers and in turn by subscribers. In the real world, what good is a network if no one can afford to use it?

8.6.4 Possible Network Configurations

Based on the requirements of IS-41 discussed in the previous section, there can be various network configurations [18]. There are essentially two types of networking requirements. One type satisfies the requirements of mobiles who travel to the adjacent systems and the other type is for those who travel to distant MSCs.

Systems that require signaling between two adjacent MSCs can be satisfied by point-to-point networks. A point-to-point network is characterized by direct link connections between a given node and all other end points with which the node has to communicate. The initial release of IS-41 required point-to-point networks. A point-to-point network is all right for a small number of nodes, but it becomes complex as the number of nodes increases. For N nodes and full connectivity, the number of links required are $N(N-1)/2$. Thus, the network becomes increasingly complex and expensive as the number of nodes increases. A variation on point-to-point network is the one which allows the messages to pass through intermediate nodes. A network of this type will require message processing at intermediate nodes, which adds additional delay to the messages. A full connectivity can, however, be achieved easily. Networks discussed in Section 8.3 are examples of point-to-point networks that can serve the adjacent, and to a limited extent nonadjacent, MSCs.

Another possible network configuration, which will satisfy the point-to-point signaling requirement, is the star network. See Figure 8.17. The network in this case has a central node through which all end-node connections are made. If the end nodes are considered to be the MSC, the connection from one MSC to another will pass through central node. This is a prevalent connection scheme for data terminals talking to the host computer, resident at the central node. The disadvantage of this configuration is that if the central node becomes unavailable, the entire network is inoperative. Thus, a configuration of this type has to be modified, that is, more redundancy for paths has to be provided before the structure can be adapted for cellular signaling networks. A variation of the star network is a distributed star. In this configuration, there are several centralized interconnected nodes. This configuration provides better connectivity as a failure of an individual node does not disable the entire network. The roamer fraud-management network discussed in Section 8.4 uses a distributed star network configuration for the roamer validation.

For satisfying the nationwide requirements of roamer signaling as required in IS-41-A/B, the point-to-point network discussed above is unsuitable. Different types of networks have to be considered. We shall discuss two candidate network types which can satisfy these requirements: circuit switched and packet switched.

A circuit-switched network allows a connection to be established between nodes on a demand basis. When node A requires connection to node B, it requests the connection. If an idle link is found between nodes, connection is established by the network and node A and B communicate. After the communication between nodes is complete either of the two nodes can ask the network for the release of the path. The network will release the paths between nodes along with the hardware at these nodes. Circuit switching is used

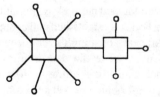

Figure 8.17 Star network.

extensively for voice communication. A typical voice terminal generates one to two calls during the peak busy hour of a day. Thus, a typical voice terminal is in use only for 10% of the busiest hour of the day. Unlike the traffic generated by voice terminals, data terminals either generate a long message with less frequency or generate a short message with high frequencies. In either case circuit switching is inefficiently utilized: More frequent messages overload the common equipment; less frequent long messages underutilize the voice trunks. Also, the problem with this scheme for cellular signaling is the delay time in establishing the path. The scheme is suitable for those cases where the duration of data transmission is far greater than the time it takes to establish the initial path through nodes. The advantage of the scheme is that once the path is established, the communication is fast as there is no slowdown due to intervening nodes. Since the requirement of signaling is in real time and short, initial path-establishment time degrades the overall performance, and thus the scheme cannot be used for cellular radio.

The second scheme that can be considered for IS-41-A/B is the packet-switched network. Here, each data message or packet has a header containing routing information. The information is used to pass the packets through intermediate nodes towards the destination. The advantage of the network is that full network connectivity can be obtained by a small number of links and intermediate equipment. The packet header makes it possible to distinguish messages and pass the packets to the correct destination node based on the header information. Since the packet transmission from node to node is controlled by the network, contention is not possible. Also, compared to circuit switching, the system can be used for a large number of short messages without establishing the path through the network for each message. The disadvantage of the scheme is that the response time is based on the traffic through nodes. If there are a large number of packets requiring transmission through the network, packet delay is increased.

There are two types of packet switching: connectionless and connection oriented. In connectionless switching there is no connection or release of virtual connections. Each packet contains all the addressing information required to route a message though the network. The advantage of connectionless packet switching is that no separate message is required for the initial connection establishment or release. The scheme is especially suitable for a large number of short messages between one transmitting node and several receiving nodes. This may be a practical requirement for cellular systems where a single MSC will have to communicate simultaneously with a large number of destination MSCs. SS7 is an example of a connectionless packet service. Unlike the connectionless network, the connection-oriented packet-switched network is similar to the circuit-switched network where the connection between nodes is established and at the end of the packet transmission the connection is released. After establishing the connection or virtual circuit, a number is assigned. This number is used while the data is transferred between the connected nodes to associate a data packet with the address information of the connected nodes. X.25 is an example of connection-oriented packet switching. The advantage of the scheme is that the actual transmission of data is fast. The disadvantage is that it takes some time at the beginning of the data transmission for connection establishment. Thus, the scheme is more efficient for the bulk transmission of data, as required for large files transmission during sessions. The scheme is also good for requiring a large number of simultaneous low-volume transmissions between nodes.

As we have seen, X.25 is the network of choice for IS-41-0 and SS7 is the choice for IS-41-A/B. We discuss below the major characteristics and the configuration of both these networks.

8.6.5 Packet-Switched Network for Signaling

With the dramatic reduction in the cost of processing power, communication networks based on the concept of packet switching offer an attractive alternative to circuit switching. Due to the bursty nature of data traffic, circuit switching is not efficiently utilized. Packet switching, on the other hand, dynamically assigns the bandwidth and thereby allows an efficient sharing of the transmission bandwidth among many active users. In packet switching, data is sent using limited size blocks called packets. The information at the transmitting end is divided into a number of packets and transmitted over the network to the destination where it is assembled to retrieve the original message. In addition, packet switching provides better connect time, reliability, economy, and flexibility. In view of their immense advantages, the TR45.2 subcommittee has aligned IS-41 with two such networks: the X.25 and the signaling network SS7. The X.25 network was originally designed to connect terminals to computers, whereas the SS7 network was designed for high-speed switching. In this section we briefly describe both these networks from the cellular applications point of view.

8.6.5.1 X.25 Signaling Network

As described earlier for the proper identification of a roamer, two databases (HLR and VLR) have to exchange data. The HLR always has the most up-to-date information about its home subscribers regarding credit worthiness, subscriber profile, and the subscriber's last known location. The VLR needs this information before the connection to the roamer can be established.

The following characteristics of X.25 should be noted: high connectivity, flexibility, accessibility, and affordability. The X.25 network allows connections to be made to a wide number of end points. Connections can also be made around the world. An X.25 network can be expanded without severely disrupting service on the existing network. In North America, X.25 connections can be made anywhere, making it a most desirable solution for signaling. This technology is simple and has also matured. We describe X.25 in the context of cellular radio. The X.25 network shown in Figure 8.18 is built using two basic building blocks: signaling points and packet switches. The signaling point in this case are the host computers and the mobile switching centers. Host computers contain the database of the HLR and the MSC interface with the VLR. The general X.25 network

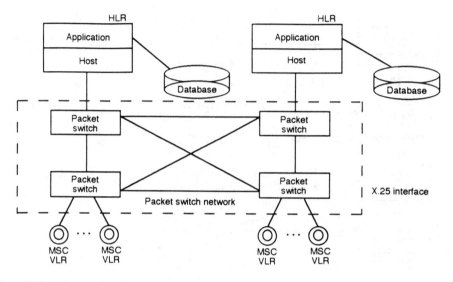

Figure 8.18 X.25 network (typical).

interface is between the data terminal equipment (DTE) and the data circuit-terminating equipment (DCE), as shown in Figure 8.19. Obviously, on one side of the network the host HLR plays the role of DTE and on the other side the MSC-VLR interface plays the role of DTE.

The X.25 interface between the DTE and DCE consists of three levels of protocols. The X.25 interface also allows the multiplexing of connections between a DTE and a number of other DTEs on the network on the same physical circuit. The three levels of the X.25 interface are the physical level, the link level, and the packet level, as shown in Figure 8.20. The physical level specifies the use of a duplex, point-to-point synchronous circuit for the physical transmission path between the DTE and the network. Recommendation X.21 specifies the electrical and procedural characteristics for digital circuits. The link level is based on the international standards organization (ISO) high-level data link

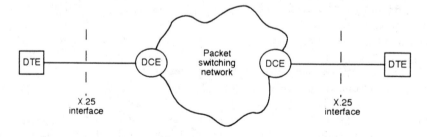

Figure 8.19 X.25 interface. After [25].

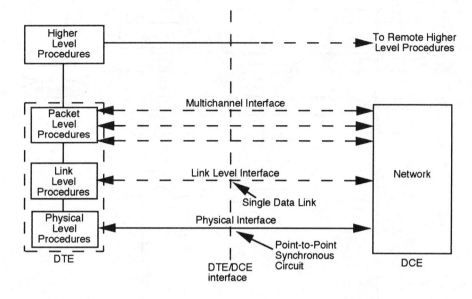

Figure 8.20 X.25 interface structure.

control (HDLC) link-level procedure. The primary responsibility of the link level is to ensure the correct exchange of data between the DTEs. Recommendation X.25 currently defines two protocols at the link level: the link access procedure (LAP) and the link access procedure balanced (LAPB). LAPB is the preferred protocol and is compatible with HDLC procedures. It differs from LAP in the mode of operation and supervisory commands. In the LAPB protocol, each data frame carries a single packet across the X.25 interface. The most significant function of the link level is to provide an error-free but variable delay link between the DTE and the network. The packet level of the interface is the one that gives X.25 its character of a virtual-circuit interface to a packet-switched service. The packet level provides facilities to establish virtual circuits and then send and receive data. Each virtual circuit may be flow controlled using a window mechanism. Recovery from errors at the interface can be effected using the clear and restart procedures.

X.25 specifies a number of formats for the packets passed between the DTE and DCE in order to establish, use, and clear virtual circuits. The format for the packet is shown in Figure 8.21. There are two types of packets: data packets and supervisory (control) packets. A single-bit C/D field indicates whether a packet is a data packet or supervisory packet. All packets contain a three-octet common header field, as shown in the figure. The first field in the common header is called the general format identifier and consists of four bits. The first bit is referred to as the Q-bit or qualifier bit. The Q-bit may be set to one or zero on data packets to give two levels of data transmission. The second bit is called the data confirmation bit or D-bit and is used to request end-to-end or local acknowledgment from the network. The third and fourth bits distinguish between

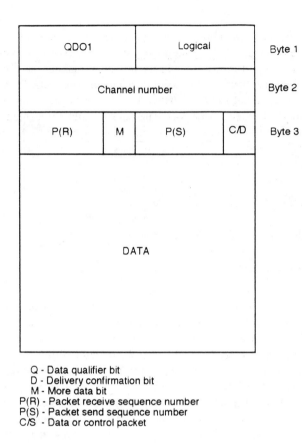

| QDO1 | Logical | Byte 1 |

Q - Data qualifier bit
D - Delivery confirmation bit
M - More data bit
P(R) - Packet receive sequence number
P(S) - Packet send sequence number
C/S - Data or control packet

Figure 8.21 Data packet format.

two possible window mechanisms for flow control. The second group of four bits provides the group number for the logical channel. In the third set of eight bits P(R) provides the packet receive sequence number, P(S) provides the send sequence number, and the designation M indicates that more data bits will follow.

8.6.5.2 Signaling System Number 7

Signaling System Number 7 (SS7) [22] was first issued by CCITT in 1980, with revisions in 1984 and 1988. SS7 is designed to be an open-ended common channel signaling standard that can be used over a variety of digital circuit-switched networks. Furthermore, SS7 is specifically designed for ISDN. The overall purpose of SS7 is to provide an internationally standardized general-purpose common channeling signaling system. The scope of SS7 is immense since it must cover all aspects of control signaling for complex digital networks, including the reliable routing and delivery of control messages and the application-oriented content of those messages. With SS7, control messages are routed through the network to perform call management (setup, maintenance, and termination) functions. These messages are short blocks of data known as packets. Though the network being controlled is a circuit-switched network, the control signaling itself is packet switched. The important characteristics of the network are high throughput, retransmission capability, flow control on data links, and retransmission speeds of up to 64 Kbps. Other characteristics, such as automatic reconfiguration, fault tolerance, and redundancy, make it quite distinct from X.25 as discussed above.

Some of the notable characteristics of SS7 networks are:

- High availability/high reliability;
- Low transit delay;
- Traffic management and fault isolation;
- Automatic congestion detection;
- Access link redundancy; and
- High connectivity

SS7 is highly reliable because it has at least one alternate path around any network node. All nodes except the service signaling point (SSP) are replicated on the network in geographically diverse locations. The SSP itself is usually redundant to ensure a high availability. Applications as shown in Figure 8.22 are duplicated. Normally the signaling transfer point (STP) routes queries to both terminating STPs. The long-term BER is 10^{-6}, though the short-term error rate may be as high as 10^{-4}. For achieving low transmission delay, the network works at relatively high speeds of 56 Kbps and 64 Kbps. The STP (packet switches are called STPs) and the signaling end points are also specified to turn messages around in a few milliseconds so that the overall query delay will be acceptable for telephone users.

The network uses signaling traffic management, signaling link management, and signaling route management. In the event of a fault, the signaling route management supports automatic procedures for informing adjacent signaling points that they must not route any more messages to the faulty nodes or to the destination address. This is done by updating the routing tables to route the messages away from the faulty nodes. Once the fault is cleared, messages are sent to that effect and normal routing is restored. In a nutshell, a single or a few node failures do not effect the operation of the network from the user's point of view. The network tries to recover automatically from congestion. Congestion is caused when more messages are requested to be sent than can be sent across a specific link. When a link becomes congested, congestion control procedures are initiated that attempt to correct the problem. This is achieved by extensive network monitoring and network controls. Simple errors are corrected by requesting retransmission of messages. A high number of errors indicates a link failure. The failed network node is removed from the network and its failure is reported to the network management system. Access link redundancy is achieved by providing two physical links between any two

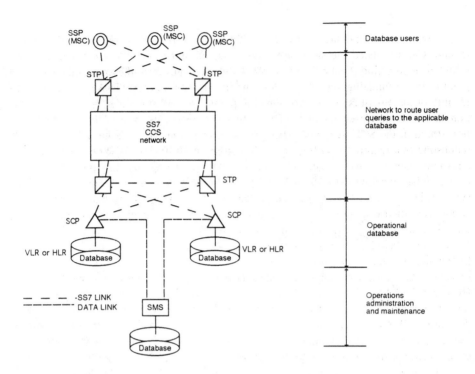

Figure 8.22 A typical SS7 network.

nodes. These links are specified to be diversely routed, that is, the routing paths are separated as far as physically and electrically possible. Links are continuously monitored for connectivity and functionality. If a link or node fails, it is detected instantly. When a link is not sending any signaling traffic, fill-in signaling units (FISUs) are sent to make sure both ends of the link are able to communicate and ready to receive and service real messages. Each SS7 message has both the forward and the backward sequence number. The sequence number is used to ensure that messages are delivered across the data link in the right order and to ensure that no message is lost. If the messages are lost, they are resent from the last known good exchange. High connectivity is due to its connectionless service with an unlimited number of possible connections.

Network Configuration

The SS7 network configuration for cellular applications is shown in Figure 8.22. The main components of the network are the service signaling point (SSP), the signaling transfer point (STP), the service control point (SCP), and the service management system (SMS).

Two types of messages flow over an SS7 network: circuit related, and database related. Circuit-related messages are used to set up and tear down a call between two switches or signaling points; STP transfers these messages from the originating signaling point to the terminating signaling point. The information in these messages includes the identity of the circuit between two signaling points, the called number, an answer indication, and a circuit release indication. Database access messages are used to access information stored in the SCP. A query message is sent from the SSP, via the STP, to the SCP requesting that required data be sent to the requesting SSP. The SCP returns the data in a response message over SS7 network, back to the originating signaling point. In the case of a cellular signaling network, the SSP includes the functions of the mobile switching center (MSC). An SSP is a telephone switching office that is able to launch SS7 queries and react to their responses. The STP is a high-speed packet switch with powerful routing facilities, including global title translation for routing packets to the proper destination. Its routing is based on link availability, node availability, and the supplied address. The SCP contains applications and databases to provide the basic intelligence for the network services. The database contains subscriber records in two different functional entities: the HLR and the VLR. The HLR is the centralized database and may be duplicated for high reliability. The VLR is included with each MSC. In the initial implementation, both these databases (HLR and VLR) may be included in the same MSC. As the network query capacity increases or the cellular carrier desires a consolidation of the subscriber's database, these functions will be migrated to different SCP platforms.

The service management system (SMS) is used to update and synchronize the databases in multiple SCPs. The service management databases are updated manually or automatically to reflect changes in the customer and network records. These changes can take place during off-peak hours. Before we leave this section, let us describe the SS7 protocol architecture.

Protocol Architecture

The SS7 architecture, as shown in Figure 8.23, consists of four levels. The lower three levels, referred to as the message transfer point (MTP), correspond to the first three layers (physical, data link, and network) of the OSI model. The first two levels of the MTP are known as the signaling data link and the signaling link. The top level of the MTP, referred to as the signaling network level or function, provides for routing data across multiple control points from the control source to the control destination. In the areas of addressing and connection-oriented service, these three layers together do not provide a complete set of the functions and services specified by OSI layers 1–3. Therefore, in the 1984 version of SS7 an additional module was added that resides in level 4, known as the singling connection control part (SCCP). The SCCP and MTP together are referred to as the network service part (NSP). The SCCP is an MTP user and therefore is in level 4 of SS7. The MTP provides a connectionless message transfer system that enables signaling

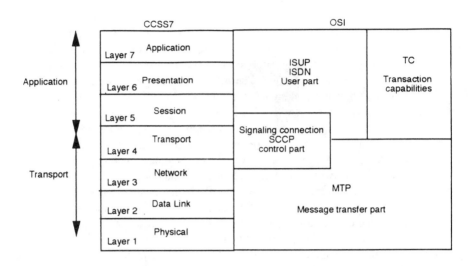

Figure 8.23 SS7 protocol architecture.

information to be transferred across the network to its desired destination. Functions included in the MTP allow for system failures to occur in the network without adversely affecting the transfer of signaling information. The SCCP provides additional functions to the MTP for both connectionless and connection-oriented network services. The transaction capabilities application part (TCAP) provides mechanisms for transaction-oriented (as opposed to connection-oriented) applications and functions. Finally, the ISDN user part (ISUP) provides for the control signaling needed in an ISDN to deal with ISDN subscriber calls and related functions. We briefly describe below each of these levels.

Message Transfer Part

The overall purpose of the MTP is to provide reliable transfer and delivery of signaling information across the signaling network and to have the ability to react and take the necessary actions in response to system and network failures to ensure that reliable transfer is maintained. Figure 8.24 illustrates the functions of the MTP, their relationship to one another and to the MTP users. These levels are now described below.

A signaling data link (level 1) is a bidirectional transmission path for signaling, consisting of two data channels operating together in opposite directions at the same data rate. It fully complies with OSI layer 1 (physical layer). This level specifies the physical, electrical, and functional characteristics of the link. For digital signaling data links, the recommended bit rate for the ANSI standard is 56 Kbps, and for the CCITT standard it is 64 Kbps. Lower bit rates may be used, but the message delay of the user parts must be taken into account. The minimum bit rate allowed for telephone call-control applications is 4.8 Kbps. In the future, bit rates higher than 64 Kbps may be required (1.544 Mbps for North America and 2.048 Mbps elsewhere).

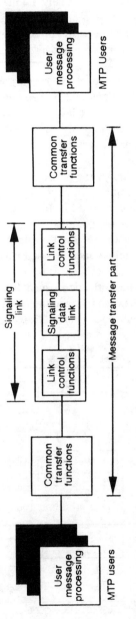

Figure 8.24 Functions of the MTP.

The signaling link level (level 2) corresponds to the data link control layer of the OSI model. Thus, its purpose is to turn a potentially unreliable physical link into a reliable data link. The SS7 signaling link level uses the same principles as the high-level data link control (HDLC) and its variants, such as LAP-B. However, the formats of signaling are different. Signaling messages are transferred over the signaling link in variable lengths called "signaling units."

The signaling network functions (level 3) correspond to the lower half of the OSI's network layer and provide the functions and procedures for the transfer of messages between signaling points, which are the nodes of the signaling network. The signaling network functions can be divided into two basic categories: signaling message handling and signaling network management. The details of these are shown in Figure 8.25. When a message comes from a level 3 user, or originates at level 3, the choice of the signaling link on which it is to be sent is made by the message routing function. When a message is received from level 2, the discrimination function is activated and it determines if it is addressed to another signaling point or to itself. If the received message is addressed to another signaling point and the receiving point has transfer capability, the message is sent to the message routing function. If the received message is addressed to the receiving signaling point, the message distribution function is activated and the message is delivered to level 4.

The purpose of the signaling network management function is to provide reconfiguration of the signaling network in the case of signaling link or signaling point failures and to control traffic in the case of congestion or blockage. The objective is that when a

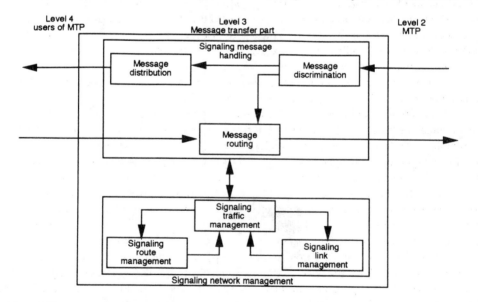

Figure 8.25 Signaling network function.

failure occurs, the reconfiguration is carried out so that the messages are not lost, duplicated, or put out of sequence, and that the message delays do not become excessive. Signaling network management consists of signaling traffic management, signaling route management, and signaling link management. We describe below the functions of each of these.

The signaling traffic management function is used to divert signaling traffic, without loss of the message or duplication, from unavailable signaling links or routes to one or more alternate signaling links or routes, or to reduce traffic in case of congestion. If unavailable, a changeover procedure is used to divert signaling traffic to one or more alternate links. When the link becomes available, a changeback procedure is used to reestablish signaling traffic on the signaling link made available. When signaling routes become unavailable or available, forced rerouting and controlled rerouting procedures are used to divert the traffic to alternative routes or to the route made available.

The signaling route management function is used to distribute information about the signaling network status in order to block or unblock signaling routes. The signaling link management function is used to restore failed signaling links and activate new signaling links. There is a basic set of signaling link management procedures, and this set of procedures is provided for any international or national signaling system. Two optional sets of signaling link management procedures are also provided, which allows for a more efficient use of signaling equipment when signaling terminal devices have switched access to signaling data links.

The signaling network layer does not provide all of the routing and addressing capabilities that the OSI model demands from the network layer. Because of this, the SCCP enhances the services of the MTP to provide the functional equivalent of OSI's network layer. The addressing capability of the MTP is limited to delivering a message to a node and uses a four-bit service indicator to distribute messages within the node. In addition to enhanced addressing capability, the SCCP provides four classes of service, two connectionless and two connection-oriented: (a) class 0, basic connectionless class; (b) class 1, sequenced (MTP) connectionless class; (c) class 2, basic connection-oriented class; and (d) class 3, flow control connection-oriented class.

Transaction capabilities (TC) refer to the set of protocols and functions used by a set of widely distributed applications in a network to communicate to one another. In SS7 terminology, TC refers to the application layer protocols, that is, TCAP plus any transport, session, and presentation layer services and protocols that support it. For all SS7 applications, TCAP directly uses the services of the SCCP, which in turn uses the services of the MTP, with the transport, session, and presentation layers being null layers. In essence, TCAP provides a set of tools in a connectionless environment that can be used by an application at one node to invoke the execution of a procedure at another node and exchange the results of such invocation.

The ISDN-UP of the SS7 protocol provides the signaling functions that are needed to support the basic bearer service as well as supplementary services for switched voice and nonvoice (data) applications in an ISDN environment. Prior to ISDN-UP, another UP, TUP, was specified that provided the signaling functions to support control of telephone calls on national and/or international connections. ISDN-UP, however, provides all the functions provided by TUP plus additional functions in support of nonvoice switched calls plus advanced ISDN and intelligent network (IN) services. In North America, ISDN-UP is used by all carriers, whereas most other countries have implemented TUP with some enhancements. TUP and ISDN-UP can coexist at the same exchange.

8.7 CONCLUSIONS

Call delivery and handoff are the two major problems faced by analog cellular subscribers today. With the introduction of digital cellular, these problems are actively being pursued and various short-term and long-term solutions have been proposed in the United States. Cellular networking will effectively extend call delivery to areas that are close as well as widely separated. In this chapter we have discussed these concepts logically. In the next chapter, we take up the problem of transporting data between cell sites and from cell sites to the MTSO. Due to the extraordinary demand for traffic, these data rates are high and the use of microwave is getting popular.

PROBLEMS

8.1 For an incoming call to a subscriber whose home office is MSC-1 (CGSA-1), should he or she be paged in its own CGSA and in the adjacent CGA simultaneously or one after another? Enumerate the necessary sequence of operation for mobile handoff between two adjacent systems.

8.2 Define the following terms with respect to cellular networking:
- Home system;
- Visited system;
- Controlling and candidate switches;
- Another MSC;
- TLDN;
- Seamless roaming;
- FMR.

8.3 Why are "shoelacing" and "trombone" effects important for successive handoffs to an adjacent system? Will the trombone effect be important if the inter-MSC voice paths have a 0 dB loss?

8.4 Compare the three schemes discussed in Section 8.3.2 for roamer registration. Can you suggest any other scheme for roamer registration?

8.5 Explain why fraud is mostly associated with roaming mobile.

8.6 A set of desirable characteristics of protocols have been discussed in Section 8.5. Can you show one-by-one that these characteristics are satisfied by the OSI protocol model?

8.7 Reorder the requirements of problem 6 according to their order of importance. Is the list complete? If incomplete, can you suggest some more requirements?

8.8 Why is the temporary telephone number assignment done for the roamer? Comment on the various advantages and disadvantages of temporary assignment as you see it.

8.9 Why can circuit switching not be used as a cellular telephone signaling network? Why is a packet-switching network more desirable in this application? List all the differences and similarities between connectionless packet switching and connection-oriented packet switching.

8.10 Assuming that a mobile subscriber X, as shown in Figure 8.26, has a call in progress with a landline subscriber Y and moves to cell B, draw the actions taken by cell sites A and B to complete the handoff. Actions should include power measurement of the mobile, command for a channel change, and the release of the original channel.

8.12 Do you agree with present and future planned improvements in the U.S. for call delivery as discussed in Section 8.6.2.1? If you do not agree with this ordering, reorder them as you see proper.

8.13 Draw the sequence of steps as shown in Section 8.6.1, assuming that X is at system B and is to be handed back to system A.

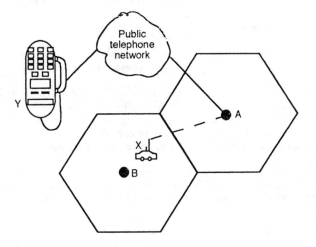

Figure 8.26 Mobile subscriber X.

REFERENCES

[1] Huff, D. "The Building of the Nationwide Cellular Network," *Cellular Business*, Aug. 1988.

[2] Fuhrmann, W., and F. R. Detecon. "Radio Access Protocol of the New GSM Land Mobile Communication Standard," *IEEE Vehicular Technology Conference*, 1988, pp. 30–37.

[3] Cheesemen, D. S., and R. Potter. "System Features-Next Generation Cellular Radio," *IEEE Vehicular Technology Conference*, 1987, pp. 152–156.

[4] Hoff, J. "Mobile Telephony in Next Decade," *IEEE Vehicular Technology Conference*, 1987, pp. 157–159.

[5] Gamst, A. "Remarks on Radio Network Planning," *IEEE Vehicular Technology Conference*, pp. 160–165.

[6] McCarthy, J. J., and G. E. Marco. "Cellular Networking Functionality and Application," *IEEE Vehicular Technology Conference*, 1987, pp. 305–311.

[7] Culp, G. D. "Cellular Intersystem Handoff: Creating Transparent Boundaries," *IEEE Vehicular Technology Conference*, 1986, pp. 304–310.

[8] Gamst, A. "A Study of Radio Network Design Strategies," *IEEE Vehicular Technology Conference*, 1986, pp. 319–328.

[9] Eckelman, S. "Minimizing Fraud," *TE & M*, Sept. 15 1990.

[10] Tsai, D., and J. Chang. "A Contention Based TDMA Technique for Mobile Data Communications," *IEEE Vehicular Technology Conference*, 1986, pp. 329–333.

[11] Thomas, R. "Performance Evaluation of the Channel Organization of the European Digital Mobile Communication System," *IEEE Vehicular Technology Conference*, 1988, pp. 38–43.

[12] Stern, H. P. "Design Issues Relevant to Developing an Integrated Voice/Data Mobile Radio System," *IEEE Vehicular Technology Conference*, 1988.

[13] Dong, Y., and A.U.H. Sheikh. "Performance Analysis of a Packetized Data Communication System on Mobile Radio Channels," *IEEE Vehicular Technology Conference*, 1988, pp. 65–70.

[14] Chaplain, D. J. "A Standard For Intersystem Operations," *Cellular Business*, June 1988.

[15] Falciasecca, G., M. Frullone, G. Riva, M. Sentinelli, and A. M. Serra. "Investigation on a Dynamic Channel Allocation for High Capacity Mobile Radio Systems," *IEEE Vehicular Technology Conference*, 1988, pp. 176–181.

[16] Chang, C. M. "Multisite Throughout of a Mobile Digital Radio Link the Uplink Channel," *IEEE Vehicular Technology Conference*, 1988, pp. 408–413.

[17] Felix, K. "Packet Switching in Digital Cellular Systems," *IEEE Vehicular Technology Conference*, 1988, pp. 414–418.

[18] Carlson, K. D. "The Promise of Seamless Coverage," *Cellular Business*, Nov. 1990.

[19] Kaczmarek, K. W. "Cellular Networking: A Carrier's Perspective," *IEEE Vehicular Technology Conference*, 1989, pp. 1–6.

[20] Kinoshita, K., N. Nakajima, I. Horikawa, and M. Kuramoto. "Considerations on the Mobile Radio System Architecture," *IEEE Vehicular Technology Conference*, 1989, pp. 708–710,.

[21] Suzuki, S., K. Funakawa, T. Utano, and A. Nakajima. "Signaling Protocol Architecture for Digital Mobile System," *IEEE Vehicular Technology Conference*, 1988, pp. 729–734.

[22] Modarressi, A. R., and R. A. Skoog. "Signaling System No. 7: A Tutorial," *IEEE Communications Magazine*, July 1990.

[23] Stallings. *Data and Computer Communications*, Macmillan: New York, 1991.

[24] Tanenbaum, A. S. *Computer Networks*, Prentice-Hall: Englewood Cliffs, NJ, 1981.

[25] Carlson, K. D. "A Data Communications Primer for IS-41." *Cellular Business,* Jan. 1991.

Chapter 9
Digital Multiplexing in Cellular Radio

9.1 INTRODUCTION

The exceptional growth in cellular telephone demand has forced operators to search for the most efficient means of collecting data at cell sites and transmitting it to the mobile telephone switching office (MTSO). What used to be point-to-point four-wire circuits for voice and signaling transmission are now being converted into T1 networks. Digital multiplexers are being used to combine and transport large numbers of telephone conversations. "Multiplex" refers to a system designed to transmit and receive two or more signals over a common transmission facility. Three alternatives exist for these: copper, fiber-optic cable, and microwave radio. Among these, fiber optic is most popular. Fiber optics offers many advantages, such as high-quality transmission, immunity to atmospheric interference, and high bandwidth capacity, making future upgrades to higher capacity easy. The major disadvantages of this technology are the need to have access to right-of-way and the need for diverse routing so that the network can be protected from cable breaks. Due to these disadvantages, microwave radio is considered more favorably in cellular applications.

The choice of microwave has several advantages. First, the microwave system is more reliable and easier to install than its counterpart multiplex lines. The ease of installation is of great advantage, especially when circumstances require the system to be online at once. The flexibility of adding channels is another factor leading to the choice of this approach. Last but not the least, the long-term cost savings are more favorable for microwave than for multiplexed cable or the fiber-optic systems. For most cellular operators, the biggest advantage of a microwave network is the fact that they have existing antenna support structures and other site facilities that have a lot of path configurations with close spacing to cell sites. Also, multiplexed equipment is well standardized in telephone companies and can easily be adopted in mobile radio surroundings. One other factor that leads to the choice of microwave radio is that the system does not require a right-of-way as the cable based systems do. In view of the above, microwave radio systems have become the primary means to interconnect cell sites for cellular systems. Most major systems in the United States , as well as many smaller systems, use microwave radio systems to carry traffic to the MTSO and also the backhaul traffic to central offices.

Early multiplex systems were frequency division multiplexed (FDM), whereby the transmission facility is divided into a number of discrete frequency bands and each channel is assigned its own band. Newer systems are digital and signals are T1 multiplexed for improved networking efficiency. Due to increased traffic load, many companies are anticipating transmission over DS-3 in the near future. A DS-3 digital backbone network also will allow the cellular telephone company to take advantage of evolving technology, including digital cell-site transmission. To be effective, a DS-3 distribution must accommodate some unique requirements of cellular industry. The important requirements include efficient handling of partially filled T1s, the ability to operate in unmanned offices, high reliability and availability, and remote alarm and surveillance capabilities. The cost of transmission facilities represents the largest recurring expense in operating a cellular telephone system. With the cost of leased T1 circuits ranging between a few hundred dollars to a few thousands dollars per month, some networks can economically justify a DS-3 system with as few as ten T1 circuits. The economic equation also considers the capability of quickly adding capacity to traffic growth. Not being able to serve customer calls due to capacity constraints in backbone facilities may lead to customer dissatisfaction and eventual loss of customers to a competitive system.

The 2-GHz frequency band is the bulk carrier of cellular traffic, followed by 6 GHz, 10 GHz, 18 GHz, and 11 GHz. The following are some of the advantages of 2-GHz band for cellular choice. Systems operating at 2 GHz can span distances up to 30–40 miles without repeaters, which covers almost all the cases of cell-site distances to the MTSO. Also, rain and other atmospheric losses are low. In spite of these advantages, the use of 2 GHz requires a large antenna size with their associated mounting and other complexities. Also, due to unexpected system growth, larger systems are changing the old 2-GHz backbone to wideband radios using other frequency bands, such as 6 GHz, 10 GHz, and 18 GHz. Even a high frequency like 23 GHz is being considered. Obviously, these higher frequencies can only be used for low rain areas.

Our discussion is divided into four sections. Section 9.2 deals with the U.S. and CCITT digital hierarchy, and the role of multiplexing in cellular systems. Sections 9.3 provides DS-1 and DS-3 applications to cellular radio. Section 9.4 provides examples of microwave applications to cellular systems in operation. Section 9.5 concludes.

9.2 GENERAL UNDERSTANDING OF MICROWAVE SYSTEMS FOR CELLULAR APPLICATIONS

Digital transmission of analog speech requires that the speech be converted into 8,000 samples per second, that is, one sample is taken every 125 μs. If each sample is converted into 8 bits, this will produce a data rate of 64 Kbps. If 24 of these channels are combined, an aggregate data rate of 24 × 64 Kbps is produced. If we add a single bit for frame synchronization, the total data rate becomes (24 × 64 + 8) Kbps = 1.544 Mbps. This data rate is the digital multiplexed standard (DS-1) of North America, which was developed originally by AT&T in the United States and has been adopted by Japan and Canada. Thus, the frame length of the T1 carrier shown in Figure 9.1 is 125 μs and carries the data at the rate of 1.544 Mbps. The number of T1 carriers (4 T1 carriers) constitutes a T2 carrier, which has a total data rate of 6.312 Mbps and provides the DS-2 standard in the United States. Other standards are obtained by combining some higher multiples of 24 channels. The North American digital hierarchy is shown in Table 9.1 and Figure 9.2(a). A similar system has been developed by CCITT for the rest of the world. The standard number of channels combined in the CCITT frame is 30, with a corresponding data rate of 2.048 Mbps. This is arrived at by combining 30 channels of data with two channels for frame alignment, alarms, and exchange signaling: (8 × 32)8,000 = 2.048 Mbps. The data rate for other levels is shown in Table 9.2. The CCITT digital hierarchy is shown in Figure 9.2(b). It should be noted that the data rate specified at each level is bounded by some limits of variation expressed in terms of parts per million. Thus, at 1.544 Mbps, one DS-1 standard can have bit variation of 200 bps over 1.544 Mbps (130 × 10^{-6} × 1.544 × 10^{6}).

Some submultiples of the channel data rate can be generated by suppressing one or more of the bits. For example, if the eighth bit (least significant bit) in a channel slot is suppressed, the data rate per channel is reduced to 56 Kbps. Similarly, by suppressing the two least significant bits one can transmit channel data at the rate of 48 Kbps.

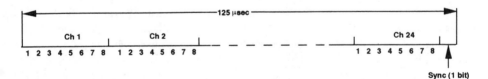

Figure 9.1 DS-1 transmission frame.

Table 9.1
North American Digital Hierarchy

Level	Signal Label	Number of Channels	Data Rate (Mbps)
1	DS-1	24	1.544 ±130 ppm
	DS-1C	48	3.152 ±30 ppm
2	DS-2	96	6.312 ±33 ppm
3	DS-3	672	44.312 ±9 ppm
4	DS-4	4,032	274.176 ±10 ppm

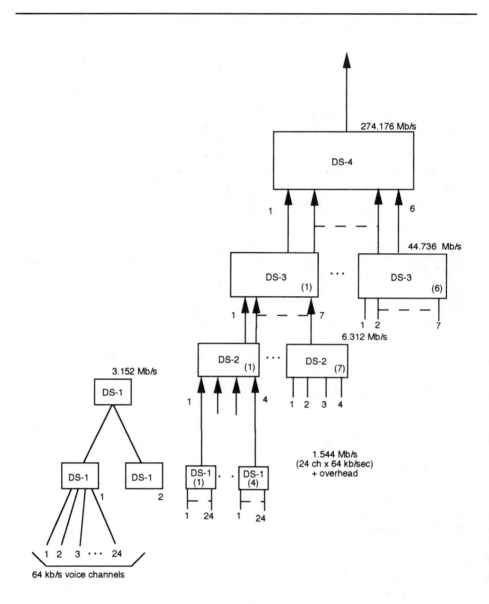

Figure 9.2(a) North American digital hierarchy.

The configuration for asynchronous multiplexing/demultiplexing is shown in Figure 9.3. There are two main functions performed in the multiplexing process [1]. The first step is to synchronize the incoming low-speed (LS) digital signals (referred as tributaries) to a common bit rate. Dummy bits are added in order to synchronize these incoming signals to a common rate. Thus, the output rate of the synchronizer is slightly higher than

Table 9.2
CCITT Digital Hierarchy

Level	Signal Label	Number of Channels	Data Rate (Mbps)
1	CEPT-1	32	2.048 ±50 ppm
2	CEPT-2	19	8.448 ±30 ppm
3	CEPT-3	480	34.368 ±9 ppm
4	CEPT-4	199	139.268 ±15 ppm
5	CEPT-5	7,680	565.148

the maximum input tributary rate allowed; for example, the bit rate of an internally synchronized DS-1 tributary is such that the multiplexed output data rate does not exceed 1.544 Mbps ± 130 ppm. Secondly, the synchronous tributaries are now combined in mux/demux block with the overhead bits into a "high-speed" digital signal by means of a parallel-to-serial conversion. Some overhead bits are added for framing information, which is necessary for the demultiplexing process to separate out the incoming bit stream from the frames. On the demultiplexing side, the incoming high-speed (HS) singular bit stream is first divided into N synchronous LS signals, which are then individually desynchronized to yield the original tributaries. As the multiplexing level increases, wider use of the overhead information, stuff control, and alarm, are necessary.

Figure 9.4 shows a common type of asynchronous high-speed digital multiplexers that are in use. The MX3 multiplexer is getting popular in cellular systems for DS-3 applications, since it offers a certain flexibility with regard to the LS tributaries. Here, various combinations of DS-1, DS-1C, and DS-2 are combined to yield the DS-3 output. The reverse is true on the receiving side for demultiplexing. At the first level of multiplexing (DS-1), 24 voice circuits may be terminated with a single T1 multiplexer, as shown in Figure 9.5(a). One multiplexing unit is required at each end for point-to-point communication. Figure 9.5(b) shows M13 multiplexers (M13 meaning the multiplexing of T1s onto a T3), which are used to combine 28 T1s for transmission over T3 or DS-3 facilities. For point-to-point communication, multiplexers are required at both ends. Because the channel bank was developed as a point-to-point multiplexer, the drop and insert multiplexer was developed to accommodate multipoint systems [16–18]. Typical setups based on DS-1 and DS-3 microwave systems for carrying data from cell sites to the MTSO are shown in Figure 9.6. Here, the first figure shows a configuration in which a T1 circuit is terminated with T1 multiplexers. Signals from three cell sites are combined before feeding the DS-1 signal to the MTSO. Point-of-presence (POP) interconnection trunks must be connected to a cellular backbone network. Here, T1 POP trunks are combined with the local cell site traffic at the drop/insert multiplexer and then the combined traffic can be carried to the MTSO. Figure 9.6(b) shows the layout of this scheme.

A back-to-back configuration based on M13 and a T1 channel bank is shown in Figure 9.6(c). A problem arises when it becomes necessary to access channels at an

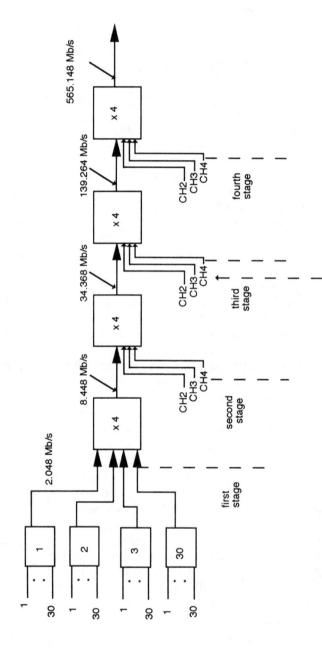

Figure 9.2(b) CCITT digital hierarchy.

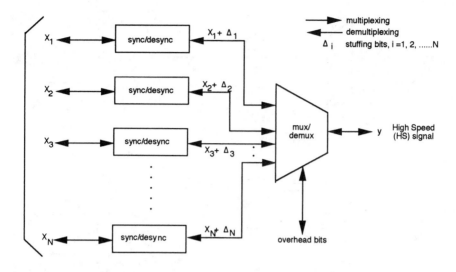

Figure 9.3 Asynchronous multiplexing/demultiplexing.

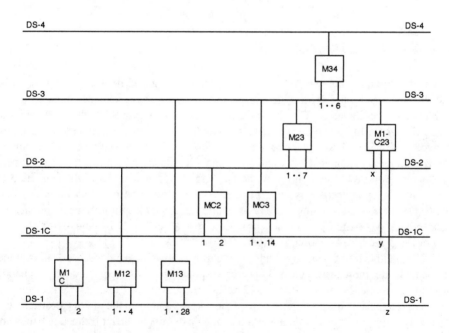

Figure 9.4 North American digital multiplexing/demultiplexing.

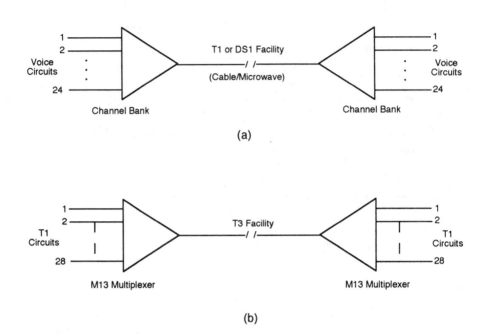

Figure 9.5 (a) DS-1 transmission; (b) DS-3 transmission.

intermediate point. This can be overcome by terminating and demultiplexing the DS-3 line into 28 DS-1 lines with a demultiplexer and using a T1 channel bank to access individual voice channels. This is again multiplexed and carried to the next cell site, where the whole thing can be repeated if desired. In the figure, the transmission to and from cell site 1 is on DS-3, and cell site 2 is an intermediate cell site where voice circuits are derived by a T1 channel bank and additional circuits are added to a T1 channel bank and then transmitted to cell site 3 by the DS-3 facility. Once again, at the cell site 3 an M13 multiplexer is used. It is an expensive configuration and can also introduce jitter due to DS-3/DS-1 conversions. Newly available DS-3 add/drop multiplexers replace the back-to-back M13 multiplexers and provide direct access of DS-1 and DS-0 channels from a DS-3 facility, as shown in Figure 9.6(d).

FCC-authorized frequency bands of operation, allowable bandwidth, and the number of channels are shown in Table 9.3. The corresponding CCIR frequency range, channel spacing in frequency bands from 2–12 GHz are shown in Table 9.4. Since cell-site distances range from 10–20 miles, between themselves and from the MTSO, cost-effective microwave equipment is chosen in certain available common-carrier frequency bands and associated FCC bandwidth allocations. Since higher frequency bands are susceptible to degradation caused by rain and other atmospheric effects, the system range decreases as the frequency increases. Low bands (2 GHz and 6 GHz) are suited for maximum path-length distances from few miles to 30–40 miles of span. The traffic capacity for 2 GHz

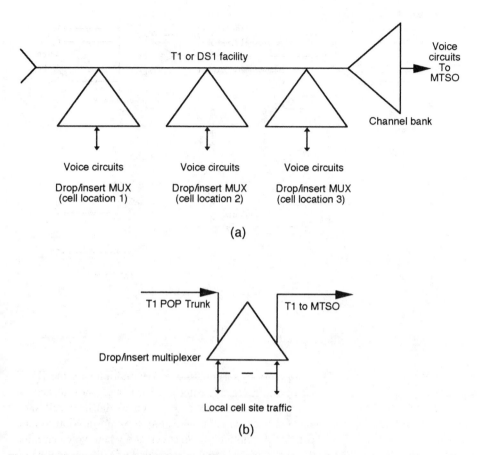

Figure 9.6 (a) Combining signals from three cell sites; (b) point-of-presence interconnection.

range from one T1 to 12 T1s and compression equipment can be used to enhance the voice channel capacity. At 10 GHz, path distances are optimized at 8–9 miles; at 18 GHz, 2–10 miles; and at 23 GHz, it is simply 1–7 miles. The use of the 10-GHz frequency is relatively new and is proving to be a good alternative to the 2-GHz band, especially because the 2-GHz band is becoming congested in large cities.

9.3 DIGITAL MULTIPLEXING IN CELLULAR TELEPHONE NETWORKS

Time division multiplexing (TDM), the action of interleaving several digital signals into a single high speed bit stream, has become one of the most widely used techniques for the efficient transmission of digital signals across a communication channel [16–18]. As the data rate increases in cellular systems, high-speed digital multiplexing becomes more important for transporting data from cell sites to the MTSO. A typical cellular network

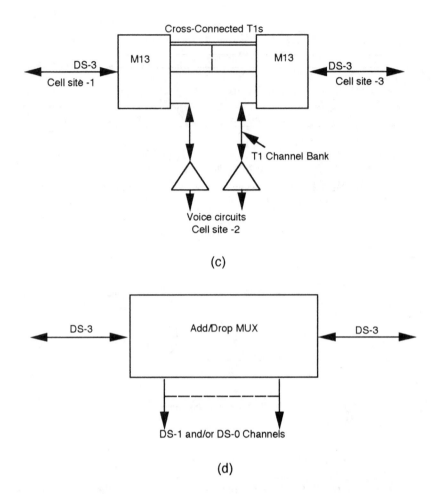

Figure 9.6 (cont.) (c) Back-to-back architecture; (d) methods of DS-3 channel access.

transporting multiplexed data from cell site 2 to the MTSO, located at the cell site (n + 2), is shown in Figure 9.7. Here, we assume that data collection from adjoining cells is at cell site 2. Similarly, the MTSO may also collect data directly from its adjoining cells in addition to the multiplexed signal from cell 2. We shall elaborate on this approach in the following section.

9.3.1 DS-1 or T1 Carrier Application in Cellular Systems

In this section we shall apply the digital multiplexing techniques discussed above, and use them at cell sites for transporting voice and signaling data economically to the MTSO.

Table 9.3

FCC Authorized Frequency Bands

Frequency Band (GHz)	Allowed Bandwidth (MHz)	Minimum Capacity (Number of channels, n)	PCM Bit Rate in Mbps (n × 64 Kbps)
2.11–2.13	3.5	96	6.144
2.16–2.18	3.5	96	6.144
3.7–4.2	9.0	1,152	73.728
5.925–6.425	30.0	1,152	73.728
10.7–11.7	40.0	1,152	73.728

Table 9.4

CCIR Frequency Allocations

Frequency Band (GHz)	Frequency Range (MHz)	Channel Spacing
2	1,700–1,900	14
4	1,900–2,100	
	2,100–2,300	
	2,500–2,700	
	1,700–2,100	29
	or	
	1,900–2,300	
4	3,700–490	29/40
6	5,925–6,425	29.65
	6,430–7,110	9/40
7	7,425–7,725	7/14
8	890–8,500	11.662
	7,725–8,275	29.65
11	10,700–11,700	40
13	12,750–13,250	14/28/35

The simplest one-carrier link application between the cell site and the MTSO is shown in Figure 9.8. Completely filled T1s result in the lowest per-channel costs, minimize the total number of lines, and streamline the system architecture. This is suitable for low traffic applications. A cell site in a busy urban center may, in fact, be served by several such circuits. When the cell site does not have adequate traffic to completely fill the T1 facility, drop and insert multiplexers (as discussed above) can be used. This type of multiplexer would be used when several cell sites are to share the T_1 facility and each site would need to access only a small number of channels. Figure 9.9 shows one such configuration where three cell sites together fill the complete T1 carrier bandwidth. This architecture is for systems serving rural areas (RSAs) and highway corridors where cell sites may require only a small number of channels. Here it is assumed that each of the three cell sites has eight channels of traffic. Leased lines or microwave can be used from

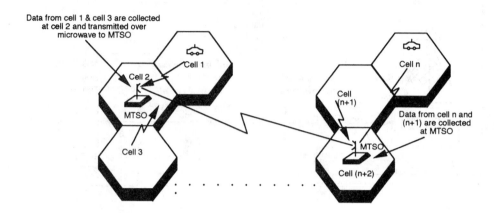

Figure 9.7 Typical cellular network.

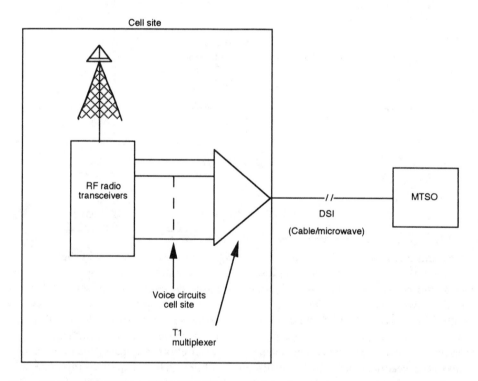

Figure 9.8 A typical MTSO to cell site DS-1 link.

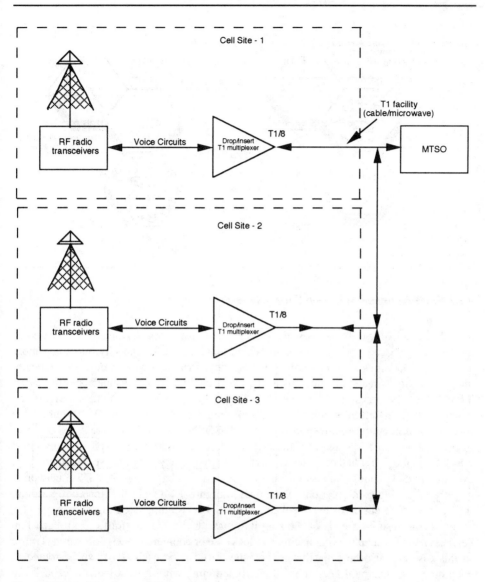

Figure 9.9 Use of drop/insert multiplexers.

cell sites 1, 2, and 3 before the T1 carrier system is completely filled. By sharing the T1 facility among several cells, T1 leased-line costs are reduced as compared to a separate T1 lines at each cell site. The star configuration uses a 1/3-filled DS-1 from cell sites 1, 2, and 3, multiplexed, which is then transported through the microwave system to the MTSO. T_1 capacities are available from different manufacturers in the frequency bands of 2 GHz, 6 GHz, 10 GHz, 18 GHz, and 23 GHz, as shown in Table 9.5

Table 9.5
T1 Capacities Available on All Radios in the Common-Carrier Frequency Bands

	Common-Carrier Frequency Band (GHz)					
Manufacturer	2	6	10	11	18	23
AT&T		56		56,84		
Avantek	4, 8	56			6, 28	1, 4
Ericsson					4	
Farinon	4, 8	56	4, 8	28, 56	4, 28	4
Fujitsu		12, 56				
MNI					28	1, 4, 28
Motorola	4				4	12, 28
NEC		56, 84		56, 84	4, 8, 28	
Rockwell	4, 8, 12	56, 84		28, 56	4, 8, 28	
Terracom		56		56	24	28

9.3.2 DS-3 Application in Cellular Systems [16]

As cellular network traffic grows, the backbone network capacity must be increased to handle the increased traffic. With an increased requirement for capacity, more and more T1 lines and facilities are used until a point is reached where it will become economically and technically attractive to implement the DS-3 facility. Voice channels and the entire T1 link must be accessed by the DS-3 facility. Figure 9.10 shows a star configuration where data from different cell sites are collected at a central point (DS-1 hub) by a microwave link and then transported to the MTSO by the DS-3 facility. A DS-3 signal operates at 44.736 Mbps and can transmit 28 T1s (equivalent to 672 voice circuits) or 7 DS-2. Here, the individual cell sites generate enough data to justify the use of a microwave system on DS-1. Thus, a hub is created around DS-1 or T1 facilities. The DS-3 networks, engineered as ring configurations, are shown in Figure 9.11(a). In the ring configuration, traffic from different cell sites is collected as DS-1 and multiplexed together into a DS-3 signal. Because both ends of the ring terminate at the MTSO, traffic from each hub can be directed around the ring in either a clockwise or counterclockwise direction. Traffic in this case can also be split, that is, half the traffic from each HUB can be routed in each direction. This will keep the cell site functioning with at least half the capacity in case of a failure in DS-3 ring path.

In Figure 9.11(b) two DS-3s are terminated at the MTSO—28 T1s from each side. If engineered properly, traffic can be routed to take full advantage of the dual DS-3 capacity. Backhaul traffic can build up as it traverses the ring toward the MTSO as long as the DS-3 bandwidth is not exceeded in any of the spans. This is easy to achieve as all the cell traffic is destined for the MTSO, and therefore does not drop at any of the intermediate cell sites. This configuration has advantages over a star configuration because in case of a failure at a hub, half the traffic can still be routed to the MTSO from the

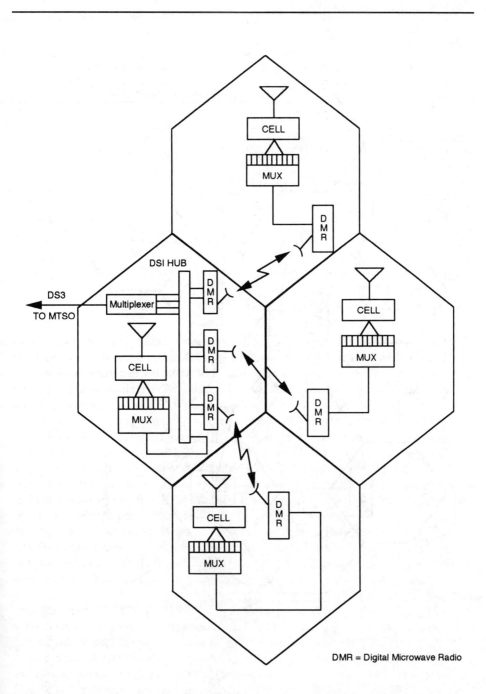

Figure 9.10 Digital microwave radio setup for cellular system. Source: [17, p. 59].

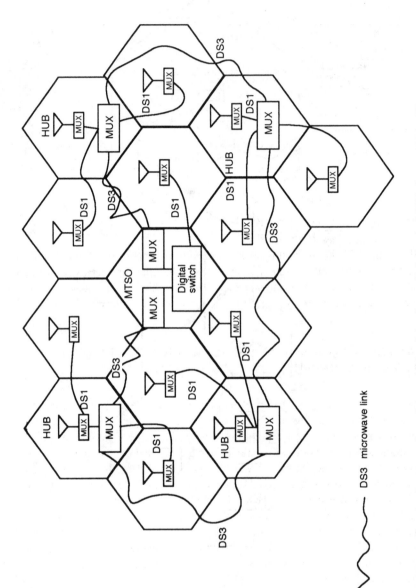

Figure 9.11(a) Ring configuration for transporting cellular multiplexed data from cell sites to the MTSO. Source: [17, p. 59].

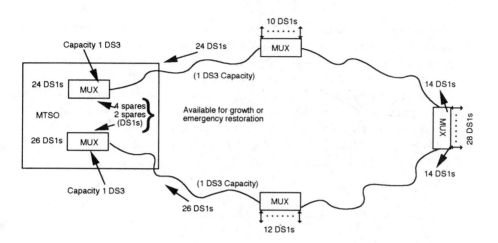

Figure 9.11(b) Traffic in two directions. Source: [17, p. 59].

other side of the ring. Although this does not prevent all traffic from being lost during a failure, it does keep the cell site functioning at least at half the capacity. Also, with remote controllable network elements located at each DS-3 hub, the system can be reconfigured in the event of a major facility outage such as a microwave failure or cable cut.

9.4 EXAMPLES OF MICROWAVE SYSTEMS FOR CELLULAR COMMUNICATIONS [17]

In this section we shall take two examples of operating microwave systems, one in Buffalo, New York; and the other in Mexico City, Mexico. The system in Buffalo is in partnership with Western Union Telegraph Corporation, Graphic Scanning Corporation, and Associated Communication Corporation. Buffalo Telephone Co., the nonwireline carrier serving more than 900 square miles in New York State, was among the first cellular companies in the United States to begin service. With a population of 1.2 million in its coverage area, Buffalo is ranked as the 25th largest metropolitan statistical area. The system consists of nine cell sites in the two-county area: Buffalo, Wheatfield, Lewiston, Elma, Clarence, Lockport, New Oregon, Eden, and Springville. Microwave systems were put in all the cell sites except at Springville, where the calling levels did not justify the use of microwave. The system operator of the Mexico cellular system is the SOS/IUSA. The microwave system for linking the cell sites to the MTSO was chosen partially due to the desired flexibility for capacity expansion as traffic grows. The other reason in Mexico was the enormous difficulty of installing cables in mountainous terrain. It was simply cost-prohibitive to connect land-leased lines in these surroundings. Maintenance of these lines would also have added a high cost.

The multiple cell-site layout of the system in Buffalo is shown in Figure 9.12. The system is built around Ericsson's AXE-10 switch in the mobile telephone switching office, located in Buffalo's Rand Building. Both 2 GHz and 10-GHz systems are in use. The 2-GHz thin-route radios provide an excellent combination of benefits for the system. As stated in the introduction, the 2-GHz system offers an advantage of serving a longer path and the common carrier frequencies are generally available in most rural parts of the country. Due to nonavailability of 2-GHz in downtown Buffalo, the 10-GHz system was selected in a number of cases. The Rand Building contains two systems operating at 10

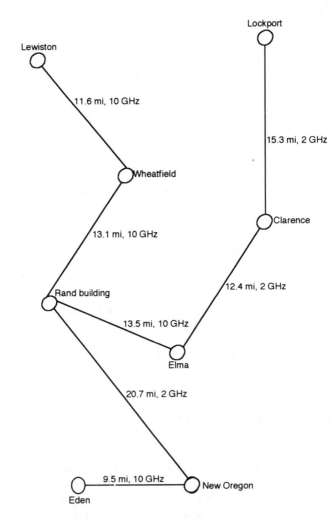

Figure 9.12 Microwave system layout in Buffalo, New York.

GHz and one system at 2 GHz. The systems serving Wheatfield and Elma are at 10 GHz and the system to the south towards New Oregon operates at 2 GHz. The systems from Wheatfield to the south towards the Rand building (MTSO) and to the north towards Lewiston are Urbanet 10 digital radio systems, as described in Table 9.5, and offer as many as three T1 circuits, or 72 PCM voice channels. The system from Elma to Clarence is at 2 GHz. In the south, the system from New Oregon to Eden operates at 10 GHz. The desired frequency of operation is 2 GHz, where both the system capacity and the desired reliability are easily met. However, in some areas of Buffalo the 2-GHz frequency band is simply not available, and this has forced the system operators to choose a frequency in the 10-GHz range. This is a common problem in metropolitan areas, where frequencies are at a premium. The problem was faced both at Wheatfield and Elma.

When the decision to improve communications service in Mexico City and outlying areas was made, the system operators (SOS/IUSA) chose microwave systems as a fast, reliable, and flexible means of communication. Because SOS/IUSA had no precise projections of consumer demand for cellular service, it wanted to leave all of its options open when it came to network expandability. Land-leased lines would have taken months of planning and installation and digging up streets to lay cable. Another factor that influenced the decision in favor of microwave is the high elevation of Mexico City. Long periods of working at a higher elevation for laying cables was considered too tiring and expensive. The use of microwave allowed the luxury of alternate access and also minimized the spare parts problem for maintenance. Each radio featured a one-for-one, hot standby configuration with independent transmitter and receiver switching to ensure uninterrupted service to system users. Today, there are approximately nine links in the network that communicate up to a radius of 60 miles in various directions outside the city. This includes the cities of Queretaro, Cuernavaca, and Toluca.

9.5 CONCLUSIONS

Due to the phenomenal growth in traffic there is an express need to provide high-capacity trunks between the cell sites and the MTSO. This can be satisfied by multiplexed microwave or wideband fiber-optic cable systems. Due to reliability, easy installation, and flexibility in adding channels, microwave is almost always preferred over fiber-optic cable, especially when the circumstances require the system to be online quickly. In this chapter we have discussed different configurations in which microwave systems are being used for cellular applications. Examples of lower capacity DS-1 and higher capacity DS-3 cellular applications have been presented.

PROBLEMS

9.1 Tabulate the advantages and disadvantages of using microwave in cellular networking instead of fiber-optic cable.

9.2 For the following approximate multiplexer data output, identify the signal level in North America or a CCITT digital hierarchy and the number of channels at the multiplexer input.

- 3 Mbps,
- 8 Mbps,
- 45 Mbps, and
- 140 Mbps.

9.3 Explain the following terms used in digital cellular multiplexing:

- Drop/insert multiplexer;
- T1 POP trunk to the MTSO;
- DS-1 link from a cell site to the MTSO;
- DS-1 HUB.

9.4 Explain why 2 GHz and 10 GHz are the most popular bands for cellular multiplexed application.

9.5 Rearrange the cellular layout of 13 cells, as shown in Figure 9.11(a), into a star network, that is, convert the ring into star from HUBs to MTSO. Let the local collection technique from cells to HUB remain the same in principle.

9.6 What signal-to-noise ratio is required to transmit a T1 carrier over 100-kHz line?

9.7 Assuming that a single bit per channel is used for signaling per sixth frame, and a single bit per frame is used for frame synchronization, compute the percent of overhead on a T1 carrier.

9.8 Discuss the reasons why 8 kHz has been established as the international sampling rate for PCM.

9.9 Twenty-four voice signals are to be multiplexed and transmitted over the telephone channel. Assuming a bandwidth efficiency of 4 bps/Hz , what is the bandwidth required for TDM using PCM?

REFERENCES

[1] Fleury, B. "Asynchronous High Speed Digital Multiplexing," *IEEE Communications Magazine*, Vol. 24, No.8, Aug. 1986, pp. 17–25.

[2] Ritchie, G. R. " SYNTRAN-A New Direction for Digital Transmission Terminals," *IEEE Communications Magazine*, Vol. 23, No. 11, Nov. 1985, pp. 20–25.

[3] Amoroso, F. "The Bandwidth of Digital Data Signals," *IEEE Communications Magazine*, Nov. 1980, pp. 13–24.

[4] Dinn, N. F. "Digital Radio: Its time has come," *IEEE Communications Magazine*, Nov. 1980, pp. 6–12.

[5] Bailey, C. C. "Digital Radio Economics," *IEEE Communications Magazine*, July 1981, pp. 6–8.

[6] Hartmann, P. R. "Digital Radio Technology: Present and Future," *IEEE Communications Magazine*, July 1981, pp. 10–14.

[7] Kohiyama, K., and O. Kurita. "Future Trends in Microwave Digital Radio: a View from Asia," *IEEE Communications Magazine*, Vol. 24, No. 2, Feb. 1987, pp. 41–45.

[8] Meyers, M. H., and V. K. Prabhu. "Future Trends in Microwave Digital Radio: a View from North America," *IEEE Communications Magazine*, Vol. 25, No. 2, Feb. 1987, pp. 46–49.

[9] Hart, G., and J. A. Steinkarmp. "Future Trends in Microwave Digital radio: a View from Europe," *IEEE Communications Magazine*, Vol. 25, No. 2, Feb. 1987, pp. 49–52.

[10] Rummler, W. D., P. C. Coutts, and M. Liniger. "Multipath Fading for Microwave Digital Radio," *IEEE Communications Magazine*, Vol. 24, No. 11, Nov. 1986, pp. 30–41.

[11] Taylor, D. P. "Telecommunications by Microwave Digital Radio," *IEEE Communications Magazine*, Vol. 24, No. 8, Aug. 1986, pp. 11–16.

[12] Ivanek, F. *Terrestrial Digital Microwave Communications*, Artech House: Norwood, MA, 1989.

[13] Smith, D. R. *Digital Transmission Systems*, Van Nostrand Reinhold Company: New York, NY., 1985.

[14] Feher, K. *Digital Communications Microwave Applications*, Prentice-Hall: Englewood Cliffs, NJ, 1981.

[15] Townsend, A.A.R. *Digital Line-Of-Sight Radio Links*, Prentice-Hall: Englewood Cliffs, NJ, 1988.

[16] Rooney, D. "A Happy Marriage Made in Buffalo," *Telephony*, Aug. 1986, pp. 40–42.

[17] Grunewald, L. E. "Cellular Companies Moving to DS-3 Networks for Backhaul," *Cellular Business*, Dec. 1988, pp. 56–60.

[18] Hamberger, H. " Microwave in Action," *Cellular Business*, July 1989.

Appendix A
Areas with Operational Cellular Systems

This appendix provides a list of the first 120 areas in the United States where cellular systems are in operation. Each area is designated by its market number, market area name, the name of the licensee (wire and nonwireline serving the area), operational date, the number of cells in the area, and the name of the switching equipment's manufacturer.

Table A.1
Operational Cellular Systems

Market Number	Area	System Operators	Number of Cells	Switching Equipment
1	New York	W-Nynex Mobile (6/15/84)	56	AT&T
		NW-Metro One (4/5/86)	36	Motorola
2	Los Angeles	W-PacTel Cellular (6/13/84)	81	AT&T
		NW-LA Cellular (3/27/87)	38	Ericsson
3	Chicago	W-Ameritech Mobile (10/13/83)	73	AT&T
		NW-Cellular One (1/3/85)	31	Ericsson
4	Philadelphia	W-Bell Atlantic Mobile (7/12/84)	38	AT&T
		NW-Metrophone (2/12/86)	32	Motorola
5	Detroit	W-Ameritech Mobile (9/21/84)	37	AT&T
		NW-Cellular One (7/30/85)	31	Ericsson
6	Boston	W-Nynex Mobile (1/1/85)	30	AT&T
		NW-Cellular One (1/1/85)	10	Motorola
7	San Francisco	W-GTE Mobilnet (4/2/85)	28	Motorola
		NW-Cellular One (9/26/86)	36	Ericsson
8	Washington	W-Bell Atlantic Mobile (4/2/84)	46	AT&T
		NW-Cellular One (12/16/83)	34	Motorola
9	Dallas	W-SW Bell Mobile (7/31/84)	41	AT&T
		NW-MetroCel (3/1/86)	28	Motorola

Table A.1 (*cont.*)

Market Number	Area	System Operators	Number of Cells	Switching Equipment
10	Houston	W-GTE Mobilnet (9/28/84)	36	Motorola
		NW-Houston Cellular (5/16/86)	32	Ericsson
11	St. Louis	W-SW Bell Mobile (7/16/84)	17	AT&T
		NW-Cyber Tel (7/16/84)	13	Motorola
12	Miami	W-BellSouth Mobility (5/25/84)	40	AT&T
		NW-Cellular One (3/7/87)	16	NTI/GE
13	Pittsburgh	W-Bell Atlantic Mobile (12/10/84)	26	AT&T
		NW-Cellular One (12/19/86)	22	Ericsson
14	Baltimore	W-Bell Atlantic Mobile (4/28/84)	46	AT&T
		NW-Cellular One (12/16/83)	34	Motorola
15	Minneapolis	W-US West (6/6/84)	14	NTI/GE
		NW-Cellular One (7/23/84)	12	NTI/GE
16	Cleveland	W-GTE Mobilnet (12/18/84)	15	Motorola
		NW-Cellular One (5/31/85)	19	NTI/GE
17	Atlanta	W-BellSouth Mobility (9/5/84)	38	AT&T
		NW-PacTel Mobile Access (2/5/87)	24	Motorola
18	San Diego	W-PacTel Cellular (8/15/85)	13	AT&T
		NW-Vector One (4/1/86)	13	Motorola
19	Denver	W-US West (7/10/84)	14	NTI/GE
		NW-Cellular One (11/21/86)	11	AT&T
20	Seattle	W-US West (7/12/84)	15	NTI/GE
		NW-Cellular One (12/12/85)	15	AT&T
21	Milwaukee	W-Ameritech Mobile (8/1/84)	12	AT&T
		NW-Cellular One (6/1/84)	9	Motorola
22	Tampa	W-GTE Mobilnet (11/30/84)	16	Motorola
		NW-Cellular One (9/25/87)	12	AT&T
23	Cincinnati	W-Ameritech Mobile (11/15/84)	18	AT&T
		NW-Cellular One (8/1/86)	14	Ericsson
24	Kansas City	W-SW Bell Mobile (8/1/14/84)	13	Motorola
		NW-Cellular One (2/14/86)	13	AT&T
25	Buffalo	W-Nynex Mobile (4/16/84)	9	Motorola
		NW-Buffalo Telephone (6/1/84)	9	Ericsson
26	Phoenix	W-US West (8/15/84)	17	AT&T
		NW-Metro Mobile CTS (3/1/86)	10	Motorola
27	San Jose	W-GTE Mobilnet (4/2/85)	28	Motorola
		NW-Cellular One (9/26/86)	36	Ericsson
28	Indianapolis	W-GTE Mobilnet (5/3/84)	7	Motorola
		NW-Cellular One (2/3/84)	9	Motorola
29	New Orleans	W-BellSouth Mobility (9/1/84)	6	Motorola
		NW-Radiofone (9/6/85)	5	Motorola
30	Portland, OR	W-GTE Mobilnet (3/5/85)	7	Motorola
		NW-Cellular One (7/12/85)	7	AT&T
31	Columbus, OH	W-Ameritech Mobile (5/30/85)	11	AT&T
		NW-Cellular One (7/1/86)	9	Ericsson

Table A.1 *(cont.)*

Market Number	Area	System Operators	Number of Cells	Switching Equipment
32	Hartford, CT	W-SNET (1/31/85)	9	AT&T
		NW-Metro Mobile CTS (10/16/87)		Motorola
33	San Antonio, TX	W-SW Bell Mobile (1/28/85)	12	AT&T
		NW-Cellular One (12/20/86)	13	AT&T
34	Rochester, NY	W-Rochester Telephone (6/4/85)	11	AT&T
		NW-Genesee Telephone (9/22/86)	7	Ericsson
35	Sacramento, CA	W-PacTel Cellular (8/29/85)	6	NEC
		NW-Cellular One (10/31/87)	5	Ericsson
36	Memphis, TN	W-BellSouth Mobility (5/1/85)	6	Motorola
		NW-Cellular One (12/30/86)	6	AT&T
37	Louisville, KY	W-BellSouth Mobility (1/3/85)	6	Motorola
		NW-Louisville Telephone (2/15/85)	5	AT&T
38	Providence, RI	W-Nynex Mobile (8/22/85)	2	AT&T
		NW-Metro Mobile CTS (7/8/87)	8	Motorola
39	Salt Lake City, UT	W-US West (1/29/85)	6	AT&T
		NW-Cellular One (12/19/86)	6	AT&T
40	Dayton, OH	W-Ameritech Mobile (5/31/85)	9	AT&T
		NW-Cellular One (8/1/86)	11	Ericsson
41	Birmingham, AL	NW-BellSouth Mobility (9/26/85)	10	Motorola
		NW-Cellular One (12/29/86)	9	AT&T
42	Bridgeport, CT	W-SNET (5/20/85)	9	AT&T
		NW-Metro Mobile CTS (11/20/87)		Motorola
43	Norfolk, VA	W-Contel Cellular (5/3/85)	18	AT&T
		NW-Cellular One (11/1/85)	6	Motorola
44	Albany, NY	W-Nynex Mobile (6/25/85)	6	NTI/GE
		NW-Albany Telephone (12/2/86)	9	Ericsson
45	Oklahoma City, OK	W-SW Bell Mobile (1/14/85)	9	AT&T
		NW-Cellular One (1/17/86)	9	AT&T
46	Nashville, TN	W-BellSouth Mobility (6/10/85)	13	Motorola
		NW-Cellular One (7/10/87)	11	AT&T
47	Greensboro, NC	W-Centel (5/15/85)	8	Motorola
		NW-Cellular One (12/27/85)	9	Motorola
48	Toledo, OH	W-United TeleSpectrum (7/25/85)	11	Motorola
		NW-Cellular One (4/15/86)	7	Ericsson
49	New Haven, CT	W-SNET (3/4/85)	6	AT&T
		NW-Metro Mobile CTS (11/20/87)		Motorola
50	Honolulu, HI	W-GTE Mobilnet (10/22/86)	11	Motorola
		NW-Honolulu Cellular (6/1/86)	11	Ericsson
51	Jacksonville, FL	W-BellSouth Mobility (6/12/85)	10	Motorola
		NW-Cellular One (8/13/87)		AT&T
52	Akron, OH	W-GTE Mobilnet (10/31/85)	15	Motorola
		NW-Cellular One (12/15/86)		NTI/GE
53	Syracuse, NY	W-Nynex Mobile (1/24/86)	3	AT&T
		NW-Cellular One (12/31/85)	8	Motorola

Table A.1 (*cont.*)

Market Number	Area	System Operators	Number of Cells	Switching Equipment
54	Gary, IN	W-Ameritech Mobile (3/11/85)	4	AT&T
		NW-Cellular One (4/21/86)	2	Ericsson
55	Worcester, MA	W-Nynex Mobile (11/18/85)	3	AT&T
		NW-Cellular One (11/18/85)	10	Motorola
56	Northeast Pennsylvania	W-Commonwealth Mobile (7/2/85)	10	NTI/GE
		NW-Cellular One (12/31/85)	5	NTI/GE
57	Tulsa, OK	W-United States Cellular (8/30/85)	13	NEC
		NW-Cellular One (5/22/86)	10	Astronet
58	Allentown, PA	W-Bell Atlantic Mobile (3/18/85)	32	AT&T
		NW-Cellular One (10/1/85)	5	NTI/GE
59	Richmond, VA	W-Contel Cellular (5/10/85)	20	AT&T
		NW-Cellular One (12/2/86)	7	NEC
60	Orlando, FL	W-BellSouth Mobility (2/27/85)	10	Astronet
		NW-Orlando Cellular (12/29/86)	8	AT&T
61	Charlotte, NC	W-Alltel (4/15/85)	10	Motorola
		NW-Metro Mobile CTS (3/1/86)	8	Motorola
62	New Brunswick, NJ	W-Nynex Mobile (1/21/87)	3	AT&T
		NW-Cellular One (11/1/86)	10	Motorola
63	Springfield, MA	W-SNET (4/10/87)	3	AT&T
		NW-Metro Mobile CTS (10/16/87)	2	Motorola
64	Grand Rapids, MI	W-Century Celluet (9/17/86)	6	NovAtel
		NW-Cellular One (6/18/86)	6	Ericsson
65	Omaha, NE	W-Centel (4/15/85)	4	Motorola
		NW-Omaha Cellular (12/23/85)	3	NTI/GE
66	Youngstown, OH	W-United TeleSpectrum (9/19/85)	4	Motorola
		NW-Cellular One (12/23/85)	5	Astronet
67	Greenville, SC	W-United TeleSpectrum (7/30/86)	4	Motorola
		NW-Metro Mobile CTS (8/18/86)		Motorola
68	Flint, MI	W-Ameritech Mobile (7/12/85)	2	AT&T
		NW-Cellular One (7/30/85)	5	Ericsson
69	Wilmington, DE	W-Bell Atlantic Mobile (3/27/85)	32	AT&T
		NW-Cellular One (6/26/86)	5	Motorola
70	Long Branch, NJ	W-Nynex Mobile (2/20/87)	3	AT&T
		NW-Cellular One (11/1/86)	5	Motorola
71	Raleigh-Durham, NC	W-United TeleSpectrum (11/11/85)	10	Motorola
		NW-Cellular One (9/16/85)	12	NTI/GE
72	W. Palm Beach, FL	W-BellSouth Mobility (5/23/85)	28	AT&T
		NW-Cellular One (3/6/87)	5	NTI/GE
73	Oxnard, CA	W-PacTel Cellular (10/30/85)	3	AT&T
		NW-Ventura Cellular (9/1/87)	6	Ericsson
74	Fresno, CA	W-Contel Cellular (5/1/86)	9	AT&T
		NW-Cellular One (10/24/87)	4	Ericsson
75	Austin, TX	W-GTE Mobilnet (9/27/85)	7	Motorola
		NW-Cellular One (12/27/85)	9	AT&T

Table A.1 (*cont.*)

Market Number	Area	System Operators	Number of Cells	Switching Equipment
76	New Bedford, MA	W-Nynex Mobile (12/9/85)	2	AT&T
		NW-Metro Mobile CTS (11/20/87)		Motorola
77	Tucson, AZ	W-US West (8/6/85)	4	NTI/GE
		NW-Metro Mobile CTS (4/1/86)	4	Motorola
78	Lansing, MI	W-Century Cellunet (9/16/87)	5	NovAtel
		NW-Cellular One (9/17/86)	7	Ericsson
79	Knoxville, TN	W-United States Cellular (7/23/85)	8	NEC
		NW-McCaw (12/21/87)	7	AT&T
80	Baton Rouge, LA	W-BellSouth Mobility (7/2/85)	6	Motorola
		NW-Cellular One (11/24/86)	5	Motorola
81	El Paso, TX	W-Contel Cellular (2/25/85)	2	AT&T
		NW-Metro Mobile CTS (5/2/86)	4	Motorola
82	Tacoma, WA	W-US West (4/18/85)	4	NTI/GE
		NW-Cellular One (12/12/85)	17	AT&T
83	Mobile, AL	W-Contel Cellular (9/3/85)	6	AT&T
		NW-Gulf Coast Cellular (6/8/87)	4	Astronet
84	Harrisburg, PA	W-United TeleSpectrum (10/18/85)	4	Motorola
		NW-Cellular One (9/18/85)	4	NTI/GE
85	Johnson City, TN	W-United TeleSpectrum (10/3/85)	6	Motorola
		NW-Cellular One (1/24/88)	5	AT&T
86	Albuquerque, NM	W-US West (8/13/85)	3	NTI/GE
		NW-Metro Mobile CTS (11/1/85)	3	Motorola
87	Canton, OH	W-GTE Mobilnet (2/25/87)	15	Motorola
		NW-Canton Cellular (2/2/87)		NTI/GE
88	Chattanooga, TN	W-BellSouth Mobility (8/1/85)	4	Motorola
		NW-Cellular One (11/20/87)		
89	Wichita, KS	W-SW Bell Mobile (2/11/85)	4	Motorola
		NW-Cellular One (1/24/86)	2	AT&T
90	Charleston, SC	W-United TeleSpectrum (9/11/85)	5	Motorola
		NW-Cellular One (1/23/87)	7	NTI/GE
91	San Juan, PR	W-Puerto Rico Telephone (8/28/86)	10	NEC
		NW-San Juan CellTel Co. (CPG 1/6/88)		
92	Little Rock, AR	W-Alltel Cellular (7/24/86)	6	Motorola
		NW-Little Rock Cellular (CPG 2/5/88)		
93	Las Vegas, NV	W-Centel Cellular (2/14/86)	5	Motorola
		NW-Amecell (CPG 8/3/87)		Ericsson
94	Saginaw-Bay, Midland, MI	W-Century Cellunet (10/21/87)	3	NovAtel
		NW-Intercont'l. Comm. (CPG 7/2/87)		
95	Columbia, SC	W-BellSouth Mobility (12/22/86)	2	Astronet
		NW-Metro Mobile (5/1/87)	3	NTI/GE
96	Fort Wayne, IN	W-GTE Mobilnet (9/16/87)	1	Motorola
		NW-Cellular One (12/2/87)	2	Ericsson
97	Bakersfield, CA	W-Contel Cellular (2/26/87)	9	AT&T
		NW-Metro Cellular (CPG 9/26/86)		Ericsson

Table A.1 (*cont.*)

Market Number	Area	System Operators	Number of Cells	Switching Equipment
98	Davenport/	W-Contel Cellular (12/3/87)	3	AT&T
	Moline, IA/IL	NW-US Cellular (12/21/87)	3	NTI/GE
99	York, PA	W-United TeleSpectrum (10/27/86)	5	Motorola
		NW-York CellTelCo. (CPG 3/30/88)		
100	Shreveport, LA	W-Century Cellunet (2/3/87)	5	NovAtel
		NW-Cellular One (7/2/87)	3	AT&T
101	Beaumont, TX	W-GTE Mobilnet (8/13/87)	3	Motorola
		NW-Defray Cellular (CPG 10/22/86)		
102	Des Moines, IA	W-US West (9/16/86)	3	NTI/GE
		NW-Us Cellular (9/11/87)	3	NTI/GE
103	Peoria, IL	W-Centel Cellular (2/17/87)	2	Motorola
		NW-Us Cellular (11/6/87)	3	NTI/GE
104	Newport News, VA	W-Contel Cellular (12/1/86)	18	AT&T
		NW-Cellular One (12/1/86)	4	Motorola
105	Lancaster, PA	W-United Telespectrum (10/27/86)	4	Motorola
		NW-Abell Cellular (CPG 9/28/87)		
106	Jackson, MS	W-Alltel Cellular Assoc. (9/23/86)	5	Motorola
		NW-Jackson CellTel Co. (CPG 12/28/87)		
107	Stockton, CA	W-PacTel Cellular (12/15/86)	2	NEC
		NW-Stockton Cell. (12/18/87)		Ericsson
108	Augusta, GA	W-Cellular Phone (6/15/87)		NTI/GE
		NW-Cellular One (5/1/87)	4	NTI/GE
109	Spokane, WA	W-US West (8/2/86)	3	NTI/GE
		NW-McCaw Cellular (CPG 12/29/87)		NovAtel
110	Huntington-	W-Indep. Cell. Network (11/23/87)	3	AT&T
	Ashland, WV	NW-Huntington Cell. (CPG 9/25/86)		Ericsson
111	Vallejo, CA	W-GTE Mobilnet (4/7/87)	54	Motorola
		NW-Cellular One (2/29/88)	4	Ericsson
112	Corpus Christi, TX	W-SW Bell Mobile (9/1/86)	2	Astronet
		NW-McCaw (CPG 11/9/87)		
113	Madison, WI	W-Ameritech Mobile (2/5/87)	4	AT&T
		NW-Cellular One (6/26/87)	4	Ericsson
114	Lakeland, FL	W-GTE Mobilnet (4/10/87)	3	Motorola
		NW-Cellular One (9/25/87)	3	AT&T
115	Utica, Rome, NY	W-Utica/Rome Cell. (10/19/87)	2	AT&T
		NW-Cellular One (2/18/87)	4	Motorola
116	Lexington-	W-BellSouth Mobility (2/5/87)	4	Astronet
	Fayette, KY	NW-Cellular One (9/8/87)	3	AT&T
117	Colorado	W-Us West (2/19/86)	2	NTI/GE
	Springs, CO	NW-Providence Cellular (CPG 7/16/87)		
118	Reading, PA	W-Bell Atlantic Mobile (3/31/86)	2	AT&T
		NW-Cell Radio Corp. (CPG 4/4/88)		
119	Evansville, IN	W-Contel Cellular (12/30/87)	3	Motorola
		NW-Cellular One (12/14/87)	2	NTI/GE

Table A.1 (*cont.*)

Market Number	Area	System Operators	Number of Cells	Switching Equipment
120	Huntsville, AL	W-BellSouth Mobility (12/18/86)	3	Astronet
		NW-Cellular One (4/16/88)	5	Ericsson

Source: *Cellular Business,* June 1988. pp. 57–65, Reprinted with permission.

Appendix B
Developing Three Basic Traffic Formulas

In this appendix we shall develop the three basic traffic formulas: Poisson, Erlang B, and Erlang C, which are listed in Table 3.8. As discussed in Section 3.5.1, for the Poisson distribution, a call entering a system with all trunks busy waits until the expiration of holding time. If a trunk becomes idle before the holding time expiration, the call is served for the remainder of the holding period. For Erlang B, blocked calls disappear and do not reappear as subsequent attempts are made. For Erlang C, blocked calls are delayed indefinitely until served. In addition to these assumptions, the availability of an unlimited number of trunks is assumed for the Poisson case. Also, the number of changes for a particular number of busy trunks to the next higher number of busy trunks is equal to the number of changes in the reverse direction. Putting it in another way, the rate of departure from the system must equal the rate of arrival, which assures the condition of statistical equilibrium. These assumptions apply to all the three formulas.

Poisson Density

In view of the above assumptions and normalizing the observation interval to unit time, we can write

$$f(0) + f(1) + \ldots + f(S) + f(S + 1) + \ldots = 1$$

or

$$\sum_{i=0}^{\infty} f(i) = 1 \tag{B.1}$$

where $f(i)$ denotes the fraction of time when i trunks are busy. Assuming λ to be the number of calls entering the system in unit time (arrival rate), then

---

417

$$\lambda \sum_{i=0}^{\infty} f(i) = \lambda = \sum_{i=0}^{\infty} \lambda f(i) \tag{B.2}$$

If A is the load carried by trunks in a unit interval, or the traffic intensity in Erlangs, then

$$E(i) = \sum_{i=1}^{\infty} if(i) = A \tag{B.3}$$

Note that $i = 0$ does not contribute to the traffic and thus is not included in the above equation. Dividing both sides of (B.3) by average holding time per unit of time (service time), $(1/\mu)$, and making use of (A.2) we obtain:

$$\mu \sum_{i=1}^{\infty} if(i) = \mu A = \lambda$$

$$= \sum_{i=0}^{\infty} \lambda f(i) \tag{B.4a}$$

$$\mu[f(1) + 2f(2) + \dots] = \lambda = \lambda[f(0) + f(1) + \dots f(S) \dots] \tag{B.4b}$$

The general term, $\{Sf[S/(1/\mu)]\}$, indicates the number of times S trunks are busy. The term $\lambda f(S - 1)$ provides the number of times the trunk groups changes state from $(S - 1)$ trunks busy to S trunks busy. Thus these two terms are same. Therefore, equating the like terms on both sides of (B.4) for statistical equilibrium we get $\lambda f(0) = \mu f(1)$, $\lambda f(1) = 2\mu f(2)$, or $f(2) = 1/2 \, (\lambda/\mu)^2 f(0)$. Thus,

$$f(S) = (1/S!) \, (\lambda/\mu)^2 f(0) \tag{B.5}$$

Substituting the values of different terms $f(i)$ in terms of $f(0)$ in (A.1), we obtain

$$f(0) = e^{-\lambda/\mu}$$

$$= e^{-A} \tag{B.6}$$

The fraction of time i trunks are busy is given by (B.5):

$$f(i) = \frac{1}{i!}\left(\frac{\lambda}{\mu}\right)^i f(0) = \frac{1}{i!}\left(\frac{\lambda}{\mu}\right)^i e^{-A} \tag{B.7}$$

The number of calls entering the system when S trunks are busy, or the number of blocked calls, is

$$\sum_{i=S}^{\infty} \lambda \frac{1}{i!} A^i e^{-A} \tag{B.8}$$

Since the probability of blocking is the ratio of the number of blocked calls to the number of offered call per unit time, λ, or

$$P_b = P(S, A)$$
$$= \sum_{i=S}^{\infty} \frac{A^i e^{-A}}{i!} \tag{B.9}$$

Thus, calls not encountering the "all trunks busy" condition or the calls served through the system (calls carried) are given by:

$$1 - P(S, A) = 1 - \sum_{i=S}^{\infty} \frac{A^i e^{-A}}{i!} \tag{B.10}$$

where S = the number of trunks in the system and A = the traffic in Erlangs.

Erlang B: Lost Calls Cleared

If in a traffic system, blocked calls immediately disappear and do not reappear as subsequent attempts, the Poisson formula is longer applicable. Proceeding as above and limiting the number of trunks to S, we obtain from (B.6)

$$f(o) = \frac{1}{\sum_{i=0}^{S} \frac{A^i}{i!}} \tag{B.11}$$

which equals the percentage of time no trunks are busy. Therefore, the percentage of time when S trunks are busy is given by:

$$f(S) = \frac{A^s/S!}{\sum_{i=0}^{S} \frac{A^i}{i!}} \tag{B.12}$$

Since the blocking probability is defined as the percentage of time when S trunks in the system are busy, (B.12) is also the blocking probability, or

$$P_b(S, A) = f(S)$$
$$= \frac{A^s/S!}{\sum_{i=0}^{S} \frac{A^i}{i!}} \tag{B.13}$$

Erlang C: Blocked Calls Held Indefinitely

The system holds the blocked calls in the theoretical trunks. If the total number of serving trunks is S, then the theoretical trunks are from $(S + 1)$ to infinity. Proceeding similar to (A.4), we get

$$\mu[f(1) + 2f(2) + \ldots + (S - 1)f(S - 1)] + S\mu f(S) + (S + 1)\mu f(S + 1) = \lambda$$

$$= \lambda f(0) + \lambda f(1) + \ldots + \lambda f(S) + \lambda f(S + 1) + \ldots \quad \text{(B.14)}$$

The term $S\mu f(S + 1)$ represents the number of calls that expire when $(S + 1)$ simultaneous calls exist. The change reflects the fact that only S servers contribute to consuming the holding time of a call. Expressing terms $f(S + 1)$, $f(S + 2)$, and so forth in terms of $f(S)$ and terms up to $(S - 1)$ in terms of $f(0)$:

$$f(0) + \frac{\lambda}{\mu}f(0) + \frac{1}{2!}\left(\frac{\lambda}{\mu}\right)^2 f(0) + \ldots + \left(\frac{1}{(S - 1)!}\right)\left(\frac{\lambda}{\mu}\right)^{S-1} f(0)$$

$$+ f(s) + \frac{\lambda}{\mu S}f(S)[1 + \left(\frac{\lambda}{\mu S}\right) + \left(\frac{\lambda}{\mu S}\right)^2 + \ldots] = 1 \quad \text{(B.15)}$$

or

$$\sum_{i=0}^{S-1} \frac{1}{i!}\left(\frac{\lambda}{\mu}\right)^i f(0) + f(S) + \frac{\lambda}{\mu S}f(S)\frac{1}{1 - (\lambda/\mu S)} = 1 \quad \text{(B.16a)}$$

or

$$\sum_{i=0}^{S-1} \frac{1}{i!}\left(\frac{\lambda}{\mu}\right)^i f(0) + f(S)\left[1 + \frac{A}{S - A}\right] = 1 \quad \text{(B.16b)}$$

Since $\lambda/\mu = A$,

$$\sum_{i=0}^{S-1} \frac{A^i}{i!}f(0) + f(S)\frac{S}{S - A} = 1 \quad \text{(B.16c)}$$

or

$$\sum_{i=0}^{S-1} \frac{A^i}{i!}f(0) + \frac{A^S}{S!}\left(\frac{S}{S - A}\right)f(0) = 1 \quad \text{(B.16d)}$$

or

$$f(0) = \cfrac{1}{\displaystyle\sum_{i=0}^{S-1} \frac{A^i}{i!} + \frac{A^S}{S!}\frac{S}{S-A}} \tag{B.17}$$

The probability that a call will be delayed is

$$\sum_{i=S}^{\infty} f(i) = f(S)\,\frac{S}{S-A}$$

$$= \frac{A^S}{S!}\frac{S}{S-A}f(0)$$

or, from equation (B.17),

$$\sum_{i=S}^{\infty} f(i) = \cfrac{\dfrac{A^S}{S!}\dfrac{S}{S-A}}{\displaystyle\sum_{i=0}^{S-1} \frac{A^i}{i!} + \frac{A^S}{S!}\left(\frac{S}{S-A}\right)} \tag{B.18}$$

Table B.1 and Figures B.1–B.4 follow.

Table B.1
Blocking Probability $P(S, A)$ for Erlang B Load ($S = 1$ to $S = 50$)

$\mathbf{A}$ in Erl

N	\multicolumn{13}{c	}{B}											
	0.01%	0.02%	0.03%	0.05%	0.1%	0.2%	0.3%	0.4%	0.5%	0.6%	0.7%	0.8%	0.9%
1	.0001	.0002	.0003	.0005	.0010	.0020	.0030	.0040	.0050	.0060	.0070	.0081	.0091
2	.0142	.0202	.0248	.0321	.0458	.0653	.0806	.0937	.105	.116	.126	.135	.144
3	.0868	.110	.127	.152	.194	.249	.289	.321	.349	.374	.397	.418	.437
4	.235	.282	.315	.362	.439	.535	.602	.656	.701	.741	.777	.810	.841
5	.452	.527	.577	.649	.762	.900	.994	1.07	1.13	1.19	1.24	1.28	1.32
6	.728	.832	.900	.996	1.15	1.33	1.45	1.54	1.62	1.69	1.75	1.81	1.86
7	1.05	1.19	1.27	1.39	1.58	1.80	1.95	2.06	2.16	2.24	2.31	2.38	2.44
8	1.42	1.58	1.69	1.83	2.05	2.31	2.48	2.62	2.73	2.83	2.91	2.99	3.06
9	1.83	2.01	2.13	2.30	2.56	2.85	3.05	3.21	3.33	3.44	3.54	3.63	3.71
10	2.26	2.47	2.61	2.80	3.09	3.43	3.65	3.82	3.96	4.08	4.19	4.29	4.38
11	2.72	2.96	3.12	3.33	3.65	4.02	4.27	4.45	4.61	4.74	4.86	4.97	5.07
12	3.21	3.47	3.65	3.88	4.23	4.64	4.90	5.11	5.28	5.43	5.55	5.67	5.78
13	3.71	4.01	4.19	4.45	4.83	5.27	5.56	5.78	5.96	6.12	6.26	6.39	6.50
14	4.24	4.56	4.76	5.03	5.45	5.92	6.23	6.47	6.66	6.83	6.98	7.12	7.24
15	4.78	5.12	5.34	5.63	6.08	6.58	6.91	7.17	7.38	7.56	7.71	7.86	7.99
16	5.34	5.70	5.94	6.25	6.72	7.26	7.61	7.88	8.10	8.29	8.46	8.61	8.75
17	5.91	6.30	6.55	6.88	7.38	7.95	8.32	8.60	8.83	9.03	9.21	9.37	9.52
18	6.50	6.91	7.17	7.52	8.05	8.64	9.03	9.33	9.58	9.79	9.98	10.1	10.3
19	7.09	7.53	7.80	8.17	8.72	9.35	9.76	10.1	10.3	10.6	10.7	10.9	11.1
20	7.70	8.16	8.44	8.83	9.41	10.1	10.5	10.8	11.1	11.3	11.5	11.7	11.9
21	8.32	8.79	9.10	9.50	10.1	10.8	11.2	11.6	11.9	12.1	12.3	12.5	12.7
22	8.95	9.44	9.76	10.2	10.8	11.5	12.0	12.3	12.6	12.9	13.1	13.3	13.5
23	9.58	10.1	10.4	10.9	11.5	12.3	12.7	13.1	13.4	13.7	13.9	14.1	14.3
24	10.2	10.8	11.1	11.6	12.2	13.0	13.5	13.9	14.2	14.5	14.7	14.9	15.1
25	10.9	11.4	11.8	12.3	13.0	13.8	14.3	14.7	15.0	15.3	15.5	15.7	15.9
26	11.5	12.1	12.5	13.0	13.7	14.5	15.1	15.5	15.8	16.1	16.3	16.6	16.8
27	12.2	12.8	13.2	13.7	14.4	15.3	15.8	16.3	16.6	16.9	17.2	17.4	17.6
28	12.9	13.5	13.9	14.4	15.2	16.1	16.6	17.1	17.4	17.7	18.0	18.2	18.4
29	13.6	14.2	14.6	15.1	15.9	16.8	17.4	17.9	18.2	18.5	18.8	19.1	19.3
30	14.2	14.9	15.3	15.9	16.7	17.6	18.2	18.7	19.0	19.4	19.6	19.9	20.1
31	14.9	15.6	16.0	16.6	17.4	18.4	19.0	19.5	19.9	20.2	20.5	20.7	21.0
32	15.6	16.3	16.8	17.3	18.2	19.2	19.8	20.3	20.7	21.0	21.3	21.6	21.8
33	16.3	17.0	17.5	18.1	19.0	20.0	20.6	21.1	21.5	21.9	22.2	22.4	22.7
34	17.0	17.8	18.2	18.8	19.7	20.8	21.4	21.9	22.3	22.7	23.0	23.3	23.5
35	17.8	18.5	19.0	19.6	20.5	21.6	22.2	22.7	23.2	23.5	23.8	24.1	24.4
36	18.5	19.2	19.7	20.3	21.3	22.4	23.1	23.6	24.0	24.4	24.7	25.0	25.3
37	19.2	20.0	20.5	21.1	22.1	23.2	23.9	24.4	24.8	25.2	25.6	25.9	26.1
38	19.9	20.7	21.2	21.9	22.9	24.0	24.7	25.2	25.7	26.1	26.4	26.7	27.0
39	20.6	21.5	22.0	22.6	23.7	24.8	25.5	26.1	26.5	26.9	27.3	27.6	27.9
40	21.4	22.2	22.7	23.4	24.4	25.6	26.3	26.9	27.4	27.8	28.1	28.5	28.7
41	22.1	23.0	23.5	24.2	25.2	26.4	27.2	27.8	28.2	28.6	29.0	29.3	29.6
42	22.8	23.7	24.2	25.0	26.0	27.2	28.0	28.6	29.1	29.5	29.9	30.2	30.5
43	23.6	24.5	25.0	25.7	26.8	28.1	28.8	29.4	29.9	30.4	30.7	31.1	31.4
44	24.3	25.2	25.8	26.5	27.6	28.9	29.7	30.3	30.8	31.2	31.6	31.9	32.3
45	25.1	26.0	26.6	27.3	28.4	29.7	30.5	31.1	31.7	32.1	32.5	32.8	33.1
46	25.8	26.8	27.3	28.1	29.3	30.5	31.4	32.0	32.5	33.0	33.4	33.7	34.0
47	26.6	27.5	28.1	28.9	30.1	31.4	32.2	32.9	33.4	33.8	34.2	34.6	34.9
48	27.3	28.3	28.9	29.7	30.9	32.2	33.1	33.7	34.2	34.7	35.1	35.5	35.8
49	28.1	29.1	29.7	30.5	31.7	33.0	33.9	34.6	35.1	35.6	36.0	36.4	36.7
50	28.9	29.9	30.5	31.3	32.5	33.9	34.8	35.4	36.0	36.5	36.9	37.2	37.6
	0.01%	0.02%	0.03%	0.05%	0.1%	0.2%	0.3%	0.4%	0.5%	0.6%	0.7%	0.8%	0.9%
N	\multicolumn{13}{c	}{B}											

Table B.1 (*cont.*)

A in Erl

								B						N
1.0%	1.2%	1.5%	2%	3%	5%	7%	10%	15%	20%	30%	40%	50%	
.0101	.0121	.0152	.0204	.0309	.0526	.0753	.111	.176	.250	.429	.667	1.00	1
.153	.168	.190	.223	.282	.381	.470	.595	.796	1.00	1.45	2.00	2.73	2
.455	.489	.535	.602	.715	.899	1.06	1.27	1.60	1.93	2.63	3.48	4.59	3
.869	.922	.992	1.09	1.26	1.52	1.75	2.05	2.50	2.95	3.89	5.02	6.50	4
1.36	1.43	1.52	1.66	1.88	2.22	2.50	2.88	3.45	4.01	5.19	6.60	8.44	5
1.91	2.00	2.11	2.28	2.54	2.96	3.30	3.76	4.44	5.11	6.51	8.19	10.4	6
2.50	2.60	2.74	2.94	3.25	3.74	4.14	4.67	5.46	6.23	7.86	9.80	12.4	7
3.13	3.25	3.40	3.63	3.99	4.54	5.00	5.60	6.50	7.37	9.21	11.4	14.3	8
3.78	3.92	4.09	4.34	4.75	5.37	5.88	6.55	7.55	8.52	10.6	13.0	16.3	9
4.46	4.61	4.81	5.08	5.53	6.22	6.78	7.51	8.62	9.68	12.0	14.7	18.3	10
5.16	5.32	5.54	5.84	6.33	7.08	7.69	8.49	9.69	10.9	13.3	16.3	20.3	11
5.88	6.05	6.29	6.61	7.14	7.95	8.61	9.47	10.8	12.0	14.7	18.0	22.2	12
6.61	6.80	7.05	7.40	7.97	8.83	9.54	10.5	11.9	13.2	16.1	19.6	24.2	13
7.35	7.56	7.82	8.20	8.80	9.73	10.5	11.5	13.0	14.4	17.5	21.2	26.2	14
8.11	8.33	8.61	9.01	9.65	10.6	11.4	12.5	14.1	15.6	18.9	22.9	28.2	15
8.88	9.11	9.41	9.83	10.5	11.5	12.4	13.5	15.2	16.8	20.3	24.5	30.2	16
9.65	9.89	10.2	10.7	11.4	12.5	13.4	14.5	16.3	18.0	21.7	26.2	32.2	17
10.4	10.7	11.0	11.5	12.2	13.4	14.3	15.5	17.4	19.2	23.1	27.8	34.2	18
11.2	11.5	11.8	12.3	13.1	14.3	15.3	16.6	18.5	20.4	24.5	29.5	36.2	19
12.0	12.3	12.7	13.2	14.0	15.2	16.3	17.6	19.6	21.6	25.9	31.2	38.2	20
12.8	13.1	13.5	14.0	14.9	16.2	17.3	18.7	20.8	22.8	27.3	32.8	40.2	21
13.7	14.0	14.3	14.9	15.8	17.1	18.2	19.7	21.9	24.1	28.7	34.5	42.1	22
14.5	14.8	15.2	15.8	16.7	18.1	19.2	20.7	23.0	25.3	30.1	36.1	44.1	23
15.3	15.6	16.0	16.6	17.6	19.0	20.2	21.8	24.2	26.5	31.6	37.8	46.1	24
16.1	16.5	16.9	17.5	18.5	20.0	21.2	22.8	25.3	27.7	33.0	39.4	48.1	25
17.0	17.3	17.8	18.4	19.4	20.9	22.2	23.9	26.4	28.9	34.4	41.1	50.1	26
17.8	18.2	18.6	19.3	20.3	21.9	23.2	24.9	27.6	30.2	35.8	42.8	52.1	27
18.6	19.0	19.5	20.2	21.2	22.9	24.2	26.0	28.7	31.4	37.2	44.4	54.1	28
19.5	19.9	20.4	21.0	22.1	23.8	25.2	27.1	29.9	32.6	38.6	46.1	56.1	29
20.3	20.7	21.2	21.9	23.1	24.8	26.2	28.1	31.0	33.8	40.0	47.7	58.1	30
21.2	21.6	22.1	22.8	24.0	25.8	27.2	29.2	32.1	35.1	41.5	49.4	60.1	31
22.0	22.5	23.0	23.7	24.9	26.7	28.2	30.2	33.3	36.3	42.9	51.1	62.1	32
22.9	23.3	23.9	24.6	25.8	27.7	29.3	31.3	34.4	37.5	44.3	52.7	64.1	33
23.8	24.2	24.8	25.5	26.8	28.7	30.3	32.4	35.6	38.8	45.7	54.4	66.1	34
24.6	25.1	25.6	26.4	27.7	29.7	31.3	33.4	36.7	40.0	47.1	56.0	68.1	35
25.5	26.0	26.5	27.3	28.6	30.7	32.3	34.5	37.9	41.2	48.6	57.7	70.1	36
26.4	26.8	27.4	28.3	29.6	31.6	33.3	35.6	39.0	42.4	50.0	59.4	72.1	37
27.3	27.7	28.3	29.2	30.5	32.6	34.4	36.6	40.2	43.7	51.4	61.0	74.1	38
28.1	28.6	29.2	30.1	31.5	33.6	35.4	37.7	41.3	44.9	52.8	62.7	76.1	39
29.0	29.5	30.1	31.0	32.4	34.6	36.4	38.8	42.5	46.1	54.2	64.4	78.1	40
29.9	30.4	31.0	31.9	33.4	35.6	37.4	39.9	43.6	47.4	55.7	66.0	80.1	41
30.8	31.3	31.9	32.8	34.3	36.6	38.4	40.9	44.8	48.6	57.1	67.7	82.1	42
31.7	32.2	32.8	33.8	35.3	37.6	39.5	42.0	45.9	49.9	58.5	69.3	84.1	43
32.5	33.1	33.7	34.7	36.2	38.6	40.5	43.1	47.1	51.1	59.9	71.0	86.1	44
33.4	34.0	34.6	35.6	37.2	39.6	41.5	44.2	48.2	52.3	61.3	72.7	88.1	45
34.3	34.9	35.6	36.5	38.1	40.5	42.6	45.2	49.4	53.6	62.8	74.3	90.1	46
35.2	35.8	36.5	37.5	39.1	41.5	43.6	46.3	50.6	54.8	64.2	76.0	92.1	47
36.1	36.7	37.4	38.4	40.0	42.5	44.6	47.4	51.7	56.0	65.6	77.7	94.1	48
37.0	37.6	38.3	39.3	41.0	43.5	45.7	48.5	52.9	57.3	67.0	79.3	96.1	49
37.9	38.5	39.2	40.3	41.9	44.5	46.7	49.6	54.0	58.5	68.5	81.0	98.1	50
1.0%	1.2%	1.5%	2%	3%	5%	7%	10%	15%	20%	30%	40%	50%	N

B

Table B.1 (*cont.*)

A in Erl

N	B												
	0.01%	0.02%	0.03%	0.05%	0.1%	0.2%	0.3%	0.4%	0.5%	0.6%	0.7%	0.8%	0.9%
50	28.9	29.9	30.5	31.3	32.5	33.9	34.8	35.4	36.0	36.5	36.9	37.2	37.6
51	29.6	30.6	31.3	32.1	33.3	34.7	35.6	36.3	36.9	37.3	37.8	38.1	38.5
52	30.4	31.4	32.0	32.9	34.2	35.6	36.5	37.2	37.7	38.2	38.6	39.0	39.4
53	31.2	32.2	32.8	33.7	35.0	36.4	37.3	38.0	38.6	39.1	39.5	39.9	40.3
54	31.9	33.0	33.6	34.5	35.8	37.2	38.2	38.9	39.5	40.0	40.4	40.8	41.2
55	32.7	33.8	34.4	35.3	36.6	38.1	39.0	39.8	40.4	40.9	41.3	41.7	42.1
56	33.5	34.6	35.2	36.1	37.5	38.9	39.9	40.6	41.2	41.7	42.2	42.6	43.0
57	34.3	35.4	36.0	36.9	38.3	39.8	40.8	41.5	42.1	42.6	43.1	43.5	43.9
58	35.1	36.2	36.8	37.8	39.1	40.6	41.6	42.4	43.0	43.5	44.0	44.4	44.8
59	35.8	37.0	37.6	38.6	40.0	41.5	42.5	43.3	43.9	44.4	44.9	45.3	45.7
60	36.6	37.8	38.5	39.4	40.8	42.4	43.4	44.1	44.8	45.3	45.8	46.2	46.6
61	37.4	38.6	39.3	40.2	41.6	43.2	44.2	45.0	45.6	46.2	46.7	47.1	47.5
62	38.2	39.4	40.1	41.0	42.5	44.1	45.1	45.9	46.5	47.1	47.6	48.0	48.4
63	39.0	40.2	40.9	41.9	43.3	44.9	46.0	46.8	47.4	48.0	48.5	48.9	49.3
64	39.8	41.0	41.7	42.7	44.2	45.8	46.8	47.6	48.3	48.9	49.4	49.8	50.2
65	40.6	41.8	42.5	43.5	45.0	46.6	47.7	48.5	49.2	49.8	50.3	50.7	51.1
66	41.4	42.6	43.3	44.4	45.8	47.5	48.6	49.4	50.1	50.7	51.2	51.6	52.0
67	42.2	43.4	44.2	45.2	46.7	48.4	49.5	50.3	51.0	51.6	52.1	52.5	53.0
68	43.0	44.2	45.0	46.0	47.5	49.2	50.3	51.2	51.9	52.5	53.0	53.4	53.9
69	43.8	45.0	45.8	46.8	48.4	50.1	51.2	52.1	52.8	53.4	53.9	54.4	54.8
70	44.6	45.8	46.6	47.7	49.2	51.0	52.1	53.0	53.7	54.3	54.8	55.3	55.7
71	45.4	46.7	47.5	48.5	50.1	51.8	53.0	53.8	54.6	55.2	55.7	56.2	56.6
72	46.2	47.5	48.3	49.4	50.9	52.7	53.9	54.7	55.5	56.1	56.6	57.1	57.5
73	47.0	48.3	49.1	50.2	51.8	53.6	54.7	55.6	56.4	57.0	57.5	58.0	58.5
74	47.8	49.1	49.9	51.0	52.7	54.5	55.6	56.5	57.3	57.9	58.4	58.9	59.4
75	48.6	49.9	50.8	51.9	53.5	55.3	56.5	57.4	58.2	58.8	59.3	59.8	60.3
76	49.4	50.8	51.6	52.7	54.4	56.2	57.4	58.3	59.1	59.7	60.3	60.8	61.2
77	50.2	51.6	52.4	53.6	55.2	57.1	58.3	59.2	60.0	60.6	61.2	61.7	62.1
78	51.1	52.4	53.3	54.4	56.1	58.0	59.2	60.1	60.9	61.5	62.1	62.6	63.1
79	51.9	53.2	54.1	55.3	56.9	58.8	60.1	61.0	61.8	62.4	63.0	63.5	64.0
80	52.7	54.1	54.9	56.1	57.8	59.7	61.0	61.9	62.7	63.3	63.9	64.4	64.9
81	53.5	54.9	55.8	56.9	58.7	60.6	61.8	62.8	63.6	64.2	64.8	65.4	65.8
82	54.3	55.7	56.6	57.8	59.5	61.5	62.7	63.7	64.5	65.2	65.7	66.3	66.8
83	55.1	56.6	57.5	58.6	60.4	62.4	63.6	64.6	65.4	66.1	66.7	67.2	67.7
84	56.0	57.4	58.3	59.5	61.3	63.2	64.5	65.5	66.3	67.0	67.6	68.1	68.6
85	56.8	58.2	59.1	60.4	62.1	64.1	65.4	66.4	67.2	67.9	68.5	69.1	69.6
86	57.6	59.1	60.0	61.2	63.0	65.0	66.3	67.3	68.1	68.8	69.4	70.0	70.5
87	58.4	59.9	60.8	62.1	63.9	65.9	67.2	68.2	69.0	69.7	70.3	70.9	71.4
88	59.3	60.8	61.7	62.9	64.7	66.8	68.1	69.1	69.9	70.6	71.3	71.8	72.3
89	60.1	61.6	62.5	63.8	65.6	67.7	69.0	70.0	70.8	71.6	72.2	72.8	73.3
90	60.9	62.4	63.4	64.6	66.5	68.6	69.9	70.9	71.8	72.5	73.1	73.7	74.2
91	61.8	63.3	64.2	65.5	67.4	69.4	70.8	71.8	72.7	73.4	74.0	74.6	75.1
92	62.6	64.1	65.1	66.3	68.2	70.3	71.7	72.7	73.6	74.3	75.0	75.5	76.1
93	63.4	65.0	65.9	67.2	69.1	71.2	72.6	73.6	74.5	75.2	75.9	76.5	77.0
94	64.2	65.8	66.8	68.1	70.0	72.1	73.5	74.5	75.4	76.2	76.8	77.4	77.9
95	65.1	66.6	67.6	68.9	70.9	73.0	74.4	75.5	76.3	77.1	77.7	78.3	78.9
96	65.9	67.5	68.5	69.8	71.7	73.9	75.3	76.4	77.2	78.0	78.7	79.3	79.8
97	66.8	68.3	69.3	70.7	72.6	74.8	76.2	77.3	78.2	78.9	79.6	80.2	80.7
98	67.6	69.2	70.2	71.5	73.5	75.7	77.1	78.2	79.1	79.8	80.5	81.1	81.7
99	68.4	70.0	71.0	72.4	74.4	76.6	78.0	79.1	80.0	80.8	81.4	82.0	82.6
100	69.3	70.9	71.9	73.2	75.2	77.5	78.9	80.0	80.9	81.7	82.4	83.0	83.5
	0.01%	0.02%	0.03%	0.05%	0.1%	0.2%	0.3%	0.4%	0.5%	0.6%	0.7%	0.8%	0.9%
N						B							

Table B.1 (*cont.*)

A in Erl

							B						N
1.0%	1.2%	1.5%	2 %	3 %	5 %	7 %	10 %	15 %	20 %	30 %	40 %	50 %	
37.9	38.5	39.2	40.3	41.9	44.5	46.7	49.6	54.0	58.5	68.5	81.0	98.1	50
38.8	39.4	40.1	41.2	42.9	45.5	47.7	50.6	55.2	59.7	69.9	82.7	100.1	51
39.7	40.3	41.0	42.1	43.9	46.5	48.8	51.7	56.3	61.0	71.3	84.3	102.1	52
40.6	41.2	42.0	43.1	44.8	47.5	49.8	52.8	57.5	62.2	72.7	86.0	104.1	53
41.5	42.1	42.9	44.0	45.8	48.5	50.8	53.9	58.7	63.5	74.2	87.6	106.1	54
42.4	43.0	43.8	44.9	46.7	49.5	51.9	55.0	59.8	64.7	75.6	89.3	108.1	55
43.3	43.9	44.7	45.9	47.7	50.5	52.9	56.1	61.0	65.9	77.0	91.0	110.1	56
44.2	44.8	45.7	46.8	48.7	51.5	53.9	57.1	62.1	67.2	78.4	92.6	112.1	57
45.1	45.8	46.6	47.8	49.6	52.6	55.0	58.2	63.3	68.4	79.8	94.3	114.1	58
46.0	46.7	47.5	48.7	50.6	53.6	56.0	59.3	64.5	69.7	81.3	96.0	116.1	59
46.9	47.6	48.4	49.6	51.6	54.6	57.1	60.4	65.6	70.9	82.7	97.6	118.1	60
47.9	48.5	49.4	50.6	52.5	55.6	58.1	61.5	66.8	72.1	84.1	99.3	120.1	61
48.8	49.4	50.3	51.5	53.5	56.6	59.1	62.6	68.0	73.4	85.5	101.0	122.1	62
49.7	50.4	51.2	52.5	54.5	57.6	60.2	63.7	69.1	74.6	87.0	102.6	124.1	63
50.6	51.3	52.2	53.4	55.4	58.6	61.2	64.8	70.3	75.9	88.4	104.3	126.1	64
51.5	52.2	53.1	54.4	56.4	59.6	62.3	65.8	71.4	77.1	89.8	106.0	128.1	65
52.4	53.1	54.0	55.3	57.4	60.6	63.3	66.9	72.6	78.3	91.2	107.6	130.1	66
53.4	54.1	55.0	56.3	58.4	61.6	64.4	68.0	73.8	79.6	92.7	109.3	132.1	67
54.3	55.0	55.9	57.2	59.3	62.6	65.4	69.1	74.9	80.8	94.1	111.0	134.1	68
55.2	55.9	56.9	58.2	60.3	63.7	66.4	70.2	76.1	82.1	95.5	112.6	136.1	69
56.1	56.8	57.8	59.1	61.3	64.7	67.5	71.3	77.3	83.3	96.9	114.3	138.1	70
57.0	57.8	58.7	60.1	62.3	65.7	68.5	72.4	78.4	84.6	98.4	115.9	140.1	71
58.0	58.7	59.7	61.0	63.2	66.7	69.6	73.5	79.6	85.8	99.8	117.6	142.1	72
58.9	59.6	60.6	62.0	64.2	67.7	70.6	74.6	80.8	87.0	101.2	119.3	144.1	73
59.8	60.6	61.6	62.9	65.2	68.7	71.7	75.6	81.9	88.3	102.7	120.9	146.1	74
60.7	61.5	62.5	63.9	66.2	69.7	72.7	76.7	83.1	89.5	104.1	122.6	148.0	75
61.7	62.4	63.4	64.9	67.2	70.8	73.8	77.8	84.2	90.8	105.5	124.3	150.0	76
62.6	63.4	64.4	65.8	68.1	71.8	74.8	78.9	85.4	92.0	106.9	125.9	152.0	77
63.5	64.3	65.3	66.8	69.1	72.8	75.9	80.0	86.6	93.3	108.4	127.6	154.0	78
64.4	65.2	66.3	67.7	70.1	73.8	76.9	81.1	87.7	94.5	109.8	129.3	156.0	79
65.4	66.2	67.2	68.7	71.1	74.8	78.0	82.2	88.9	95.7	111.2	130.9	158.0	80
66.3	67.1	68.2	69.6	72.1	75.8	79.0	83.3	90.1	97.0	112.6	132.6	160.0	81
67.2	68.0	69.1	70.6	73.0	76.9	80.1	84.4	91.2	98.2	114.1	134.3	162.0	82
68.2	69.0	70.1	71.6	74.0	77.9	81.1	85.5	92.4	99.5	115.5	135.9	164.0	83
69.1	69.9	71.0	72.5	75.0	78.9	82.2	86.6	93.6	100.7	116.9	137.6	166.0	84
70.0	70.9	71.9	73.5	76.0	79.9	83.2	87.7	94.7	102.0	118.3	139.3	168.0	85
70.9	71.8	72.9	74.5	77.0	80.9	84.3	88.8	95.9	103.2	119.8	140.9	170.0	86
71.9	72.7	73.8	75.4	78.0	82.0	85.3	89.9	97.1	104.5	121.2	142.6	172.0	87
72.8	73.7	74.8	76.4	79.0	83.0	86.4	91.0	98.2	105.7	122.6	144.3	174.0	88
73.7	74.6	75.7	77.3	79.9	84.0	87.4	92.1	99.4	106.9	124.0	145.9	176.0	89
74.7	75.6	76.7	78.3	80.9	85.0	88.5	93.1	100.6	108.2	125.5	147.6	178.0	90
75.6	76.5	77.6	79.3	81.9	86.0	89.5	94.2	101.7	109.4	126.9	149.3	180.0	91
76.6	77.4	78.6	80.2	82.9	87.1	90.6	95.3	102.9	110.7	128.3	150.9	182.0	92
77.5	78.4	79.6	81.2	83.9	88.1	91.6	96.4	104.1	111.9	129.7	152.6	184.0	93
78.4	79.3	80.5	82.2	84.9	89.1	92.7	97.5	105.3	113.2	131.2	154.3	186.0	94
79.4	80.3	81.5	83.1	85.8	90.1	93.7	98.6	106.4	114.4	132.6	155.9	188.0	95
80.3	81.2	82.4	84.1	86.8	91.1	94.8	99.7	107.6	115.7	134.0	157.6	190.0	96
81.2	82.2	83.4	85.1	87.8	92.2	95.8	100.8	108.8	116.9	135.5	159.3	192.0	97
82.2	83.1	84.3	86.0	88.8	93.2	96.9	101.9	109.9	118.2	136.9	160.9	194.0	98
83.1	84.1	85.3	87.0	89.8	94.2	97.9	103.0	111.1	119.4	138.3	162.6	196.0	99
84.1	85.0	86.2	88.0	90.8	95.2	99.0	104.1	112.3	120.6	139.7	164.3	198.0	100
1.0%	1.2%	1.5%	2 %	3 %	5 %	7 %	10 %	15 %	20 %	30 %	40 %	50 %	

B

N

Table B.1 (*cont.*)

A in Erl

N							B							
	0.01%	0.02%	0.03%	0.05%	0.1%	0.2%	0.3%	0.4%	0.5%	0.6%	0.7%	0.8%	0.9%
100	69.3	70.9	71.9	73.2	75.2	77.5	78.9	80.0	80.9	81.7	82.4	83.0	83.5
102	70.9	72.6	73.6	75.0	77.0	79.3	80.7	81.8	82.7	83.5	84.2	84.8	85.4
104	72.6	74.3	75.3	76.7	78.8	81.1	82.5	83.7	84.6	85.4	86.1	86.7	87.3
106	74.3	76.0	77.1	78.5	80.5	82.8	84.3	85.5	86.4	87.2	87.9	88.6	89.2
108	76.0	77.7	78.8	80.2	82.3	84.6	86.2	87.3	88.3	89.1	89.8	90.5	91.1
110	77.7	79.4	80.5	81.9	84.1	86.4	88.0	89.2	90.1	90.9	91.7	92.3	92.9
112	79.4	81.1	82.2	83.7	85.8	88.3	89.8	91.0	92.0	92.8	93.5	94.2	94.8
114	81.1	82.9	84.0	85.4	87.6	90.1	91.6	92.8	93.8	94.7	95.4	96.1	96.7
116	82.8	84.6	85.7	87.2	89.4	91.9	93.5	94.7	95.7	96.5	97.3	98.0	98.6
118	84.5	86.3	87.4	89.0	91.2	93.7	95.3	96.5	97.5	98.4	99.2	99.9	100.5
120	86.2	88.0	89.2	90.7	93.0	95.5	97.1	98.4	99.4	100.3	101.0	101.7	102.4
122	87.9	89.8	90.9	92.5	94.7	97.3	98.9	100.2	101.2	102.1	102.9	103.6	104.3
124	89.6	91.5	92.7	94.2	96.5	99.1	100.8	102.1	103.1	104.0	104.8	105.5	106.2
126	91.3	93.2	94.4	96.0	98.3	100.9	102.6	103.9	105.0	105.9	106.7	107.4	108.1
128	93.1	95.0	96.2	97.8	100.1	102.7	104.5	105.8	106.8	107.7	108.5	109.3	109.9
130	94.8	96.7	97.9	99.5	101.9	104.6	106.3	107.6	108.7	109.6	110.4	111.2	111.8
132	96.5	98.5	99.7	101.3	103.7	106.4	108.1	109.5	110.5	111.5	112.3	113.1	113.7
134	98.2	100.2	101.4	103.1	105.5	108.2	110.0	111.3	112.4	113.4	114.2	115.0	115.6
136	100.0	101.9	103.2	104.9	107.3	110.0	111.8	113.2	114.3	115.2	116.1	116.8	117.5
138	101.7	103.7	105.0	106.6	109.1	111.9	113.7	115.0	116.2	117.1	118.0	118.7	119.4
140	103.4	105.4	106.7	108.4	110.9	113.7	115.5	116.9	118.0	119.0	119.9	120.6	121.4
142	105.1	107.2	108.5	110.2	112.7	115.5	117.4	118.7	119.9	120.9	121.8	122.5	123.3
144	106.9	109.0	110.2	112.0	114.5	117.4	119.2	120.6	121.8	122.8	123.6	124.4	125.2
146	108.6	110.7	112.0	113.8	116.3	119.2	121.1	122.5	123.6	124.6	125.5	126.3	127.1
148	110.4	112.5	113.8	115.5	118.1	121.0	122.9	124.3	125.5	126.5	127.4	128.2	129.0
150	112.1	114.2	115.6	117.3	119.9	122.9	124.8	126.2	127.4	128.4	129.3	130.1	130.9
152	113.8	116.0	117.3	119.1	121.8	124.7	126.6	128.1	129.3	130.3	131.2	132.0	132.8
154	115.6	117.8	119.1	120.9	123.6	126.5	128.5	129.9	131.2	132.2	133.1	133.9	134.7
156	117.3	119.5	120.9	122.7	125.4	128.4	130.3	131.8	133.0	134.1	135.0	135.9	136.6
158	119.1	121.3	122.7	124.5	127.2	130.2	132.2	133.7	134.9	136.0	136.9	137.8	138.5
160	120.8	123.1	124.4	126.3	129.0	132.1	134.0	135.6	136.8	137.9	138.8	139.7	140.4
162	122.6	124.8	126.2	128.1	130.8	133.9	135.9	137.4	138.7	139.8	140.7	141.6	142.4
164	124.3	126.6	128.0	129.9	132.7	135.8	137.8	139.3	140.6	141.7	142.6	143.5	144.3
166	126.1	128.4	129.8	131.7	134.5	137.6	139.6	141.2	142.5	143.5	144.5	145.4	146.2
168	127.9	130.2	131.6	133.5	136.3	139.4	141.5	143.1	144.3	145.4	146.4	147.3	148.1
170	129.6	131.9	133.4	135.3	138.1	141.3	143.4	144.9	146.2	147.3	148.3	149.2	150.0
172	131.4	133.7	135.2	137.1	139.9	143.1	145.2	146.8	148.1	149.2	150.2	151.1	151.9
174	133.1	135.5	136.9	138.9	141.8	145.0	147.1	148.7	150.0	151.1	152.1	153.0	153.9
176	134.9	137.3	138.7	140.7	143.6	146.9	149.0	150.6	151.9	153.0	154.0	155.0	155.8
178	136.7	139.0	140.5	142.5	145.4	148.7	150.8	152.4	153.8	154.9	156.0	156.9	157.7
180	138.4	140.8	142.3	144.3	147.3	150.6	152.7	154.3	155.7	156.8	157.9	158.8	159.6
182	140.2	142.6	144.1	146.1	149.1	152.4	154.6	156.2	157.6	158.7	159.8	160.7	161.6
184	142.0	144.4	145.9	147.9	150.9	154.3	156.4	158.1	159.5	160.6	161.7	162.6	163.5
186	143.7	146.2	147.7	149.8	152.8	156.1	158.3	160.0	161.4	162.5	163.6	164.5	165.4
188	145.5	148.0	149.5	151.6	154.6	158.0	160.2	161.9	163.3	164.4	165.5	166.5	167.3
190	147.3	149.8	151.3	153.4	156.4	159.8	162.1	163.8	165.2	166.4	167.4	168.4	169.3
192	149.1	151.6	153.1	155.2	158.3	161.7	163.9	165.6	167.0	168.3	169.3	170.3	171.2
194	150.8	153.4	154.9	157.0	160.1	163.6	165.8	167.5	168.9	170.2	171.2	172.2	173.1
196	152.6	155.2	156.7	158.8	161.9	165.4	167.7	169.4	170.8	172.1	173.2	174.1	175.0
198	154.4	156.9	158.5	160.7	163.8	167.3	169.6	171.3	172.7	174.0	175.1	176.1	177.0
200	156.2	158.7	160.3	162.5	165.6	169.2	171.4	173.2	174.6	175.9	177.0	178.0	178.9
	0.01%	0.02%	0.03%	0.05%	0.1%	0.2%	0.3%	0.4%	0.5%	0.6%	0.7%	0.8%	0.9%
N							B							

Table B.1 *(cont.)*

A in Erl

1.0%	1.2%	1.5%	2%	3%	5%	7%	10%	15%	20%	30%	40%	50%	N
84.1	85.0	86.2	88.0	90.8	95.2	99.0	104.1	112.3	120.6	139.7	164.3	198.0	100
85.9	86.9	88.1	89.9	92.8	97.3	101.1	106.3	114.6	123.1	142.6	167.6	202.0	102
87.8	88.8	90.1	91.9	94.8	99.3	103.2	108.5	116.9	125.6	145.4	170.9	206.0	104
89.7	90.7	92.0	93.8	96.7	101.4	105.3	110.7	119.3	128.1	148.3	174.2	210.0	106
91.6	92.6	93.9	95.7	98.7	103.4	107.4	112.9	121.6	130.6	151.1	177.6	214.0	108
93.5	94.5	95.8	97.7	100.7	105.5	109.5	115.1	124.0	133.1	154.0	180.9	218.0	110
95.4	96.4	97.7	99.6	102.7	107.5	111.7	117.3	126.3	135.6	156.9	184.2	222.0	112
97.3	98.3	99.7	101.6	104.7	109.6	113.8	119.5	128.6	138.1	159.7	187.6	226.0	114
99.2	100.2	101.6	103.5	106.7	111.7	115.9	121.7	131.0	140.6	162.6	190.9	230.0	116
101.1	102.1	103.5	105.5	108.7	113.7	118.0	123.9	133.3	143.1	165.4	194.2	234.0	118
103.0	104.0	105.4	107.4	110.7	115.8	120.1	126.1	135.7	145.6	168.3	197.6	238.0	120
104.9	105.9	107.4	109.4	112.6	117.8	122.2	128.3	138.0	148.1	171.1	200.9	242.0	122
106.8	107.9	109.3	111.3	114.6	119.9	124.4	130.5	140.3	150.6	174.0	204.2	246.0	124
108.7	109.8	111.2	113.3	116.6	121.9	126.5	132.7	142.7	153.0	176.8	207.6	250.0	126
110.6	111.7	113.2	115.2	118.6	124.0	128.6	134.9	145.0	155.5	179.7	210.9	254.0	128
112.5	113.6	115.1	117.2	120.6	126.1	130.7	137.1	147.4	158.0	182.5	214.2	258.0	130
114.4	115.5	117.0	119.1	122.6	128.1	132.8	139.3	149.7	160.5	185.4	217.6	262.0	132
116.3	117.4	119.0	121.1	124.6	130.2	134.9	141.5	152.0	163.0	188.3	220.9	266.0	134
118.2	119.4	120.9	123.1	126.6	132.3	137.1	143.7	154.4	165.5	191.1	224.2	270.0	136
120.1	121.3	122.8	125.0	128.6	134.3	139.2	145.9	156.7	168.0	194.0	227.6	274.0	138
122.0	123.2	124.8	127.0	130.6	136.4	141.3	148.1	159.1	170.5	196.8	230.9	278.0	140
123.9	125.1	126.7	128.9	132.6	138.4	143.4	150.3	161.4	173.0	199.7	234.2	282.0	142
125.8	127.0	128.6	130.9	134.6	140.5	145.6	152.5	163.8	175.5	202.5	237.6	286.0	144
127.7	129.0	130.6	132.9	136.6	142.6	147.7	154.7	166.1	178.0	205.4	240.9	290.0	146
129.7	130.9	132.5	134.8	138.6	144.6	149.8	156.9	168.5	180.5	208.2	244.2	294.0	148
131.6	132.8	134.5	136.8	140.6	146.7	151.9	159.1	170.8	183.0	211.1	247.6	298.0	150
133.5	134.8	136.4	138.8	142.6	148.8	154.0	161.3	173.1	185.5	214.0	250.9	302.0	152
135.4	136.7	138.4	140.7	144.6	150.8	156.2	163.5	175.5	188.0	216.8	254.2	306.0	154
137.3	138.6	140.3	142.7	146.6	152.9	158.3	165.7	177.8	190.5	219.7	257.6	310.0	156
139.2	140.5	142.3	144.7	148.6	155.0	160.4	167.9	180.2	193.0	222.5	260.9	314.0	158
141.2	142.5	144.2	146.6	150.6	157.0	162.5	170.2	182.5	195.5	225.4	264.2	318.0	160
143.1	144.4	146.1	148.6	152.7	159.1	164.7	172.4	184.9	198.0	228.2	267.6	322.0	162
145.0	146.3	148.1	150.6	154.7	161.2	166.8	174.6	187.2	200.4	231.1	270.9	326.0	164
146.9	148.3	150.0	152.6	156.7	163.3	168.9	176.8	189.6	202.9	233.9	274.2	330.0	166
148.9	150.2	152.0	154.5	158.7	165.3	171.0	179.0	191.9	205.4	236.8	277.6	334.0	168
150.8	152.1	153.9	156.5	160.7	167.4	173.2	181.2	194.2	207.9	239.7	280.9	338.0	170
152.7	154.1	155.9	158.5	162.7	169.5	175.3	183.4	196.6	210.4	242.5	284.2	342.0	172
154.6	156.0	157.8	160.4	164.7	171.5	177.4	185.6	198.9	212.9	245.4	287.6	346.0	174
156.6	158.0	159.8	162.4	166.7	173.6	179.6	187.8	201.3	215.4	248.2	290.9	350.0	176
158.5	159.9	161.8	164.4	168.7	175.7	181.7	190.0	203.6	217.9	251.1	294.2	354.0	178
160.4	161.8	163.7	166.4	170.7	177.8	183.8	192.2	206.0	220.4	253.9	297.5	358.0	180
162.3	163.8	165.7	168.3	172.8	179.8	185.9	194.4	208.3	222.9	256.8	300.9	362.0	182
164.3	165.7	167.6	170.3	174.8	181.9	188.1	196.6	210.7	225.4	259.6	304.2	366.0	184
166.2	167.7	169.6	172.3	176.8	184.0	190.2	198.9	213.0	227.9	262.5	307.5	370.0	186
168.1	169.6	171.5	174.3	178.8	186.1	192.3	201.1	215.4	230.4	265.4	310.9	374.0	188
170.1	171.5	173.5	176.3	180.8	188.1	194.5	203.3	217.7	232.9	268.2	314.2	378.0	190
172.0	173.5	175.4	178.2	182.8	190.2	196.6	205.5	220.1	235.4	271.1	317.5	382.0	192
173.9	175.4	177.4	180.2	184.8	192.3	198.7	207.7	222.4	237.9	273.9	320.9	386.0	194
175.9	177.4	179.4	182.2	186.9	194.4	200.8	209.9	224.8	240.4	276.8	324.2	390.0	196
177.8	179.3	181.3	184.2	188.9	196.4	203.0	212.1	227.1	242.9	279.6	327.5	394.0	198
179.7	181.3	183.3	186.2	190.9	198.5	205.1	214.3	229.4	245.4	282.5	330.9	398.0	200
1.0%	1.2%	1.5%	2%	3%	5%	7%	10%	15%	20%	30%	40%	50%	N

B

Source: Siemens Aktiengesellschaft, *Telephone Traffic Theory and Chart—Part I*, 1970.

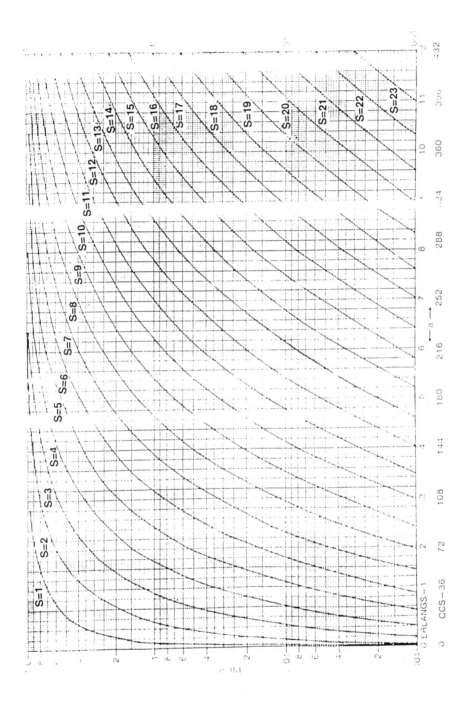

Figure B.1 Blocking probability $P(S, A)$ for Poisson load ($S = 1$ to $S = 23$).

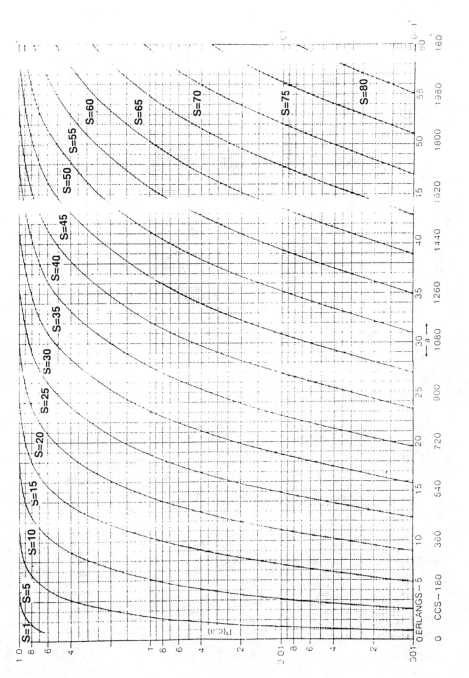

Figure B.2 Blocking probability $P(S, A)$ for Poisson load ($S = 1$ to $S = 80$).

430

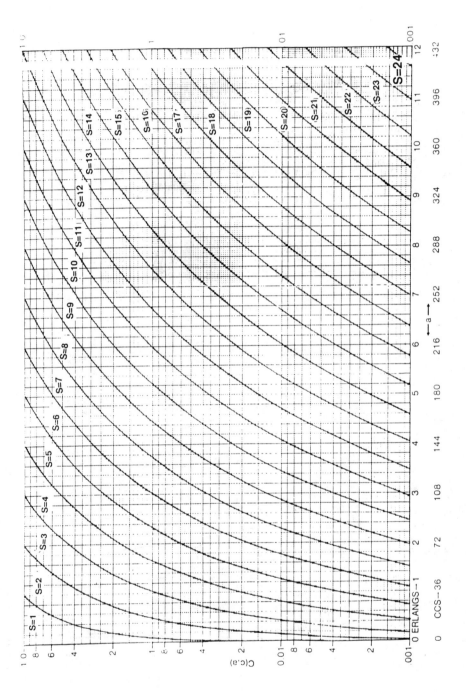

Figure B.3 Blocking probability $P(S, A)$ for Erlang C load ($S = 1$ to $S = 24$).

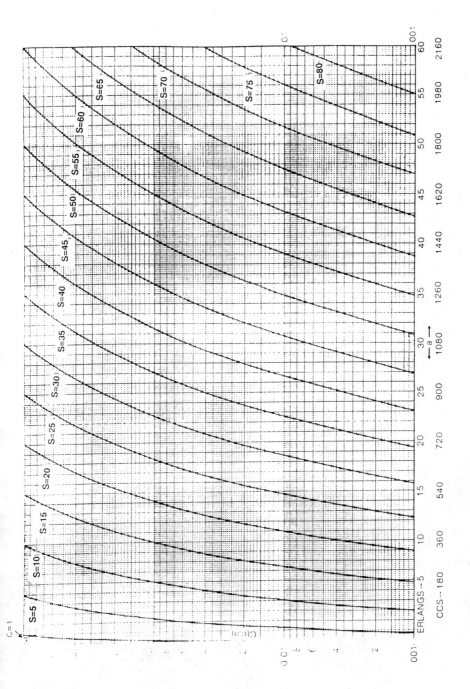

Figure B.4 Blocking probability $P(S, A)$ for Erlang C load ($S = 1$ to $S = 80$).

Glossary

Access	When initiating a call at the mobile phone, the function of gaining access to the telephone network consists of informing the cellular system of the mobile's presence, supplying the system with the mobile's identification and dialed digits, and waiting for the proper channel designation from the cellular system.
Access channel	A control channel used by a mobile station to access a system to obtain service.
Alerting	On command from the MTSO, the serving cell site transmits a data message over the voice channel to an alerting device in the mobile telephone which signals the customer that there is an incoming call.
AMPS	Advanced mobile phone system, a name coined by Bell Laboratories to cover their version of a cellular phone system.
AMPS	Advanced mobile phone service, the analog standard for mobile telecommunications
Analog access channel	An analog control channel used by a mobile station to access a system to obtain service.
Analog color code	An analog system (see SAT) transmitted by a land station on a voice channel and used to detect the capture of a mobile station by an interfering land station and/or the capture of a land station by an interfering mobile station.
Analog paging channel	A forward analog control channel that is used to page mobile stations and send orders.

Analog voice channel	A channel on which a voice conversation occurs and on which brief digital messages may be sent from a base station or from a mobile station to a base station.
ARTIS	Automatic transmitter identification system
AT&T	American Telephone and Telegraph
Authentication	A procedure used by base stations to validate a mobile station's identity at system access.
BS	Base station, a station in the domestic public cellular radio telecommunications service, other than a mobile station, used for radio communications with mobile stations.
BCH	Bose-Chaudhuri-Hocquenghem code.
BHCA	Busy hour call attempts
Blank-and-burst	Signaling over a channel, whereby the voice is blanked briefly and a burst of data is sent. The data may be sent in the land-to-mobile or the mobile-to-land direction.
Blocked calls cleared system	A system whereby calls that find all servers busy leave the system and have no effect upon it, also known as lost calls clear system. The Erlang B formula is also used in this case.
BPSK	Binary phase shift keying is where the phase of the transmitting carrier changes by 180° whenever the logical value of the binary data changes.
BSC	Base station controller. Functions as a control and communication interface between the RF equipment and the MSC.
Busy hour	The busy hour is the uninterrupted period of 60 minutes during which the traffic is at the maximum.
Busy hour call attempts	The average number of times cellular users attempt to originate calls during the hour of the day which contains most traffic.
CAI	Common air interface

CCIR	International Consultative Committee on radio, the branch of the International Telecommunications Union that coordinates international use of the radio spectrum.
CCITT	International Consultative Committee for Telephone and Telegraph
CDMA	Code division multiple access, a modulation and multiple access scheme based on spread-spectrum techniques
CEPT	European Conference of Post and Telecommunications Administration
Cell	A geographical region within which calls are expected to be served by a particular cell site. A hexagonal cell has advantages over other regular geometrical shapes.
Cell site	Installation containing radio and control equipment in order to complete the talking path to the mobile unit.
Cell splitting	The process that handles increased traffic demand by dividing larger cells into smaller cells.
Cellular geographic service area	CGSA. The area a system operator is authorized to cover.
Channel	Refers to a pair of frequencies used for a cellular radio talk path. One is used for cell site to mobile transmission while the other is used for mobile to cell site transmission. The bandwidth of the total (full duplex) channel is 60 kHz.
Channel assignment	(a) The process of specifying which voice channels are to be used at each cell site face in mobile service area; (b) a channel so assigned.
Channel reuse	Refers to the simultaneous use of a signal voice channel for multiple conversations in separate cells such that the mutual interference is below a specified level.
Channel reuse pattern	Pattern in which assignments are repeated in a mobile service area.
Channel set	A group of channels which are generally assigned into cell site.

CIR	Carrier-to-interference ratio
Clear key	Control unit key, which clears dialed number display.
CMRS	Cellular mobile radio telephone communication system (or service)
CNR	Carrier-to-noise ratio
Cochannel interference	Radio interference between channels using the same frequency. In cellular radio, cochannel interference is of concern is among cells using the same channel set.
Coded digital verification color code	CDVCC. A 12-bit data field containing the 8-bit DVCC and 4 protection bits, sent in each time slot to and from mobile stations and base stations. It is used to indicate that the correct rather than cochannel data is being decoded.
Compandor	A compressor and expander of the audio signal. In the line transceiver, the compressor is part of the transmitter and the expander is in the receiver. As applied in the cellular telephone system, its use improves the signal-to-noise ratio and the quality of received audio.
Compandor bypass	A bypass test point provided to permit simultaneous disabling of both the compressor and the expander circuit within the transceiver unit.
Compatibility	The Federal Communications Commission (FCC) in the United States requires that cellular telephone systems be compatible throughout the country, so that a given mobile subscriber can obtain service anywhere in the U.S., irrespective of the service-providing organization.
Compressor	This stage must include the compressor portion of a 2:1 syllabic compandor. For every 2-dB change in input level to a 2:1 compressor within its operating ranges, the change in output level is a nominal 1 dB. The compressor must have a nominal attack time of 3 ms and a nominal recovery time of 13.5 ms.
CPU	Central processor unit
CRS	Cellular radio system

CTIA	Cellular Telecommunications Industry Association.
DDD	Direct distance dialing
Digit receivers	Receivers in the mobile telephone switching office (MTSO) used to collect dialed digits from mobile phone units.
Digital color code	DCC. A digital signal transmitted by a land station on a forward control channel that is used to detect the capture of a land station by an interfering mobile station.
Digital section	That part of the mobile transceiver logic circuit that performs all necessary logic operations required for cellular telephone call processing.
Disconnect	During a normal disconnect when the land party goes on-hook first, the MTSO instructs the mobile via a data message to tune back to the control channel. This clears the associated voice channel for the next call and puts the mobile in the correct state to initiate or receive the next call. If the mobile disconnects first, it autonomously returns to the control channel and the MTSO turns off the associated cell site transmitter by sending a data message.
Discontinuous transmission	DTX. A mode of operation in which a mobile-station transmitter autonomously switches between two transmitter power levels while the mobile station is in the conversation state on an analog voice channel or a digital traffic channel.
Display brightness dimmer	The NITE key dims the brightness of the dialed number display for the comfort of the user during night operations.
DOC	Department of Communication. Provides licensing for Canadian cellular systems.
DPLMRS	Domestic public land mobile radio service.
DPM	Digital phase modulation where the shaped data signals are applied directly to a phase modulator.
DSI	Digital speech interpolation. A feature of E-TDMA that takes advantage of the pauses in speech by assigning the active speech segments from many conversations to a smaller number of radio frequency channels.

DTI	Department of Trade and Industry (U.K.).
DTMF signaling	Dual tone multifrequency signaling, which is widely used in telephone dialing. It is the same type of signaling used in touch-tone, which is Bell Telephone's registered service mark for DTMF signaling. Two tones are used simultaneously for each digit and in the coding scheme two tones out of seven frequencies dial the available digits.
DYNA T.A.C.	Motorola's registered trademark
EBCH	Equated busy hour call. A traffic unit.
End key	Control unit key, which terminates a call.
Electronic lock	An "electronic lock" is provided to prevent unauthorized use of the mobile telephone. The electronic lock is activated by operating the lock key and is unlocked by entering a combination of any three or four digits, which maybe determined and programmed at the time of installation.
EMX	Electronic mobile exchange
Erlang	A unit of telephone traffic measurement named after the Danish engineer and mathematician, A. K. Erlang. Traffic in Erlangs equals the average number of concurrent calls. One Erlang is regarded as an average of one call of one hour duration.
ERP	Effective radiated power.
ETACS	Extended TACS, with more bandwidth.
E-TDMA	Extended-TDMA. Developed by Hughes Network Systems; it extends the capacity of TDMA through the use of digital speech interpolation (DSI). For typical speech activity factors, Hughes claims that the E-TDMA can achieve a network subscriber capacity gain of up to 16 times that of a single analog channel.
ETSI	European Telecommunications Standards Institute. Located in Nice, France, the European Community's organization for standards

EO	End office.
Expander	This stage must include the expander portion of a 2:1 syllabic compandor. For every 1-dB change in input level to a 2:1 expander, the change in output level is a nominal 2 dB.
FCA	Fixed channel assignment.
FCC	Federal Communication Commission.
FDMA	Frequency division multiple access; a modulation and multiple access scheme in while each channel is assigned a specific frequency band in the total signal.
FM	Frequency modulation. A form of nonlinear angle modulation.
Flash request	A message sent on a voice channel from a mobile station to a land station indicating that a user desires to invoke special processing.
Forward setup channel	A setup channel from a land station to a mobile station. Also termed a forward control channel (FCC).
Forward analog voice channel	FVC. An analog voice channel used from a base station to a mobile station.
FSK	Frequency shift keying, where the instantaneous carrier frequency is dependent on the value of the input symbol. If the phase is continuous at symbol boundaries, continuous phase FSK (CPFSK) results.
FFSK	Fast frequency shift keying is an adaptation of offset phase shift keying (OQPSK) in which baseband pulses are sinusoidal instead of rectangular.
FVC	Forward voice channel
GHz	Gigahertz; a thousand megahertz
GMSK	Gaussian minimum shift keying is CPFSK with a Gaussian-shaped premodulation filter. It has a narrower bandwidth than an equivalent unfiltered CPFSK signal.

GMX	Group multiplexer.
Group identification	A subset of the most significant bits of the system identification (SID) that is used to identify a group of cellular systems.
GOS	Grade of service.
GSM	Group Special Mobile. Stands for the next generation digital cellular system to be deployed in Europe.
Handoff	The process by which a deteriorating transmission path between a cell site and a mobile unit is rerouted to a different cell site on a different channel in order to obtain a stronger signal. Handoff is one of the distinguishing features of CMRS.
HCA	Hybrid channel assignment.
HCS	Hundred call seconds per hour. A traffic unit.
HLR	Home location register.
I and Q signals	Inphase and quadrature signals are vector components of the base-band information carrying signal.
IBT	Illinois Bell Telephone Company
Improved mobile telephone service	IMTS. An automatic mobile telephone service introduced in 1964 by the Bell System.
IMTS	Improved mobile telephone service.
Initialization	Whenever the mobile equipment is turned on, various mobile equipment parameters and system variables are set and tested to be ready for performing functions required to communicate with a cell site. Such a process is referred to as initialization.
Interleaving	A process whereby data, in either bit or symbol form, are distributed over a time frame in order to disperse error bursts at the receiver after deinterleving has occurred.

IS/54	National digital standard for telecommunications in the U.S. It is the EIA/TIA-approved standard for mobiles and base stations.
ISDN	Integrated service digital network.
ISO	International Organization for Standardization. The global standards-making body.
Kbps	Kilobits per second
Land-originated call	A call to a mobile unit in which the calling party is located at a telephone connected through public switched telephone network.
Land station	A station in the domestic public cellular radio telecommunication service, other than a mobile station used for radio communications with mobile stations.
Limiter	A stage in the receiver section of the mobile (FM) transceiver. Its purpose is to reduce amplitude variations in the incoming signal.
Location receiver	A receiver used exclusively for gathering mobile location.
Lock switch	A control unit switch, which activates an electronic lock to prevent unauthorized use of the mobile telephone. It is activated by the LOCK key and is unlocked by entering a combination of any three or four digits, which may be determined and programmed at the time of installation.
MIN	Mobile identification number
Mobile assisted handoff	MAHO. A process where a mobile in digital mode, under direction from a base station, measures the signal quality of specified RF channels. These measurements are forwarded to the base station upon request to assist in the handoff process.
Mobile identification number	MIN. The 34-bit number that is a digital representation of the 10-digit directory telephone number assigned to a mobile station.

Mobile service area	MSA. A basic coverage region in which the cellular service is made available. Generally referred to as CGSA (cellular geographic service area) for cellular phone service.
Mobile station	A station in the domestic public cellular radio telecommunications service intended to be used while in motion or during halts at unspecified points. It is assumed that mobile station includes portable unit (e.g., handheld ''personal'' units) as well units installed in vehicles.
Mobile-telephone switching office	MTSO. A switching and control entity that serves as the central coordinator and controller for the cellular system.
Mobile-terminated call	Call terminated in a mobile unit.
Mobile unit	Radiotelephones that are free to move (e.g., car telephones, hand-held units) as opposed to cell-site or fixed-station units.
MSA	Mobile service area.
MSC	Mobile switching center. Used in the digital cellular system, the MSC provides connections between the mobile subscribers and the PSTN, other mobile switches, and other mobile subscribers.
MSK	Minimum shift keying is CPFSK with a modulation index of 0.5.
MTSO	Mobile telephone switching office. It is a term used in the analog cellular system.
MTX	Mobile telephone exchange.
NADC	North American digital cellular. American digital cellular standard.
NAM	Number assignment module.
NAM socket	A 16-contact socket used as a mounting device for the NAM. To facilitate field changeability, the NAM socket is located so the NAM can be removed or installed with the removal of the top cover.

NCS	Network control switch. Also MTSO (mobile telephone switching office).
NEC	Nippon Electric Company
NF	Noise figure. Provides an estimate of signal degradation due to noise generated within the receiver.
NMT	Nordic mobile telephone.
NOI	Notice of inquiry issued by FCC.
NPA	Numbering plan area.
NRZ	Nonreturn to zero code.
NVM	Nonvolatile memory.
OHD	Overhead message type field.
OLC	Overload control.
Off-air call setup	The technique assigns channels only after both the calling and the called parties answer.
On-off sense line	The on-off sense line provides a binary input to the transceiver unit, which controls the application of DC power to the mobile equipment.
Order message	A message sent from the cell site to a mobile unit directing the mobile to perform some specific action.
Order response message	A message sent from a mobile unit to a cell site indicating that an order message has been received.
OSI	Open systems interconnection. A seven-layer model developed by ISO that standardizes data communications.
OMC	The operations and maintenance center. The OMC provides software upgrades and network configuration capabilities.

Overload protection	The mobile interconnecting cable has fuses for the main power line (10 amperes, quick acting) and the ignition sense line (4 amperes, quick acting). The mobile equipment has current-limiting devices in all power lines, as necessary, to prevent damages by faults, unwanted grounds, or overloads.
Paging	The act of seeking a mobile station when an incoming call has been placed on it.
Paging channel	A forward control channel that is used to page mobile stations and send orders.
Page reply	The mobile telephone responds to the cell site it selected over the reverse control channel. The selected cell site then reports the page reply to the MTSO over its dedicated landline data link.
PCN	Personal communications network. A microcell-based radio telephone network, providing two-way mobile telephony, with at least slow-speed handoffs from cell to cell. At the present, PCN is more an idea than a service, since there is no consensus yet on what features it should include, let alone how to provide them.
PCS	Personal communications services; the collection of services and features that a PCN is expected to provide.
Personal identification number	PIN. A secret number managed by the system operator for each subscriber.
Portable mobile unit	A mobile unit that is not constrained to operate exclusively from a vehicular installation.
Pre-emphasis	The preemphasis characteristics at the mobile transmitter must have a nominal +6 dB/octave response between 300 and 3,000 Hz.
Preorigination dialing	With preorigination dialing, the dialing sequence takes place before the mobile unit's first communication with the local system. A mobile customer dials the telephone number of the party being contacted into the unit's memory. The customer then initiates the communication by depressing the SEND key.
PM	Phase modulation. A form of nonlinear angle modulation.

PRCS	Personal radio communications service General Electric System.
PSN	Public switch network.
PSTN	Public switched telephone network.
QPSK	Quaternary phase shift keying. Allows two bits per symbol of transmission.
RACE	Research into advanced communications for Europe; a long-term development effort to design the European communications infrastructure for the next century.
Radio common carrier	RCC. The non-wireline carrier, which is granted a competitive share of the cellular telephone service.
Recall key	Control unit key, which recalls telephone numbers in memory.
Received signal strength indication	RSSI. A voltage level proportional to the logarithm of the received signal strength is provided to the logic circuitry and the RSSI test point.
Registration	The steps by which a mobile station identifies itself to a land station as being active in the system at the time the message is sent to the land station.
Release	When the user desires to terminate a call (mobile initiated release) or when a release request is received (forced release), the mobile equipment enters release procedures.
Release request	A message sent from a mobile station indicating that the user desires to disconnect the call.
RELP	Residual excited linear predictive, an algorithm used in speech vocoding.
Reuse	The reusing of the same radio channels in two or more cells of a single cellular system. Reuse requires sufficient cell separation to provide an adequate signal-to-noise ratio.
Reverse blank-and-burst	Sending blank-and-burst data from mobile to land.

Reverse setup (control) channel	The setup (control) channel used from a mobile unit to a cell site.
Reverse voice channel	RVC. The voice channel used from a mobile unit to a cell site.
Roamer	A mobile unit that operates in a cellular service area other than the one from which service is subscribed.
RSSI	Received signal strength indication.
RVC	Reverse voice channel.
SACCH	Slow associated control channel in the GSM (European) digital cellular system.
SAT	Supervisory audio tone (5,970, 6,000, 6,030 Hz).
SCC	SAT color code.
SCPC	Single channel per carrier.
Secondary paging channels	In addition to the primary paging channels, a supplementary set of analog control channels developed specifically for IS-54 compatible mobile stations. Such channels are used to page mobile stations and send orders.
Serial number	SN. A 32-bit binary number that identifies the mobile equipment. It differs from the NAM in that it is factory set and installed and it is not alterable in the field without rendering the mobile equipment inoperable. The circuitry that provides the SN is well buried within the equipment and is of a permanent nature. Any attempt at field alteration requires extensive disassembly and is supposed to result in damage to the equipment.
Setup channel	A channel used for the transmission of digital control information from a cell site to a mobile unit or from a mobile unit to a cell site (same as control channel).

Setup radios	Radios that transmit only data are used in the initial phase of "setting up" the call prior to establishing a voice path for communications. They are for the general (shared) use of the cell site in communicating with all mobiles within its cell. In addition, setup radios also transmit overhead messages to ensure that idle mobiles within the cell coverage zone are ready and able to communicate should a call be initiated for or from a mobile.
SID	System identification.
Sidetone circuit	Contained in the base set. It consists of a 10-dB attenuator, and a 20-dB volume control for the handset receiver-audio sidetone amplifier. The sidetone circuit takes the signal at the handset transmitter interface and develops the acoustic output 12 dB lower than the acoustic input over the 300–3,000 Hz band.
Signaling tone	A 10-KHz tone transmitted by a mobile unit on a voice channel to: 1) confirm orders, 2) signal flash requests, and 3) signal release requests.
Slow associated control channel	SACCH. A continuous channel used for signaling message exchanges between the base station and the mobile station. A fixed number of bits are allocated to a SACCH in each TDMA slot.
SMSA	Standard metropolitan statistical area.
SN	Serial number.
ST	Signaling tone.
Supervisory audio tone	Tone frequencies are: 5,970, 6,000, and 6,030 Hz, transmitted by cell site and transponded by mobile unit.
System identification	SID. A digital identification associated with a cellular system; each system is assigned a unique number.
TACS	Total access communication system.
TASI	Time assignment speech interpolation. Dynamically switches active voice signals to idle channels. This technique enhances the capacity of the system.

TDMA	Time division multiple access. A technique whereby stations communicate with each other on the basis of nonoverlapping time-sequenced transmissions through a common medium.
TIA	Telecommunications Industry Association.
Traffic carried	For systems in statistical equilibrium, the traffic carried is equal to the mean number of busy servers.
Traffic channel	That portion of the digital information transmitted between the base station and the mobile station, or between the mobile station and the base station, that is used to the transport of user and signaling information.
Transceiver	A transmitter-receiver pair.
TSP	Tandem switching point. An intermediate switch used to establish a connection.
UC	Unit call. A traffic unit.
Ultrahigh frequency	UHF. A 300 MHz to 3 GHz cellular radio system belongs to UHF range.
USD	U.S. digital. The standard being developed by TIA for cellular telephone system in the United States.
VAD	Voice activity detector. The circuit determines when a speaker is not talking and interrupts transmission .
VLR	Visitor location register
Voice channel	A channel on which a voice conversation occurs and on which brief digital messages may be sent from a cell site to a mobile unit (station) or from a mobile unit (station) to a cell site.
VSELP	Vector sum excited linear predictive. TIA's IS-54 voice coding standard.
WARC	World Administrative Radio Conference.

WCC   Wireline common carrier. Local telephone company providing cellular service.

X.25   Connection-oriented packet-switching network.

About the Author

Asha K. Mehrotra received his B.S., M.S., and Ph.D. degrees in electrical engineering from Bengal Engineering College, Sibpur, India; Nova Scotia Technical College, Halifax, Canada; and Polytechnic Institute of New York (formerly Brooklyn Polytechnic) in 1961, 1967, and 1981, respectively. He is presently working as a member of the technical staff at the Analytic Sciences Corporation (TASC), Reston, Virginia, where he deals with a large range of communication system engineering problems. In the past he has worked on the MILSTAR satellite project at MITRE as a member of the technical staff and with TDRSS satellite at Space Communication Company as a manager of the systems analysis group. His past experience includes areas of HF communication, switching systems, and telephone transmission systems. His hardware experience spans the design of central office equipment, PBXs, and facsimile systems. In the past, Dr. Mehrotra has taught graduate courses in computer science and electrical engineering at Virginia Polytechnic Institute and State University and George Mason University. He has been an adjunct professorial lecturer at George Washington University for the past 13 years, where he teaches graduate courses in communication engineering. He is a member of the IEEE Communication and Computer Society.

Index

The Artech House Telecommunications Library

Vinton G. Cerf, Series Editor

Advanced Technology for Road Transport: IVHS and ATT, Ian Catling, editor

Advances in Computer Communications and Networking, Wesley W. Chu, editor

Advances in Computer Systems Security, Rein Turn, editor

Analysis and Synthesis of Logic Systems, Daniel Mange

Asynchronous Transfer Mode Networks: Performance Issues, Raif O. Onvural

A Bibliography of Telecommunications and Socio-Economic Development,
   Heather E. Hudson

Broadband: Business Services, Technologies, and Strategic Impact, David Wright

Broadband Network Analysis and Design, Daniel Minoli

Broadband Telecommunications Technology, Byeong Lee, Minho Kang, and Jonghee Lee

Cellular Radio: Analog and Digital Systems, Asha Mehrotra

Cellular Radio Systems, D. M. Balston and R. C. V. Macario, editors

*Client/Server Computing: Architecture, Applications, and Distributed Systems
   Management*, Bruce Elbert and Bobby Martyna

Codes for Error Control and Synchronization, Djimitri Wiggert

Communication Satellites in the Geostationary Orbit, Donald M. Jansky and
   Michel C. Jeruchim

Communications Directory, Manus Egan, editor

The Complete Guide to Buying a Telephone System, Paul Daubitz

Computer Telephone Integration, Rob Walters

The Corporate Cabling Guide, Mark W. McElroy

Corporate Networks: The Strategic Use of Telecommunications, Thomas Valovic

Current Advances in LANs, MANs, and ISDN, B. G. Kim, editor

Design and Prospects for the ISDN, G. Dicenet

Digital Cellular Radio, George Calhoun

Digital Hardware Testing: Transistor-Level Fault Modeling and Testing, Rochit
   Rajsuman, editor

Transmission Performance of Evolving Telecommunications Networks, John Gruber and
    Godfrey Williams

Troposcatter Radio Links, G. Roda

UNIX Internetworking, Uday O. Pabrai

Virtual Networks: A Buyer's Guide, Daniel D. Briere

Voice Processing, Second Edition, Walt Tetschner

Voice Teletraffic System Engineering, James R. Boucher

Wireless Access and the Local Telephone Network, George Calhoun

Wireless LAN Systems, A. Santamaría and F. J. Lopez-Hernandez

Writing Disaster Recovery Plans for Telecommunications Networks and LANs, Leo A.
    Wrobel

X Window System User's Guide, Uday O. Pabrai

For further information on these and other Artech House titles, contact:

Artech House
685 Canton Street
Norwood, MA 01602
617-769-9750
Fax: 617-762-9230
Telex: 951-659
email: artech@world.std.com

Artech House
Portland House, Stag Place
London SW1E 5XA England
+44 (0) 71-973-8077
Fax: +44 (0) 71-630-0166
Telex: 951-659